Home Ranges of Animals

Home Ranges of Animals

Cognitive Maps, Research Approaches, and Insights

Roger A. Powell

Department of Applied Ecology, North Carolina State University

OXFORD
UNIVERSITY PRESS

OXFORD
UNIVERSITY PRESS

Great Clarendon Street, Oxford, OX2 6DP,
United Kingdom

Oxford University Press is a department of the University of Oxford.
It furthers the University's objective of excellence in research, scholarship,
and education by publishing worldwide. Oxford is a registered trade mark of
Oxford University Press in the UK and in certain other countries.

Published in the United States of America by Oxford University Press
198 Madison Avenue, New York, NY 10016, United States of America

British Library Cataloguing in Publication Data
Data available

Library of Congress Control Number: 2025938045

ISBN 9780192858337

DOI: 10.1093/9780191949142.001.0001

Printed and bound by
CPI Group (UK) Ltd., Croydon, CR0 4YY

In memory of Monte Lloyd and Don Hayne, who showed me good science and good writing. They left me with a lifelong challenge to follow their examples.

Foreword

Whether you are a seasoned scientist with long experience in animal ecology or you are a young student observing for the first time the movements of your study animals, the very first question that will likely come to your mind is about the size, form, and composition of the spatial range used by your animals. Indeed, the home range is often one the first and most basic pieces of information that we biologists seek and produce when studying animal ecology. For many decades, we have been using the concept without asking too many questions on its real meaning, protected as we have been by Burt's (1943) reassuring definition that gave us the illusion of objectivity and robust science.

I thought back, as I wrote this Foreword, to my first attempt, ages ago, to map the home range of my first radio-collared wolf in central Italy. At that time, I thought that simply putting all radio-locations on a map would obviously and automatically produce the home range. Nonetheless, as the radio-locations accumulated on the map, it was also obvious that several assumptions and decisions had to be made to make sense of those "clouds" of points on the map. No possible statistical manipulation of the points was ever fully satisfying; they were always a partial and imperfect representation of reality. Something was obviously missing in my clouds of points.

The answer slowly unveiled as my research team and I proceeded in our field work following on foot and by car our collared wolves through their nightly movements, trying to be as near as possible to them and in continuous touch with the beeps of the radio. The activities and movements of the wolves took a completely new meaning when we could "read" the details of those movements in relation to the local spatial and temporal context, just as perceived by the wolves. Suddenly, the selection of one path over another made sense because we could hear a local disturbance that no radio-telemetry data could ever reveal: the hesitation to enter a village made sense as an unexpected noise broke the silence of the night; the choice of a daily resting site made sense in view of the local presence of tourists. In a nutshell, we biologists understood that the way we looked at home ranges as clouds of points was inevitably wrong, no matter how perversely we tortured them through analytical exercises, because we reflected too roughly and indirectly the animals' views of their homes. By following the animals at close range for nights and nights, we were closer to understanding the events that shaped the animals' points of view, though still imperfectly. But, we were closer to the wolves' perceptions. Roger Powell, in this magnificent book, would more appropriately say that we were closer to understanding the cognitive map of the wolves.

Comprehending a cognitive map requires much more than collecting just a series of locations and correlating them with environmental features such as landscape, prey or predator presence, conspecific competition, etc. Comprehending means adding the deep understanding of the animals' behaviors, personalities and, why not, emotions. It is not easy, nor often feasible, as animals can rarely be studied at close range, as if we were literally with them for hours and days. This book goes a long way towards teaching us how to shift our approach to the study of animals' homes, overturning a human-centered approach and moving to an animal-centered one. This change will benefit from a bit of new technology but also, and equally important, from a good deal of empathy and love and participation with the animals we study. It is not an option; it is essential if we wish to move from describing the distribution of where the animals have been found (i.e. the clouds of points) to understanding why the animals are there and how they keep moving to the various parts of their homes. Moving from a correlational to a causational result should be the ultimate prize for the animal ecologist.

Today's GPS technologies allow us to collect tons of locations from the comfort of our offices far away from the bearers of the GPS collars. The enormous quantity of very precise locations, however, is not a substitute for the quality of fewer points with information on the "why". As well explained in this book, imagination, creativity, and curiosity are necessary tools to build and explore models of home ranges that can help us in generating hypotheses to be tested objectively with our clouds of locations and movement tracks. But models are useful only when based on deep knowledge of natural history, as shown in this unique book filled with countless anecdotes and stories of movements, home ranges, and behaviors of free-ranging animals (with a bias for mustelids and bears which the author has been following for decades). Home ranges will reveal all their secrets if approached with the combination of firm scientific methods, robust analytical skills, and ingenuous love for the animals.

Luigi Boitani

Contents

CHAPTER 1

An Introduction to Home Ranges

Female black bear 260 did not want to climb up and over Little Pisgah Ridge. This deduction struck me as I looked at a map of bear 260's wanderings (Figure 1.1). She walked around the base of Little Pisgah Ridge, sticking to the more level ground but longer distance along Big Creek and Bee Tree Creek. She understood the trade-offs. North Carolina's Blue Ridge Mountains have rugged topography across short distances and ridges have steep sides. Over the course of 1 km, Little Pisgah Ridge drops 750 m from the top of Little Pisgah Mountain to the convergence of Bee Tree and Big Creeks. Although capable of traversing the toughest terrains the mountains have to offer, bears do not purposefully waste energy tackling the steepest slopes when easier routes can take them where they want to go. In fact, steep slopes often define the peripheries of bears' home ranges (Powell and Mitchell 1998).

Home Ranges of Animals. Roger A. Powell, Oxford University Press. © Roger A. Powell (2025). DOI: 10.1093/9780191949142.003.0001

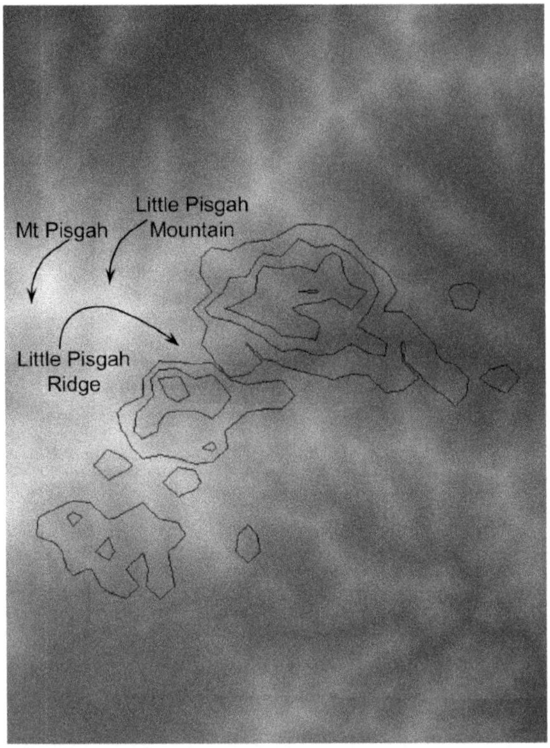

Figure 1.1 Bear 61's familiarity distribution has a notch showing how she walked around the steep sides of Little Pisgah Ridge instead of hiking up and over the steep ridge. High elevations are white, low elevations dark grey.

When I first noted that 260 avoided Little Pisgah Ridge, I understood her behavior completely. I had hiked Little Pisgah Ridge myself countless times. From any approach, Little Pisgah Ridge is exhausting. Even the ridgeline is steep enough that hiking up leads to sweat and heavy breathing, while hiking down leads to aching quadriceps from braking with every step. The sides of the ridge are steeper. Bear 260 understood the basic topography of her home range. Travelling across a landscape is energetically expensive, setting a premium on being able to minimize travel energy while still directing moves toward important resources. While the reasons for avoiding a steep ridge were obvious, I am convinced that 260 understood myriad tiny details across her whole home range as well. Undoubtedly, all animals do.

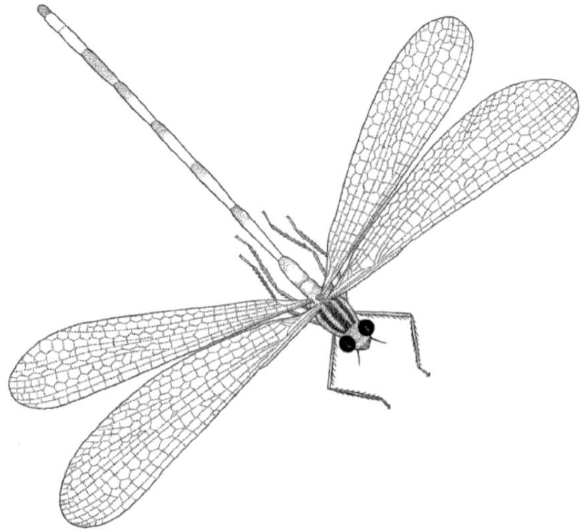

1.1 Introducing Home Ranges

This chapter provides an introduction to home ranges, including the standard definition and discussion of some of the problems with that definition. It then discusses why a biologist should study home ranges and what the challenges are for studying home ranges. Next the chapter introduces the scientific method, discusses how to use it for best research results, and introduces the use of models in ecology and behavior. This book uses many terms that are not always used consistently in the biological literature and, therefore, I define my use of some important terms and explain how I use them. Lastly, the chapter introduces the following chapters, explaining my logic for their order.

During most of their lives, many animals confine their movements on the landscapes where they live and, thereby, develop home ranges (their homes). Thus, a home range is an emergent property of an animal's confined movements (Börger et al. 2008, Spencer 2012, Kenkro and Giugglioli 2019). Even migration movements lead to building a home range but the home range will have 2 sections connected by the migration route or routes. Exploratory movements that end back at an animal's home range may not contribute to the animal's home range but have the potential to do so should the animal be forced to abandon her[1] home range and establish a new one. The animal may remember her exploratory movements and return to where she had explored to seek the new home range or to acquire a resource not readily available elsewhere. Likewise, early dispersal movements do not usually contribute to an animal's home range but late dispersal movements, those movements where the animal begins to confine her movements because she has found a potential home range, may contribute.

Aldo Leopold (1949: 78) wrote: "The wild things that live on my farm are reluctant to tell me, in so many words, how much of my township is included within their daily or nightly beats. I am curious about this, for it gives me the ratio between the size of their universe and the size of mine, and it conveniently begs the much more important question, who is the more thoroughly acquainted with the world in which he lives?"

Like many other biologists (e.g. Grimm 2023, Macdonald 2021, Vucetich 2021), including me, Leopold wanted to conceive of what it is to be a cottontail rabbit ducking into a brush pile to elude a red-tailed hawk, or to be the first field sparrow to sing at 03:35 of an April morning. He wanted to feel how they understand their environments, which is wanting to know their home ranges from their perspectives. How an animal is acquainted with her environment, from her point of view, is her home range.

Understanding the importance of animals' perspectives of their home ranges is my first goal for readers of this book. Trying to understand how animals conceive of their home ranges provides insights into why animals behave as they do. Understanding how brains make cognitive maps of environments and situations, of scheduling activities, and of interacting with neighbors and strangers can help us understand what home ranges are from animals' perspectives. Cognitive maps are not Euclidian maps, they are not like maps on paper or on a computer monitor. Brains' cognitive maps are more like organizing functions, taking knowledge remembered about environments, about changing components of environments, about social interactions and other aspects of everyday life, and about using that knowledge to plan future actions, including future thinking. Animals do not conceive of their home ranges as we conceive maps. To understand animals that are subjects of research and animals whose populations are managed, biologists benefit immeasurably from understanding the perspectives of those animals.

My second goal is to present different approaches, especially modelling approaches, for developing hypotheses that have potential, when tested, to provide new insights into how animals view their home ranges. The different approaches complement each other by providing different perspectives on home ranges. I hope, thereby, to give biologists tools to understand more deeply what home ranges are to the animals that they study.

My third goal is to present interesting information on animals and their home ranges. Although some sections of the book have become technical, nonetheless, I hope that a reader with a good biological background who does not do research related to home ranges will find the biology interesting enough to make reading the book worthwhile.

The goals for the book complement each other. Sharing the world with animals, learning about them, conserving them, and managing them is exciting. These actions can best be done by understanding *why* animals do what they do. Different individual animals, animals in different populations, and animals of different species use the landscapes where they live differently. A single-size-fits-all approach to living on a landscape can never work as well as can approaches tailor-fit to the needs of animals. Different animals behave differently to meet those needs. Having at least some understanding of what animals' home ranges are to them should precede attempts to quantify home ranges for research and management. Understanding what animals' home ranges are allows biologists to adapt their hypotheses and their research or management to be most efficient and effective, and to choose the most pertinent ways to quantify home

[1] English lacks a singular personal pronoun that does not specify sex for living beings. Our only choices are "she" or "he." I do not apply "it" to animals. Consequently, in this book, I switch back and forth arbitrarily between "she" and "he" when I need a personal pronoun for an animal. Kimmerer (2017) explained that the Potowatomi language has such a non-sex-specific pronoun, "ki," which I have wanted to use in this book but decided that introducing the word might distract readers.

ranges. Of course, the more hypotheses a biologist tests, the more she learns about how her study animals think, which allows her to generate more hypotheses that will lead to more insights.

Sometimes, of course, management and conservation decisions must be made when the biology of the animals is not well understood. Those situations make gaining an understanding of the animals' biology even more important.

1.2 A Home Range Overview

Through my research, observations, and experiences, I want to gain a concept of what it is to be a weasel[2] (or a vireo or a bumblebee) and thereby to understand something about how such animals conceive of their home ranges. Although I understood female bear 260's avoidance of Little Pisgah Ridge's steep slopes, animals' knowledge of their environments usually differs tremendously from the knowledge that biologists gain of those environments. At the other end of North Carolina from 260's mountains, on the coastal plain, black bears take advantage of shelter provided by habitats called pocosins. A pocosin is a mix of deciduous and conifer bog shrub-forest woven intricately, densely, and impenetrably together with thorny vines. In the 1970s, Joe Hamilton studied black bears in coastal North Carolina, using radio telemetry to follow the bears (Hamilton 1978, Landers et al. 1979). He drove roads that wove through the coastal habitats and stopped to take compass readings of the directions of the strongest signals from his radio-collared bears. He became familiar with pocosins but not in the way a bear does. Biologists have trouble conceiving of slipping gracefully through pocosins, or, for that matter, moving through them at all. When a person approaches a pocosin, he or she does not see a way to enter, only a wall of tangled branches, vines, and thorns extending from the ground to above head height. Black bears, in contrast, are intimately familiar with all pocosins within their home ranges, know all the routes through those pocosins, and know how to use pocosins to avoid people. One of Joe's bears died in a pocosin just 400 m from a dirt road. Joe did not enter that pocosin at bear height (his knee height) nor did he look through the pocosin from the bear's perspective to see potential routes. Instead, armed with tools to cut shrubs and vines, Joe struggled for 8 hours to reach the bear. Joe spoke to me of pocosins with reverence, yet he understood pocosins differently than do black bears.

Darwin[3] (1871: 213) noted that "no one with an unbiased mind can study any living creature, however humble, without being struck with enthusiasm at its marvellous structure and properties." I can not help but believe that, had Darwin seen how black bears traverse pocosins so easily, he would have wanted to know how bears do it. I have followed black bears closely on foot in the Blue Ridge Mountains and have gaped when one slips through a seemingly impenetrable curtain of *Smilax*, a vine with extremely effective thorns. I wanted to know how she did it.

Home ranges can be as small or smaller than 100 m^2 for some wasps and dragonflies (Dunkel 2000, Eaton 2021) and 100 m^2 for some small mammals (Hayne 1950, Kelt and Van Vuren 2001). For carnivorans (members of the mammalian Order Carnivora), which have large home ranges for their body sizes, home ranges can be as small as 1 ha for least weasels (Lockie 1966, Nyholm 1959). Yet, home ranges can be very large, 100s of km^2 for wolverines (Vangen et al. 2001) to 1000s of km^2 for polar bears (Amstrup et al. 2000), and up to 10,000 km^2 for albatrosses (black-footed albatross, Laysan albatross, shy albatross; Gutowsky et al. 2015, Mason et al. 2018). In general, home range size varies with

[2] I use the North American convention for common names of weasels, stoats, long-tailed weasels, etc. I use "weasel" to refer to all of them as a group. For individual species, I use their accepted common names.

[3] One gains the impression that all evolutionary biologists must pay homage to Darwin and cite him at least once.

animal body size, resource density, and competition (Jetz et al. 2004).

Home ranges vary in diverse ways other than size. Animals of the same species in different parts of the species range can have home ranges in dense forest, in savannahs, or on vast prairies; they can have home ranges at high elevations or home ranges near sea level. Female black bears in boreal forests with low productivity maintain non-overlapping home ranges while those black bears that live further south may tolerate extensive home range overlap (Powell et al. 1997, 2022). Animals of different species in the same ecosystem use home ranges in vastly different manners. North American black bears forage for ants, berries, and carrion, capture small mammals on the forest floor when the opportunity arises, and during spring female black bears prey on ungulate newborns. North American red squirrels on the same landscapes have 3-dimensional home ranges, foraging on the ground and in trees, and caching seeds, dried mushrooms, and other food underground, as well as travelling under ground, across the ground, through the trees, and even through the air. Ravens on those same landscapes incorporate into their home ranges the space above the forests and the air within the forests. Black bears can scan the forest and the air for smells that are sometimes kilometers away; ravens can scan areas kilometers afield visually. Red squirrels, black bears, and ravens all listen keenly.

Home ranges may be shared by mates, by families or extended families, or by groups of conspecifics, or not shared at all. Why do animals maintain home ranges? Simply being familiar with the critical resources in an area makes using those resources efficient, raising survival and reproductive output. The benefits exceed the costs, thereby increasing an animal's fitness. Being familiar with the critical resources has strong enough benefits that animals stay in their home ranges even during short periods when resource productivity is less than the costs of maintaining the home range. During those periods, animals may lose weight or experience somewhat higher mortality rates. Animals may make short-term changes in their home ranges to increase efficiency of resource use. Some animals avoid predictably negative resource balances by hibernating, aestivating, or migrating. Sometimes, an animal leaves his home range temporarily to find a mate, or to find food when food is scarce in the home range. And sometimes the habitats within a home range are destroyed by fire, by logging, or by some other major event. At these times, an animal must find a new place to live and establish a new home range. In the long term, a negative resource balance leads to death or to seeking a new home range.

1.3 A Home Range Definition

Most contemporary discussions of animals' home ranges begin with William Burt's (1943: 351) definition of a home range:

> ". . . that area traversed by the individual in its normal activities of food gathering, mating, and caring for young. Occasional sallies outside the area, perhaps exploratory in nature, should not be considered part of the home range."

Burt knew that biologists before him had understood that individuals confine their movements and do not wander across entire landscapes. Even Darwin (1859: 45) referred to "individuals of the same species inhabiting the same confined locality."

Figure 1.2 Burt's definition of a home excluded "occasional sallies," referring to travels that Burt considered to be outside an animal's home range for exploring or for other 1-time visits. His definition does not acknowledge that animals remember explorations and incorporate them into their cognitive maps.

The earliest reference, beyond Darwin's, that I have found of mammals setting up residence was by Elliot Coues (1877) in his volume on the Mustelidae for his report to the North American Northern Boundary Commission of 1873–1876. Coues noted that short-tailed weasels that take up residence on farms usually provide more benefits for the farmer by killing mice and other rodents that eat grain than the costs of chickens killed by the weasels (pp 131–132). Rich (1888) reported an observation of a puma who resided in "his mountains to the east." Anthony (1928), in his field guide to North American mammals, referred to making long-term observations of individual local mammal residents. Ernest Seton Thompson[4] (1898) wrote the story of a wolf and its pack that confined their movements to the headwaters of the Currumpaw River in New Mexico. Seton (1909) appears to have been the first to use the term "home range" and, by the 1930s, the term appears to have been in general use for mammals (Murie 1936, Hamilton 1937, Dalke and Sime 1938, Blair 1940, Haugen 1942) and soon after for other diverse animals (Stickel and Cope 1947, Stebbins 1948, Gerking 1953, Frank 1964).

[4] Before he became Ernest Thompson Seton.

Therefore, Burt (1943) did not introduce a new concept but, instead, provided a clear definition of a home range when biologists needed such a definition. His definition implies 3 behavioral characteristics of animals that lead to home ranges (Horne et al. 2019). First, home ranges emerge from movements limited to a confined area, resulting in resource patches being revisited. Second, those movements provide animals with access to resources: to forage, to find mates, to avoid or to escape predators, and to avoid competitors. Lastly, home ranges exclude exploratory "sallies" (Figure 1.2).

Burt's definition of a home range accommodates a home range having areas used in diverse ways for diverse behaviors related to diverse resources. The definition could be understood to push biologists to use all the tools available to gain the best understanding of those resources and behaviors. Different research and management goals will require diverse approaches to research beyond statistical approaches.

Burt considered a home range to be an area of ground. But to animals, their home ranges are the complex amalgam of memories and knowledge of resource locations and conditions, the changing of conditions, the locations and identities of neighbors, and the risk of predators. A biologist can draw maps of where she finds study animals but the maps are not *the animals'* home ranges, just the biologist's understanding of the places. Once a biologist gains a concept of a study animal's home range, she may state that "This is where I located animal *X*" but will not state "This is animal *X*'s home range."

1.4 Why Study Home Ranges?

Home ranges affect and reflect all biology of free-living, individual animals. Home ranges are complex mental concepts that animals use to plan foraging movements, to decide what resource to exploit next, or to decide which way to run from a predator. Biologists can gain a concept of how animals conceive their home ranges by doing research into those home ranges, by studying how animals use their home ranges, and by designing research that answers questions about why animals do what they do where they live.

Home ranges dictate the habitats, foods, and other resources available to each individual animal. Juxtaposition of home ranges affects competition with conspecifics and with members of other species and affects understanding that competition is also part of home ranges. Home range locations on a landscape affect exposure to predators and dictate predator avoidance measures. Perhaps most consequentially, home ranges affect access to potential mates and, thereby, reproductive output and direct effects on fitness. *No behavior of a resident animal makes sense without the context of that animal's home range.* Thus, to study individual behaviors in the field, one must understand something about the animals' home ranges.

Even when studying animals in the lab, a biologist must know the effects on behaviors of having such a sterile home range and must know the wild conditions to which behaviors are adapted. Studying animals in the lab does not constitute a null condition but, instead, a highly aberrant treatment to which animals may behave naturally or aberrantly (Smith 2023). Laboratory rodents raised and maintained in different labs but under the same housing protocols often behave differently when given standardized tests because of small differences in how the protocols were followed. Those small differences (like having a friendly lab technician who talks to his animals) represent slightly different home ranges.

Biologists and wildlife managers often target the study of populations, not individual animals. Nonetheless, home ranges can not be ignored. Populations are an amalgam of local individuals who depend on their home ranges. Population sizes and distributions depend on abundances and distributions of foods and other resources, which are distributed differentially among individuals' home ranges. Study of animal populations and population dynamics, therefore, requires knowledge of the characteristics of home ranges within populations. Only with such knowledge can a biologist or manager predict how specific changes in environmental conditions will lead to specific responses by populations.

A worrisome reason to study home ranges is that human effects on landscapes can lead to poor choices by animals (Berger-Tal and Saltz 2019, Merkle et al. 2022). Animals know their home ranges. They know distributions of resources and how best to use those resources. Site fidelity is strong because the cost of having to find a new home range and having to learn all about a new home range has many dangers. Nonetheless, humans can cause animals' home ranges to decrease in quality sometimes in ways that the animals can not detect immediately. For example, humans may make a subtle environmental change, one not detected or understood by a resident, that affects predators' access to the site. Such a change could become an ecological trap, where prey animals have high mortality and high population turnover while not perceiving the cause. The animals' cues for what constitutes a good site have become uncoupled from the actual quality of the site (Robertson et al. 2013, Merkle et al. 2022). If human-caused changes do not appear to animals to be as consequential as they really are, and as they will be in the long run, animals may mistakenly delay abandoning their home ranges. Be'er Sheva fringe-fingered lizards in Israel's Negev desert, for example, continued to use areas despite higher rates of predation after the experimental addition of structures that facilitated predator activity (Hawlena et al. 2010). In these situations, biologists and wildlife managers will need to recognize problems that animals have not evolved to perceive and will need to take action. Merkle et al. (2022) proposed the term "fidelity-induced ecological trap" to specify site fidelity as a mechanism underlying an ecological trap.

1.5 Challenges for the Study of Home Ranges

Because home ranges are complex mental concepts, biologists can only gain ideas of how animals conceive their home ranges by doing research into home ranges. Developing hypotheses and designing research aimed at understanding how animals' minds work is difficult and requires imagination, creativity, and clear thinking. Why do animals do what they do in the environments of their home ranges?

Technology advances steadily, which definitely aids research. Animal locations, movements, foraging, predator avoidance, and reproduction can all be studied in greater and greater detail as technology advances, incorporating land- and satellite-based telemetry, remote DNA sampling, remotely triggered cameras, scat dogs, machine learning to identify topics for hypotheses, and more. To study home ranges, most biologists quantify location data for an individual study animal by building a digital map, generally called a "utilization distribution," that describes the animal's densities of locations across a landscape, usually including smoothing to account for location error. Thus, these utilization distributions are descriptive distributions. Assuming representative sampling, a "descriptive utilization distribution" can be considered to be a probability distribution for how the animal distributed time spent in different places during the period of study (Van Winkle 1975). Alternatively, it can be considered the probability of a biologist finding an animal in any given place at any random time.

Unfortunately, "utilization" is an inappropriate term for these distributions. First, we know that animals seldom utilize anything within their home ranges. To utilize means to use something with unconventional thought or in a novel or unconventional way. I know of no evidence for animals always to use resources in unconventional ways. Second, "utilization" can always be replaced by "use" with little to no loss of meaning, replacing 5 syllables with 1. And third, almost all the time, "utilization distributions" are built from location data or other data that show where animals have travelled. Being in a given place, however, does not provide any information about whether an animal "used" that place in any way other than to have been there.

In this book, I use "familiarity distribution" instead of "utilization distribution" for several reasons. First, we know that by being in a place an animal has become familiar with that place, at least to some degree. Second, we know that the animal perceives the environment around himself and, therefore, has some familiarity with that surrounding environment. The more times the animal visits a given place, the more familiar he becomes with that place and its nearby surroundings. The further away the surroundings, the less familiar the animal is with them. Which leads to: fourth, "familiarity" is appropriate for using smoothing techniques, such as kernel smoothing and Brownian bridge smoothing, to build distributions. Places farther from actual locations are less familiar. And, fifth, "familiarity" does not imply anything about "use" other than presence. Unfortunately, "familiarity" and "utilization" both have 5 syllables, though "familiarity" does seem to slip off the tongue a bit more smoothly.

Development of technologies for collecting location data and other behavioral data, and development of techniques to calculate familiarity distributions that describe familiarity with environments, have far outpaced theoretical explorations of how biological processes affect the distributions of animals' movements (Powell and Mitchell 2012). The paucity of conceptual underpinnings for home range behavior means that biologists generally build familiarity distributions to "estimate" "home ranges" with little understanding of what they are actually estimating, leading to myriad potential problems (Houle et al. 2011, Powell and Mitchell 2012). Having the technological cart before the conceptual horse has confused biologists, who try to understand the statistical descriptions of location or movement data while not understanding the biological basis for the behaviors that generated the data. If a biologist wants only to show where study animals have been located, he should simply show locations as dots on a map.

On the flip side, biologists who study behavior by direct observation (e.g., Bekoff and Wells 1981; Altmann1998; Byers and Kitchen 1988, Byers 2003) seldom explore how behaviors build home ranges. The consequence is that analyzing ever more location data with ever better statistical procedures is misguided without a better understanding of what, exactly, needs to be estimated. I find this situation similar to giving more sophisticated tools to the blind men describing the elephant.

Contemporary approaches for building familiarity distributions are descriptive because they describe where animals have been located. A great pitfall of using models in science is to confuse a model with the phenomenon that the model was built to represent (Petroski 2023). Familiarity distributions are not home ranges and are not descriptions of how animals view their environments, nor are they even complete descriptions of the places with which study animals are familiar. Familiarity distributions are usually presented as contours on a map of a landscape, describing the sum of an animal's movements over a sampling period. For an animal with a stable home range over a long enough time period, a familiarity distribution will estimate how the animal distributed her time across the landscape to meet 100% of her requirements. Changes in food availability, the use of the landscape by competitors, and the arrival of a new predator or food may cause the animal to change her use of the landscape. If the time spent in different places rises or falls enough over time, the topographies of successive familiarity distributions change and the outlines may expand or contract, quantifying how the animal changed her use of her environment to meet her requirements. The consequence is that, to understand the boundaries and area of a familiarity distribution, one must understand what an animal does within the boundaries. Don Hayne noted this point in 1949, as did Aldo Leopold.

Biologists acknowledge that the lives of animals are far, far more fascinating and complex than what data and analyses can illuminate. Biologists who try to use informed imagination will understand better what living another animal's life is like (Yamada and Besom 2013). Like Nagel (1974: 439), "I want to know what it is like for a bat to be a bat." Nagel went on to acknowledge "Yet if I try to imagine this, I am restricted to the resources of my own mind, and those resources are inadequate to the task." Nevertheless, biologists can use good research design and the best tools available to strive toward understanding how animals think. What is it like to be a red-backed vole moving through sub-nivean tunnels? Or a fisher attacking a porcupine? Or a grey jay knowing where he has cached hundreds of food packages? Or a fledgling peregrine learning to land on a branch?

Biologists can gain insights from unexpected situations. When my wife was working on a children's book about the development of a brood of peregrine falcons, she spent hours watching live, remote videos of peregrine chicks. That summer she also struggled to learn how to maneuver a dinghy with a small outboard motor up to a dock so that the boat snuggled smoothly to the dock facing out into the lake. As she watched a just-fledged peregrine struggle to land on a branch she realized that both landing on a branch and landing at the dock require complex maneuvers to reach just the right place where the bird or motor must stall, allowing a safe landing (KleinHeerenbrink et al. 2022).

Biologists may never find completely objective methods for field research and analyses that yield unbiased, biologically significant information about animals' home ranges (Powell 1987). Nonetheless, my goal as a biologist is to make those methods and analyses less inadequate by suggesting ways to consider home ranges and to gain insights into animals' lives. In addition, with imagination and creativity, subjective observations can suggest outstanding hypotheses for testing.

A goal most biologists have for studying home ranges is to define boundaries, Don Hayne's (1949) warning notwithstanding. It is true that even crudely estimated areas for home ranges provide some insights and that some animals do maintain home ranges with distinct boundaries: especially animals with non-overlapping home ranges and animals that use physical features of the landscape to dictate boundaries (Reid and Weatherhead 1988, Powell and Mitchell 1998). Extensive overlap of boundaries might suggest that animals do not experience limiting resources. If, however, use of specific resource patches within the boundaries shows little overlap, then the familiarity distributions may have been calculated on the wrong scale.

In contrast, however, many animals' home ranges may have no identifiable boundaries that biologists can discern. Animals with non-exclusive home ranges are likely to spend the majority of their lives far from the peripheries of their home ranges, making the peripheries of most home ranges diffuse and, thus, mathematically undefined (Gautestad and Mysterud 1993, 1995). Ambiguous boundaries do not reduce the importance of a home range to its animal. And, just as an island with a fractal, infinite shoreline has no mathematically defined area, that island still has a finite limit to its area. Therefore, areas of animals' home ranges can be estimated even if boundaries are diffuse.

Nonetheless, to define boundaries, biologists must estimate some radius of familiarity around each location point. This radius might be quantified by how far an animal can see, smell, hear, or otherwise have familiarity with its environment. With no information on sensory perception, location error is a passable fall-back measure. If biologists do not estimate some radius of familiarity, a familiarity distribution is reduced, *reductio absurdum,* to the places where an animal actually placed his feet or to his flight path. Foot placement or flight path are reasonable units for analyses of movements but are on the wrong scale for analyses of the extent of familiarity that an animal has with his environment. Thus, the boundaries that biologists calculate may mean absolutely nothing to animals themselves and, if so, area (size) may be an inappropriate statistic for those animals' familiarity distributions.

Despite these myriad problems with home range area, Chapter 12 presents some approaches to estimating the area within an animal's home range that is needed to meet his requirements.

1.6 The Scientific Method and Research Design

Aldo Leopold noted: "Like people, my animals [the animals on his farm] frequently disclose by their actions what they decline to divulge in words." Leopold was an advocate of observation and of well-designed research.

More so than with anything else, productive research starts with excitement, curiosity, imagination, and creativity. I am in awe of the richness, diversity, and complexity of life on this earth. Being able to do research on beautiful, wonderful animals has been a reward in and of itself. Having the research be productive so that I learn new "stuff" about animals is an even greater reward[5]. Designing and doing productive research leads to more and more new knowledge about the subjects of my research, stimulating my curiosity to learn more. Imagination and creativity stemming from what I and all the other researchers have learned leads to my next step with research.

Productive research requires good design based on relevant, non-trivial, testable a priori hypotheses (Popper 1934/1959, Hayne 1978, Hurlbert 1984, Reynolds-Hogland and Mitchell 2007a, Epling and Powers 2022, Skipper 2024a) that, in turn, are based on the synthesis of observations, previous research, and relevant theory. Good research seeks to learn *why* animals do what they do and follows the scientific method (Figure 1.3).

The conceptual bulk of research design is done before setting foot on a research field site. Observations direct the development of hypotheses. Hypotheses dictate data that must be collected to test the hypotheses. Field research must be planned to collect the data needed to test the research hypotheses. Hypotheses and field data direct what analyses are needed to test the hypotheses. Consequently, good research design includes, up front, all aspects of the research up to considering the results, which can not be known before field research and analyses are completed.

Observations—The scientific method begins with observations that are not readily explainable by the known biology of organisms of interest and that otherwise raise questions. Here is curiosity. Observations include direct observations of wild or captive animals, the results of theoretical modelling, and the results of previous research. Observations constitute the foundation for all the steps of the scientific method that follow.

Hypotheses—Hypotheses are stimulated by the questions raised by observations. Curiosity, imagination, and creativity lead to the best hypotheses. Though hypotheses are exciting, they must be stated bluntly so that they can be proven to be false if they are false; for example: "Members of the larger sex of a species characterized by large sexual size dimorphism have home ranges that incorporate proportionally more food resources than do the home ranges of members of the smaller sex." Hypotheses do not have weasel words, such as "suggest," "should be," "might be," because those weasel words prevent clear rejection of a hypothesis. Biologists do not need to believe, or have a hunch, that their hypotheses are true. Often, clarity and ease of rejection depend on wording and the best wording for a hypothesis might be what a biologist figures is probably false but that can be rejected cleanly. Having favorite hypotheses and trying to "prove" them true is not good science. Developing hypotheses after data are collected can lead to false acceptance (Epling and Powers 2022). Hypotheses stated in the negative (". . . have home ranges that do not incorporate more food . . .") can be harder to reject than those stated in the positive.

Biological hypotheses are relevant to why animals behave as they do and are based on biological knowledge (Brosi and Biber 2009). When a priori information is robust, biologists should generate biological null and alternate hypotheses to guide the development of research design. The hypothesis about sexual size dimorphism used as an example in the previous paragraph is a biological null

[5] Most of my research has required live-trapping. Handling animals, inspecting and feeling each one closely, and smelling them to evaluate their health are all intrusive but have allowed me to know the animals personally, in a way, despite most having been drugged. From the animals' perspectives, each had a negative experience, potentially terrifying for the carnivorans who were constrained and had their personal spaces invaded when drugged, and for the small mammals. Although some small rodents and weasels became trap-happy after their initial captures, the fishers' and black bears' experiences left them deathly afraid of me. I had strong emotional ties to the animals that were not reciprocated. That result is not rewarding.

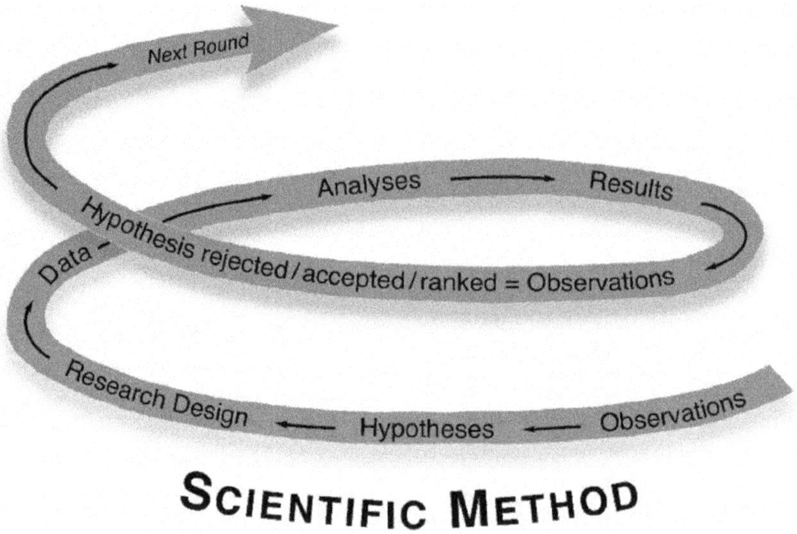

Figure 1.3 The scientific method envisioned as a spiral starting with observations that lead to hypotheses, which require good research design to test; data collected and analyzed lead to results, which are the observations for the next loop around the spiral.

hypothesis. It states what is predicted by the present state of knowledge about sexual size dimorphic species. Even though the statistical null hypothesis of "no difference" is standard for most statistical software, the statistical null is not the appropriate null hypothesis when a difference is expected to exist.

Because a home range is an emergent property of movements, hypotheses related to home ranges may differ from hypotheses related strictly to movements (contra White and Garrott 1990). Home range hypotheses should be used to understand how individuals establish their home ranges on a landscape with respect to the predicted distributions of resources and neighbors, how individuals might evaluate landscapes, and other questions that extend beyond simply the composite of all movements. Home range hypotheses include how animals use information about their home ranges to plan future movements. Hypotheses related to movements and individual locations should address (individually or in combinations) travel, short-term responses of individuals to competitors and predators, short-term changes in specific resources, and other topics that concern locations related to specific movements. Being hierarchically related, home ranges and movements lend themselves to using hierarchy theory to investigate how biological properties emerge across different spatial and temporal scales (Reynolds-Hogland and Mitchell 2007a).

Data Collection—The a priori hypotheses, and the data required to test the hypotheses, both the type of data and sample sizes, dictate the data to be collected. Being enamored with new technology need not lead to collecting the most pertinent data. Much research on animals is now done remotely, using some system of telemetry, cameras, remote DNA, or combinations of these and other methods. Direct observations, however, should be used whenever possible. Direct observations allow continuous data collection, follow study animals within the context of their landscapes and the context of other animals, allow multiple types of data to be collected simultaneously, allow observations of behaviors not under study, allow study of cooperative, dominance, and other within-group behaviors, and lend themselves to serendipity. Tracking animals in snow shows complete movement paths including behaviors related to very small-scale aspects of habitat and other animals and lacks only precise timing of behaviors, which can usually be obtained in other ways. Drones with video may explode the possibilities of direct observations.

Often, however, direct observations limit data collection to 1 or a few individuals at once, while telemetry allows data collection on many animals at once. Choice of a telemetry system must be matched to research questions and to the hypotheses being tested. For example, using relatively inexpensive very high frequency (VHF) transmitters on many animals is usually appropriate for answering population-level questions, while using more expensive Global Positioning System (GPS) collars with accelerometers and using collars with video cameras on a few animals are able to provide better data to answer questions about the behaviors of individuals. Similarly, environmental DNA data are most appropriate for analyses of a species presence or absence while hair snares, which capture hair for individual identification, provide data more appropriate for individual analyses. As expensive technologies become cheaper and miniaturized, new trade-offs will undoubtedly develop and old ones disappear.

Hypotheses related to movements and home ranges require information about the landscapes on which the animals move and establish home ranges. These hypotheses relate, for example, to such topics as how far and when an animal moves, how the local habitat quality affects timing of movements, whether food densities cause the animal to linger or to continue travel, whether local habitat features facilitate foraging, and whether a local competitor depresses food availability (as defined by Charnov et al. 1976). Telemetry systems that can download data on weather and local environments will facilitate having critical data for testing hypotheses (Kays et al. 2015). In addition, incorporating machine learning into telemetry capabilities and into analyzing field data has the potential to provide more insight into behavior of study animals (Rutz et al. 2023).

Data Analyses—Statistical analyses used should be those able best to reject hypotheses directly or to rank hypotheses (Wasserstein et al. 2019). Sober's (2008) review of the major statistical approaches is probably still the best overview. Biological hypotheses are appropriate for traditional parametric and non-parametric statistics, Bayesian, likelihood, and information-theoretic approaches. Use biological nulls for statistical tests (Brosi and Biber 2009). Using biological null hypotheses often results in exchanging Type II for Type I error with the statistical null hypothesis; alternately one can use calculations of power, B, for Type I tests instead of using the chosen p limit.

Thus, if the best biological information available predicts a particular outcome, then that outcome should be the null hypothesis for a statistical test because it is the expected outcome, even if it is not the outcome associated with "no effect." This approach is consistent with "Morgan's Canon," even though many biologists are under the mistaken impression that Morgan's Canon dictates use of statistical null hypotheses. Morgan actually argued against highly anthropomorphized interpretations of animal behavior and not against appropriate, biological interpretations and biological null hypotheses (Morgan 1894, Kimler 2000).

Results—The results of statistical analyses provide observations that generate new hypotheses (Figure 1.3). Within the breadth possible given the data, the results of research provide new information and the potential for new interpretations. Hypotheses can be rejected or temporarily accepted.

Finally, good research design allows serendipity and a good biologist is always alert to the unexpected and the unknown (Foresman 2012, Catania 2020). The most exciting words in science are not "These results are just what I predicted" but, instead, "That's weird!"

Consider an example. Large sexual dimorphism in body size is the norm for most mammal species that have a long evolutionary history of males competing for mates and not contributing to the raising of young. The home ranges of such males have appropriately more food resources and, therefore, larger outlines than do the home ranges of smaller females. Such sexual dimorphism in body size and home ranges is pronounced for many species within the mammalian family Mustelidae. Were a new mustelid discovered this year and, at first, only females live-trapped or documented, one would hypothesize that the yet-to-be-discovered males would be larger than the females and have larger familiarity distributions. That would be the biological null hypothesis for statistical tests because large sexual dimorphism is expected.

Now, imagine that a few adult males are collected and imagine that they are *not* larger than females. If the appropriate statistical analysis rejects the biological null hypothesis of large sexual dimorphism, then myriad, exciting hypotheses come to mind regarding why, such as: Sexual selection on males is not large for this species, or The habitat is so unproductive that monogamy and long-term pair bonds are required to raise young. In contrast, what would the result be for this scenario using the encrusted, traditional statistics and the null hypothesis of "no effect"? The result would be failure to reject the statistical null hypothesis of "no effect." That result has 2 potentially dire problems. First, accepting the statistical null hypothesis is generally not exciting and could cause one to miss the excitement of having found a mustelid without sexual dimorphism, and all the associated hypotheses to test. Second, had the result been caused by a small sample size and insufficient power, then one might believe incorrectly that the new species is not sexually dimorphic when, actually, no one yet knows.

1.7 Models in Biology

1.7.1 Types of Models

This book deals extensively with models. The term "model" in biology refers to a representation of a natural phenomenon or system (Lincoln et al. 1982). The best model describes a given situation precisely but without complication that causes one to lose intuition (Aspect 2024). Models come in different flavors, each able to contribute to a research meal in different ways. An organismal model is used as a representative or a stand-in for other organisms that are not as easy to study. Lab rats, lab mice, zebra finches, African clawed frogs, and *Drosphila* are classic examples, though many other species have been used as models and what constitutes a good model organism has been debated (e.g. Bolker 2014).

A descriptive model is built to match a data set and describes past actions and conditions. Descriptive models can be equations, graphs, verbal descriptions, or other descriptions. They can not predict anything, but 2 or more descriptive models can be compared to learn which describes a data set better. Descriptive models are often used as the observations to start the scientific method.

A mechanistic model describes the effects of particular rules of behavior (Adams 2001) and is used to build a mechanistic familiarity distribution, or mechanistic distributions (Chapter 6). These models describe hypothesized or established patterns of behavior and ecological conditions and are used to test hypotheses related to what behaviors or ecological conditions are most important to target animals.

A statistical model states a hypothesis to be tested or ranked in the specific format of a specific statistical test.

A theoretical model (different variations of which can be called conceptual, functional, mathematical, mechanistic, etc.) provides an explanation, or hypothesized explanation, for how or why something works. These models are built on known or assumed or hypothesized biology. They facilitate logical extensions from known relationships, which then lead to new hypotheses. The best theoretical models provide insight into *why* organisms do what they do and have the potential to produce unanticipated results. Theoretical models make biologists specify assumptions and conditions. If hypotheses from theoretical models are rejected, the theoretical models have highlighted what biologists do not know but thought that they did. Theoretical models come in a continuum of abstraction and clarity of logic that are sometimes split into 3 levels:

Verbal theoretical model: based on logic that is only as clear as the wording used to present the model;

Graphical theoretical model: based on logic made clearer than for a verbal model and better defined by showing relationships graphically;

Mathematical theoretical model: based on logic that is clear and well-defined, using equations to present relationships clearly and to make assumptions clear.

1.7.2 Using Models

All models are wrong but some can be very useful (Box 1976: 792).[6] They are especially useful for

[6] Although George E. P. Box did state that "All models are wrong but some are useful" and he did write "all models are wrong" in his 1976 publication, the concept predates him by decades, if not centuries. Even Picasso (1923) made a similar statement about art.

generating hypotheses to test. If animals behave as predicted by models, then the general understanding of the constraints and assumptions and the mathematical shortcuts built into the models are reasonably accurate. This is a good result because it allows biologists to explore the biology of the animals further, establishing new research on the newly established understanding of the target animals' biology. When animals do *not* behave as predicted by models, this is a good result. Biologists now know that the general understanding of the biology of those animals, or the mathematical shortcuts taken, are not accurate and biologists can not build on the now-known-to-be-false understanding, but must go back to re-examine what they thought they knew.

Models can be judged on many criteria, such as generality, realism, precision, flexibility, and adaptability. Looking at generality, realism, and precision, Levins (1966) proved that no model can maximize all 3. A model can, at best, maximize 2. General and realistic make a good combination. Likewise, precise and realistic can be good, though the combination must be narrowly defined and narrowly applied. One can, however, string many precise and realistic models together by systematically varying parameter values, thereby gaining some idea of generality. This is the logic behind sensitivity and elasticity analyses. By definition, a general model simply can not be precise and no one wants a model that might be unrealistic.

Compared to the physical sciences, biology can appear quite "sloppy." Even when environmental and habitat information is complete, individual animals of the same species differ in how they behave in a given environment and habitat. And, a given individual animal has different needs at different times. Nutritional requirements, for example, change often with the latest meal, with available foods, with energy expenditure since the last meal, with needs for different nutrients at different times of day, and with the last time sleeping. Thus, biology is not really sloppy as much as it is hampered by lots of unknowns, including unknown unknowns. Nonetheless, exploring possibilities with theoretical models allows one to play "what if?" games and to explore and to compare biological systems that have different rules. I am the last person who would discourage biologists from building and playing with theoretical models to learn what hypotheses the models might suggest. Modellers must remember, however, that theoretical models can easily exceed biological reality and precision and can relate to biological unknowns. Modellers and field biologists must maintain communication. When I was building models for energy expenditure by animals (part of my PhD research), Monte Lloyd noted that any model with an exponent can be made to do whatever one wants, whether that matches biology or not. Models with multiple exponents and other complexity, require substantial justification. Consequently, applying Ockham's Razor, my approach is to use the simplest models that capture the essence of the biology I wish to understand. I favor general and realistic theoretical models.

For models that deal with behavior, both learning in the short term and evolution of behavior in the long term, model builders often assume that if a particular behavior is the optimal or evolutionarily stable behavior in a model then that behavior can evolve (Fawcett et al. 2013, McNamara and Leimar 2020). Evolved psychological mechanisms, however, can prevent animals from learning some behaviors, while competing, adaptive constraints can prevent some behaviors from evolving. Thus, not all modelled behaviors can evolve. In addition, modelling behaviors without the context of animals' whole lives can generate unrepresentative hypotheses (Houston and McNamara 2005). When behavioral hypotheses generated by models are rejected, new biology can be learned by exploring potential constraints on behavior.

Models are misused or misunderstood when they are stated to describe what animals *must* do. When optimal foraging models were originally filling the ecological literature, the models were lambasted in the North American wildlife literature because such models did not describe accurately what all wildlife do (using the US definition of "wildlife" as animals, especially vertebrates, that are managed for hunting, conservation, preservation, or observation). Optimal foraging models, however, are not meant to predict how all animals behave. Think of optimal foraging models as null hypotheses for how animals will behave given the known biological constraints and mathematical assumptions built into the models.

Similarly, Steve Hubbell's (2001) "unified neutral theory of biodiversity and biogeography" was criticized harshly because his theory did not describe all patterns of biodiversity and biogeography, despite Hubbell's using "neutral theory" in the title of his book. His neutral theory was a null hypothesis for biodiversity and biogeography absent natural selection and, thus, allowed biologists to identify when selection was consequential.

Unfortunately, many modellers *do* state that their models predict what animals *should* do, using the same sort of shorthand that most evolutionary biologists use when stating that "such-and-such evolved to do so-and-so." Evolutionary biologists know that natural selection is not directed, just as modellers know that models are not all-knowing. Models are best discussed without such shorthand (as is evolution).

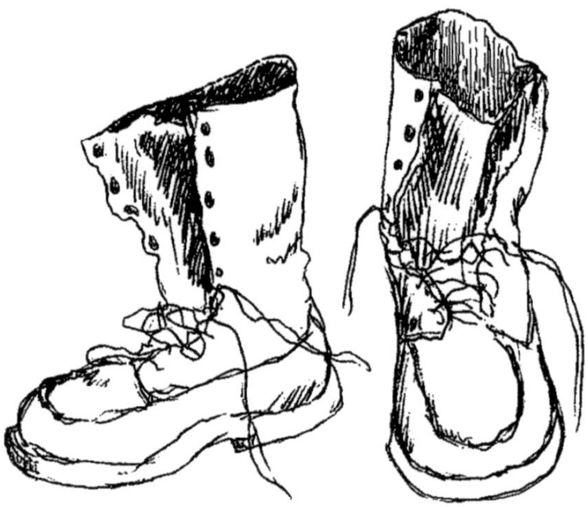

1.8 Some Definitions and Background

In the vein of clarity from that last sentence, I present here some definitions and some of my preferences that have been built into this book.

Abbreviations—I have avoided abbreviations. Asking readers to remember abbreviations is not necessarily a big request, I know, but I avoid abbreviations for 2 reasons. First, a reader who does not read this entire book but picks it up just to read, say, a single chapter that has been recommended, should not have to search through the book to find where an abbreviation was defined. Even a definition supplied earlier in a chapter can be hard to find when searching back for it. Even an appendix with definitions of abbreviations can be hard to find, especially for a quick reader who does not realize that the appendix exists.

Second, readers who have just read something else that required them to remember abbreviations, perhaps some with the same abbreviations I would have used, might easily forget my abbreviations. I once reviewed a manuscript that asked readers to remember abbreviations for 3 types of forest, one of which was mixed deciduous forest, abbreviated as MDF. Before reading that manuscript, however, I had been reading a woodworking magazine and all I could think when I read "MDF" was "medium-density fiberboard." I do not want readers to be confused like that. Spelling out whole names is worth the space used to be clear.

Anthropomorphism—Gaining ideas for hypotheses requires trying to think like animals and using some knowledge about humans that can be applicable to other animals. This is not anthropomorphizing. Humans have many characteristics in common with other animals, especially with other mammals and with birds, which have complex lives and brains (e.g. Griffin 1976, Bekoff 2004, Bastos et al. 2024, Andrews et al. 2025). Of course, humans differ from other animals in many ways, yet biologists should avoid anthropectomy, which is baselessly assuming other animals do *not* share certain characteristics with humans. Using pertinent knowledge of human biology allows biologists to develop good hypotheses to test. The results of the tests will show whether assumptions were appropriate. Using human biology in this way is consistent with Morgan's Cannon (Morgan 1894, Kimler 2000) and was promoted as long ago as by Theophrastus in *Historia Plantarum*.

Home Range: Familiarity and Mechanistic Distributions—When I use "home range" I mean it as from an animal's perspective. I use "familiarity distribution" to refer to the maps that biologists calculate from distributions of locations or similar data. Familiarity distributions are descriptive, though a familiarity distribution can be considered to be a probability distribution describing how familiar a study animal was with its surroundings or a probability distribution for how the animal distributed time spent in different places during a period of study.

A "mechanistic utilization distribution" is a map of hypothesized use of an environment based on biological knowledge, usually knowledge that has been modelled. Because mechanistic distributions represent hypothesized *use* of resources, using the term "utilization" is acceptable, in contrast to familiarity distributions: mechanistic utilization distributions are not based on animal location data. Mechanistic utilization distributions can be tested to learn whether animals move as hypothesized. Given that "utilization distributions" is an awkward phrase without adding "mechanistic," I use "mechanistic distribution" for the rest of the book.

Home Range: Fundamental and Realized—The concepts of fundamental and realized home ranges parallel the concepts of fundamental and realized niches (Hutchinson 1957). The fundamental niche is the diversity of resources that members of a species or population require or can use and the diversity of environmental conditions that they can tolerate; the actual diversity of resources and the environment used by each individual constitutes the realized niche (Hutchinson 1957, Chase and Liebold 2003, Carscadden et al. 2020). As species and populations evolve, they develop their fundamental niches.

Likewise, as an individual animal matures and leads her (or his) life, she develops her fundamental home range, which consists of the areas of land with which she has become familiar and that she maintains updated or usable in her cognitive map (Tolman 1948, O'Keefe and Nadel 1978, Eichenbaum 2017). An animal's realized home range is that part of her fundamental home range that she uses during a particular period of time. Given that an animal's realized home range can not be larger than her fundamental home range, the realized home range will be well bounded by the animal's movements. Those movements, however, would be the result of the particular conditions and of the animal's condition on the days when she made her most extreme movements while being studied. The animal knows the area beyond these arbitrary movements of those given days and, thus, the particular boundary of a realized home range has no meaning without knowing the state of the individual and the state of her resources.

Most research, even relatively long-term research, quantifies realized home ranges. In this book, discussion related to data sets and to familiarity distributions emphasizes realized home ranges. Discussion of home ranges as cognitive maps and of an animal's knowledge of her environment emphasizes fundamental home ranges.

Hypotheses, Rejecting and Accepting—My preference is not to state that a hypothesis can be accepted: I prefer accepted "temporarily" because future research may show that a hypothesis has been accepted because a key component was not incorporated into the research, causing a hypothesis that should have been rejected to be accepted. I also understand that many biologists understand that hypotheses can not be accepted unequivocally but find "accepted temporarily" to be awkward or unnecessarily wordy. I can accept this disagreement but prefer the accurate wordiness.

Patch—From an ecological perspective, a patch is an area on a landscape that has some characteristics important to a specific animal and these characteristics distinguish the patch from the surrounding area (Wiens 1995). Generally, this is how I use "patch." This conceptually clear definition can be a challenge to apply in the field, where "patches" contain multiple resources that vary in local, individual distributions and that often lack clear edges (Mitchell and Powell 2002). Most of the time in this book, patches and their resources are generally clear. For simulation models, however, one can define patches as cells in a grid, each characterized as having a resource value to the model animals. Where contiguous cells have similar resource values, the result is an ecological patch, thereby retaining the biological function in the models. I have made clear when I use "patch" to mean a grid cell for a model.

Resource—I define a resource as an item or a set of items or conditions at a site on a landscape that affects an animal's fitness: mates, food, attributes of the environment that facilitate availability of food, shelter, travel routes, escape routes, and also attributes of the environment that facilitate vigilance and avoidance of danger and that reduce wasted time, if that time has a cost. Consequently, I define other members of a social group to be resources if they facilitate foraging for a target individual, or are vigilant and warn other group members, or are part of a selfish herd (Hamilton 1971). Resources are diverse, often complex, and very often do not correlate well at all with human-derived vegetation "types" (of the sort regularly used to study "habitat" selection).

Space, Environments, and Landscapes—Heft (2013) noted that the term "space" is promiscuously and often misleadingly used to mean an "environment." Conceptually, "space" is a term for an abstraction: 3 associated dimensions not tied to any particular real place. Using the term "space" often leads to ignoring critical qualities of environments that are important for living organisms (Gibson 1979). That is, by using "space" in expressions such as "space use" or "use of space," we fool ourselves into thinking (often subconsciously) that real animals are disjunct from the very real environments in which they live.

Thus, "space" and "environment" are not synonymous (Heft 2013). Cartesian space is empty except for defined shapes, while environments are filled with sundry living organisms and physical features important to animal residents. Unlike objective shapes in space conceived in isolation, items in an environment are perceived by animals in relation to other items in the environment. Cartesian space is limitless but environments are limited by animals' abilities to perceive and to remember. Objects in space are static and do not change, while other animals in an environment can move. Even those organisms and physical features that stay in place (plants, streams, etc.) do not remain static. Plants grow and die, physical features erode, water flows and melts and freezes. In general, only 1 space exists but our world is filled with multitudinous environments and, because different organisms perceive their environments differently, a single place on earth constitutes different environments to different animals and plants.

To keep animals and their home ranges within the context of their environments, I avoid the terms "space use" and "use of space" in this book. I refer to "use of environments," "use of landscapes" or "environmental use." I do, however, use "space" as a verb to refer to how animals space themselves on a landscape.

My Background—Since 1970, I have tracked fishers and martens and weasels in snow, have live-trapped mammals of dozens of species, have learned diverse technologies, and trained red-tailed hawks to attack fake weasels. I have also held a wild fisher by his tail as he attempted to run away, had a weasel urinate on the top of my head, and from less than 3 m watched a female black bear nurse her cubs in an open den under a dense canopy of mountain laurel near the top of Little Pisgah Mountain. Consequently, I have gained some insights into animals' daily beats and *why* they have them. Yet, still, my insights seem trivial compared to the vast diversity of living organisms and the insights biologists can gain from them.

I admit to being an unapologetic, whole-hearted biologist—a field biologist. I have been called a quantitative biologist but I am certainly not a statistician. Animal behavior, animal ecology, and evolution are the subjects that drive my curiosity. Statistics are diagnostic tools that provide insights into behavior, ecology, and evolution but other tools often provide more consequential insights, tools such as direct observation, snow tracking, remote cameras, live-trapping, telemetry, measuring metabolic rates (in lab and field), habitat analyses, hand raising orphaned mammals, training captive wild mammals and birds for experiments, DNA analyses, models, and taking a deep smell within the fur of every mammal I have handled. Healthy animals smell wonderful. I use whatever tools appear appropriate for resolving "why" questions related to behavior, ecology, and the evolution of animals.

1.9 Organization of This Book—Chapter Summaries

Chapter 1 An Introduction to Home Ranges—This chapter introduces the concept of a home range, briefly describes some of the diversity of home ranges, discusses why animals maintain home ranges, and presents a brief history of the origin of the concept, including Burt's (1943) definition of a home range. Reasons for studying home ranges include learning how animals share or compete for resources, learning how animals use habitats related to resources, and applying such information for conservation and wildlife management. Challenges for studying home ranges include lack of understanding of what familiarity distributions describe, limits for what biologists can learn about home ranges from location data, and what, if anything, biologists can learn from studying home range areas. Research design is critical for testing hypotheses and using models is a good way to generate hypotheses to guide research.

Chapter 2 Territories—A territory as an area occupied mostly exclusively by an animal or a group of animals by means of repulsion of conspecifics through overt defense or advertisement or by means of mutual avoidance. An animal or a group of animals maintains a territory when a resource is in short supply and the territory may be all, or just a part, of the animal's entire home range. Territories may be maintained year-round or only for shorter periods. Territory size tends to vary positively with body size and negatively with productivity of the limiting resource but territory size can be difficult to measure, depending on characteristics of boundaries. Maintaining territories is a rare example of non-human animals maintaining private property.

Chapter 3 Cognitive Maps—Building on the first 2 chapters, this chapter explores how animals perceive their home ranges. The hippocampus and associated brain structures organize (map) relationships among memories, including memories of places, plans of future movements, planned use of memories, and contributions to an expansive suite of cognitive operations that stretch the concept of memory. This organization is an animal's cognitive map leading to my definition of a home range: "That part of an animal's cognitive map that the animal keeps up-to-date; includes places not visited recently but that are held in memory for future, specific use." All complex animals have brain areas equivalent to the mammalian hippocampus that are key to problem solving, to planning, to memory, as well as to wayfinding. Animals use visual thinking, auditory thinking, olfactory thinking, and thinking in other senses. Thus, understanding animal's cognitive maps is to understand their home ranges. Burt's definition of a home range does not address an animal's perspective of its home range and, therefore, is inadequate.

Chapter 4 Movement—Animals' home ranges, their cognitive maps, are built initially from experiences gained by moving across landscapes. I define movement terms, describe associated movements, and present ways to quantify movements. Kriging, Brownian bridges, and other approaches can be used to fill gaps in movement (location) data, but can only be filled when successive locations are autocorrelated. Identifying dispersal moves is important for much research but 2 methods designed to identify dispersal have complications.

Chapter 5 Information, Movements, and Home Ranges—The optimal time for sequential foraging bouts within a resource patch balance food renewal rates with the loss of information about food renewal caused by other animals foraging in that patch. Models of linear (1-dimensional) and 2-dimensional landscapes predict that animals will sometimes maintain, without aggression, non-overlapping home ranges so as to maximize information about the status of resource patches. Long-term data on black bears and fishers demonstrate that predictions developed in 1 dimension apply on 2-dimensional landscapes.

Chapter 6 Mechanistic Familiarity Distributions —The term "mechanistic" is used in this chapter to refer to mechanisms hypothesized to be important to the fitness of study animals, such as foraging behaviors, predator avoidance, and interactions with conspecifics. Models of the mechanisms are used to build mechanistic distributions that can be tested for their abilities to predict animals' movements, thereby testing biologists' understandings of the ecology, behavior, and other aspects of the biology of study animals. The chapter introduces how to build testable, or predictive, familiarity distributions, or mechanistic distributions, and presents examples, beginning with simple, mechanistic distributions and progressing to more complicated mechanistic distributions used to test multiple hypotheses. The mechanisms used to build mechanistic distributions can incorporate movements governed by information (Chapter 5) and behaviors, as modelled in Chapters 7–9. I build a mechanistic distribution incorporating 6 behavior models for black bears and test the mechanistic distributions with data for bears in the Blue Ridge Mountains of North Carolina, USA. The results highlight individual differences among the bears.

Chapter 7 Economic and Cost-Benefit Models and Home Ranges—Cost–benefit models of territoriality are based on the foundation that benefits must exceed the costs of maintaining a territory.

Benefits and costs are modelled mostly using daily energy expended and energy gained from food. Initial tests used animals that defended territories using vocalizations, visual displays, chasing when necessary, and occasionally fighting. When built to incorporate habitat patchiness, the models predict that a condition can appear for which shrinking of a home range can not be allowed because that would cause loss of patches needed during some times of year. When this condition exists, home range overlap occurs. I apply one of the early but still most useful models to my long-term data sets on black bears in the Blue Ridge Mountains, highlighting population differences among bears.

Chapter 8 Foraging Optimally for a Home Range—Optimal foraging models generate the hypothesis that animals "forage," or explore, to choose resource patches optimally for inclusion in their home ranges. Linear programming can predict the optimal proportions in a home range of patches with different resources. Resource-maximizing models predict that resource productivity has little effect on home range area while area-minimizing models predict that home range areas change with productivity. Black bears in North Carolina, USA, maintain area-minimizing home ranges, as do wolf packs in Montana, USA. The chapter ends with an overview of applications of optimal foraging models applied to understanding establishment and maintenance of home ranges.

Chapter 9 Game Theory and Home Ranges—This chapter introduces game theory, Nash equilibria, and evolutionarily stable strategies. Most game theoretic approaches to home ranges model territories. Territory owners are predicted to cooperate to minimize altercations. Nonetheless, conditions exist that can cause territory invasion. Game theoretic models predict the conditions for animals to adopt landmarks as territory boundaries.

Chapter 10 Familiarity Distributions: Infrequently Asked Questions—I review the history of quantifying home ranges and then discuss several concepts that need to be addressed by a biologist before attempting to build familiarity distributions: location and other data error, autocorrelation, what location data do and do not represent, what familiarity distributions actually show, details about smoothing data, whether some data should be excluded before calculating a familiarity distribution, and concepts of the area of a familiarity distribution. These topics are critical for how conservation and wildlife management actions can use animal location data.

Chapter 11 Building Familiarity Distributions—I present and compare the most common statistical approaches for building familiarity distributions, starting with minimum convex polygons and location counts on a grid. Then I cover kernel, bridge and local convex hull estimators, their strengths, and how to deal with some of their weaknesses.

Chapter 12 Familiarity Distributions: Overlap, Cores, Activity and Sensory Kernels, Areas—Overlap of familiarity distributions can be quantified for both static and dynamic overlap. Cores of familiarity distributions are used disproportionately heavily and preferentially and must be determined by animals themselves, otherwise cores can be calculated for animals that move randomly. I present 2 objective ways to identify what animals consider to be their cores. Understanding animals' perceptions of their environments can be incorporated into familiarity distributions. Examples include varying kernel band widths; weighting locations using kernels that differ for visual, smell, and hearing abilities; shaping kernels to fit wind direction and speed and to fit topography. Combining evenness of familiarity distributions with area provides a better understanding of the amount of a landscape that an animal actually uses. Finally, I present research I have done on my own home range and discuss insights from that research.

Chapter 13 A Doorway to New Cogitations on Home Ranges—This last chapter starts with a review of my goals for this book, especially to convince readers that thinking of home ranges as animals' cognitive maps can lead to research that increases understanding of animals' ecology, behavior, and evolution. The bulk of the book has dealt with complementary approaches for developing research hypotheses that, when tested, lead to productive research on animals' home ranges. Machine learning has much potential to contribute to understanding of animal behavior and ecology but also has important constraints. Final comments review the importance of home ranges and cognitive maps.

CHAPTER 2

Territories

Marcel had heard that an Egyptian nightjar had been sighted in far southern Israel, toward the Gulf of Aden, where we had planned to explore some wadis over the weekend. So, we added finding the nightjar to our plan for one evening. At an unpaved road, we left our rental car and hiked through a date palm orchard to a clearing about the size of a football field and covered with coarse grasses and herbaceous vegetation. As dusk settled, we walked slowly around the clearing, pausing, listening, and

Home Ranges of Animals. Roger A. Powell, Oxford University Press. © Roger A. Powell (2025). DOI: 10.1093/9780191949142.003.0002

watching. No nightjar. We found ourselves at the far side of the clearing when darkness had really settled. No nightjar. So we stood and listened and absorbed the early night. From the center of the clearing, no more than 25 m away, brazen, raucous howls broke from a family of jackals. They overwhelmed the night with their glorious advertisement of their territory.

2.1 Definition

Territories constitute a special subset of home ranges. Territorial behavior by animals was originally conceived as an extension of the concept of territorial defense by humans (Howard 1907–1914, 1920, Burt 1943) and by the 1930s a territory was defined as "any defended area" (Noble 1939). The definition now widely, but not universally, accepted is an area with exclusive use not requiring defense though often defended or advertised. Requiring defense fails to include the possibility that territories can be maintained by mutual avoidance caused by neighbors who make availability of resources unpredictable (exploitation competition; Stamps and Krishnan 1999, Spencer 2012, Wakefield et al. 2013, Riotte-Lambert et al. 2015, Ellison et al. 2019, Collet et al. 2022) or by animals preferring to use areas that are in their cognitive maps and that they have used before (Lourie et al. 2021).

Consequently, here is my definition of a territory: A territory is all or part of a realized home range to which its animal (or its group of animals) has exclusive access and use due to repulsion of conspecifics through overt defense, advertisement, or mutual avoidance[1].

An animal maintains a territory only when she has a limiting resource: a resource that is in short supply and that limits reproduction or survival and, consequently, individual fitness (Brown 1964, 1969). Only at intermediate levels of productivity is defending a territory cost effective (Carpenter and MacMillen 1976, Hixon 1980, Powell et al. 1997). When productivity of the limiting resource is high enough that the resource can be shared, then the costs of territorial behavior are an unneeded expense (Brown 1969, Carpenter and MacMillen 1976, Hixon 1980, Stamps and Krishnan 1999, Spencer 2012). When productivity is very low, then the costs of territorial behavior are not returned through exclusive access to that limiting resource and an animal must be nomadic, that is, must seek another site for his home range.

[1] This distinction between a home range and a territory is not always followed by biologists with British backgrounds, who may use "home range" and "territory" interchangeably.

An individual or social group might maintain exclusive access only to those parts of their home range with the most important resources, as exhibited by some coyotes (no overlap at den sites and central, areas of good habitat; Person and Hirth 1991) and eastern chipmunks (Getty 1981). Individuals of some species maintain territories only in parts of their species' ranges but tolerate home range overlap in other places (Rogers 1977, 1987, Garshelis and Pelton 1980, 1981, Crabtree 1989, Person and Hirth 1991, Powell et al. 1997, 2022, Solomon et al. 2004, Martin and Martin. 2007, Robertson 2016). Territorial behavior by female black bears and prairie voles can be predicted from variation in the local productivity of food (Powell et al. 1997, 2022, Robertson 2016). Thus, territorial behavior is not necessarily a species characteristic. If, however, environmental conditions leading to territorial behavior remain stable for long periods, flexibility of behavior can be lost as behaviors that depend on territoriality are selected and become fixed.

For territorial individuals, territory size tends to vary positively with body size and inversely with the productivity of the limiting resource (Harestad and Bunnell 1979, Powell et al. 1997, Jetz et al. 2004, Seigle-Ferrand et al. 2021, Séchaud et al. 2022), though not necessarily if territory holders are energy maximizers (maximize energy density in territory irrespective of requirements) and food is the limiting resource (Ebersole 1980). Within a breeding territory, multiple resources have the potential to be limiting, including a mate or mates, food, display sites, refuges from predators, reproductive sites (nest or den), and tunnel systems. Wolff (1989, 1993) warned that food may appear superficially to be the limiting resource when the limiting resource is actually offspring. Many vertebrates and even some invertebrates maintain breeding territories but are not territorial during the rest of the year (e.g. multitudinous birds, many mammalian ungulates, anoles, fence lizards, dragonflies, damselflies, wasps, and crickets; Alexander 1961, Waage 1988, Haenel et al. 2003, Bush and Simberloff 2018, Kamath and Losos 2018, Aromaa et al. 2019, Eaton 2021).

2.2 Territory Variations

Most mammals that maintain territories do so year-round and food is identified most often as the limiting resource (Sandell 1989, Rogers 1977 1987, Palomares 1994, Powell et al. 1997), though for solitary carnivorans the limiting resource for males may be females during the breeding season (Erlinge and Sandell 1986, Sandell 1989, Powell et al. 1997). Members of many species defend territories as mated pairs (barbets, Soma and Brumm 2020; Siberian cranes, Klenova et al. 2020; coyotes, Person and Hirth 1991) and many ant colonies defend territories (e.g. red harvester ant, Adler and Gordon 2003). Members of still other species defend territories as extended family groups (red-cockaded woodpeckers, Walters et al. 1988 1992; scrub jays,

Woolfenden and Fitzpatrick 1984; white-browed sparrow-weavers, York et al. 2019; beavers, Jenkins and Busher 1979; Eurasian red foxes, Macdonald 1979; meerkats, Madden et al. 2009; wolves, Mech and Boitani 2003; woodland voles, Powell and Fried 1992). Sometimes the extended family groups contain non-family members; most often such a non-family member has replaced a member of the parental pair who has died (Mech 1970, Rood 1986, Kruuk 1989, Madden et al. 2009).

When youngsters delay dispersal and help to raise their younger siblings, the social system is called cooperative breeding. For cooperatively breeding animals in general, theoretical (Powell 1989) and empirical research (Jenkins and Busher 1979, Woolfenden and Fitzpatrick 1984, Walters et al. 1988 1992, Powell and Fried 1992) suggest that the limiting resource is usually some aspect of breeding and living space, such as European badgers' setts (Doncaster and Woodroffe 1993, Kruuk 1989), woodland voles' tunnel systems (Powell and Fried 1992), and red-cockaded woodpeckers' sleeping trees (Walters et al. 1988, Carter et al. 1989). Whether individuals, mated pairs, or families maintain territories appears to depend not just on the productivity of limiting resources but also on predictability and fine-grained vs coarse-grained patchiness (Bekoff and Wells 1981, Macdonald 1981 1983, Kruuk and Parish 1982, Macdonald and Carr 1989, Powell 1989, Doncaster and Macdonald 1992, Brandell et al. 2020, Twining et al. 2024).

Although members of many species maintain individual or group territories against all conspecifics, many solitary carnivorans maintain territories only among members of the same sex and tolerate territory overlap with one or more members of the opposite sex: called intrasexual territoriality (Powell 1979a 1993 1994, Sandell 1989, Johnson et al 2000, Facka and Powell 2021). For members of species with intrasexual territoriality, generally: 1) sexual dimorphism in body size is large; 2) males are polygynous while females are selectively polyandrous; 3) females raise young without help from males; and 4) the large body sizes and large territories of males may be considered a cost of reproduction (Seaman 1993, Powell et al. 1997, Yamaguchi and Macdonald 2003). Some animals affect food supplies through resource depression (i.e. have rapidly renewing food resources such as prey that become wary when they perceive a predator and later relax to become available again; Charnov et al. 1976). For these animals, the cost of intrasexual territoriality appears to be minor because the limiting resource renews. The cost of intrasexual territoriality is imposed on females by males (Powell 1993 1994, Facka and Powell 2021). Male ungulates of many species defend breeding territories from other males while females generally occupy a male's territory but take no part in territory maintenance (Owen-Smith 1977, Byers et al. 2010).

2.3 Advertisement

Animals advertise territories using diverse behaviors, including vocalizations and other sounds, scent marking, visual displays, environmental displays, and backup displays (Johnstone 1996); animals with understudied senses (electric, magnetic, chemotaction) undoubtedly use those senses for advertisement, too. Advertisements are safer and more economical than is defense with tooth and claw and advertisements are evolutionarily stable (Maynard Smith 1976, Lewis and Murray 1993; Chapter 9). Theoretically, a strategy of attempting to take over a territory and forcing defense by the owner loses to strategies that do not force defense, such as by using non-combative advertisement (Morris and MacEachern 2013). Calls and overt displays are immediately transient and disappear. Environmental displays include scent marks, hind leg scrapes by wolves, bears' scrapes and bites on trees, and Eurasian badgers' setts. Most environmental displays change over time. Individual identification, dominance or social status, and time since last visit may all fade from scent marks and environmental displays but at different rates.

The assumption that scent marks, calls, and displays actually constitute "defense" has not been tested. An alternative explanation of these behaviors is that they advertise resource depletion or depression, thereby communicating that an advertised area has had its resources depressed or depleted by the territory owner (Spencer 1992 2012;

Chapter 5). Neighbors understand that foraging in an advertised area will not be productive and, therefore, the neighbors avoid foraging there. If environmental conditions do not change over evolutionary time, selection could favor behaviors such as scent marking, vocalizing, and other advertising behaviors that promote minimal home range overlap. Consequently, such advertisement of territories need not be territory defense but, simply, communication with known neighbors about resource availability, which keeps neighbors out. The hypothesis is that advertisement and subsequent avoidance of poor foraging sites caused by resource loss or depression can maintain territories within a population (Spencer 2012). For North American red squirrels, survival decreases with the turnover of neighbors (Siracusa 2019). A hypothesis is that, because new neighbors do not know their social neighborhood, they do not know whose territorial communications can be trusted.

2.4 Territory Size

Biologists intent on measuring home range areas might hope that territorial behavior would eliminate the problem of home range boundaries being vague or meaningless. Sometimes that is the case, as with North American red squirrels (Smith 1978, 1981) and kinkajous (Kays et al. in press). Yet, for members of other species or under different conditions, territories may not yield clear boundaries. Wolves respond to the scent marks of neighboring wolf packs by marking at high rates. The dominant male of a pack (usually the father of subordinate pack members) may intrude 100 m into the territory of a neighboring pack to scent mark (Peters and Mech 1975, Ausband et al. 2013). The neighbors reciprocate. Consequently, the territory boundaries become fuzzy spaces a few hundred meters wide with an amalgam of scent marks of at least 2 packs, and without a distinct boundary. The same holds true for coyote packs. Boundaries can be so wide and fuzzy that lone wolves and coyotes can move through the boundary areas without much danger from resident packs (Peters and Mech 1975, Mahoney and Young 2017). Distinct boundaries of territories may at times be no easier to identify than are boundaries of undefended home ranges. Nonetheless, apparent overlap of territories seen in telemetry data may not be real but, instead, be due to location error or due to tracking animal movements on the wrong scale (Kays et al. in press).

Sometimes, however, a buffer zone can develop between 2 territories, such as between 2 wolf packs or between indigenous human groups (Hoskinson and Mech 1976, Mech 1977a, Botkin 1995, Kay 2007). At those times, the buffer acts as a refugium, allowing prey populations to grow because of reduced use of the buffer by neighboring predators. Hamelin and Lewis (2010) suggested that buffers can develop when the relative cost of a conflict with a neighbor in the buffer zone exceeds the cost of extra foraging pressure on resources in the core region of the territory. This hypothesis deserves to be tested.

2.5 Population Regulation

The ultimate regulator of a population of territorial animals is the limiting resource that stimulates territorial behavior (Nevoux et al. 2011). Although population regulation through territoriality has received extensive theoretical attention (Brown 1969, Fretwell and Lucas 1970, Watson and Moss 1970, Maynard Smith 1974, López-Sepulcre and Kokko 2005, Wang and Grimm 2007, Varga et al. 2020), the general conclusion of such theory is that territoriality can only regulate populations proximally. The most common limiting resource is food and, for territorial individuals, territory size tends to vary inversely with food availability (Harestad and Bunnell 1979, Ebersole 1980, Hixon 1980, Schoener 1983, Saitoh 1991, Powers and McKee 1994, Powell et al. 1997, Fuller et al. 2003, Jetz et al. 2004, McRoberts and Mech 2014, Heuck et al. 2017, Grant et al. 2017, Chambert et al. 2020, Webber et al. 2023).

2.6 Private Property

Maintaining territories is a rare example of non-human animals, or groups of animals, maintaining private property (Klopfer 1969, Klopfer and Rubenstein 1977, Strassmann and Queller 2014, Sherratt and Mesterson-Gibbons 2015). Other examples of private property, though less consistent, include nest structures, cached food, mates at certain times of year, and sometimes the resources and partners in mutualisms, none of which have the same potential for long-term property rights as do territories. Territories generally orient around land, which has more permanence than do other examples of property rights. Resources worth privatizing have high value, will be useful to their owner in the future, are durable and defensible (Strassmann and Queller 2014), and, therefore, are likely to be sought by other animals and to be the subject of confrontations or contests among animals (Chapter 9). Hinsch and Komdeur (2017) went so far as to state that modelling of territories should be done using game theoretic approaches only, a position that I would venture is short-sighted and grossly limiting. Much is to be gained from modelling territories using as many approaches as possible.

CHAPTER 3

Cognitive Maps and the Definition of a Home Range

Hurtling down Avent Ferry Road, wind whistling through my bike helmet, going so fast that I can not keep up pedalling, I aim toward the causeway over Lake Johnson and then toward the steep uphill grade that I want to crest without having to downshift. I survey the landscape automatically, instantaneously without thinking. I note not many cars in the parking lot for the trail around the lake and, for an instant, think about biking that trail, though it would add 2 km to my route. I also see that no hikers are coming from the woods on the other side of the road. Crossing the causeway, I note that the parallel foot bridge has few people fishing, and remember my assessment earlier that fish populations were down. As I pass the next parking lot, I cross the road to bike on the separate bike path. Going up the hill I hear a dog barking in an apartment on the right side of the road and, briefly, think that I could reach the wooded trail around Lake Johnson, had I wanted to, by cutting through the woods to my left, across the high school playing field and across 2 parking lots. I see a car coming toward me on the road and then hear a car coming from behind. The 2 cars should pass each other about where they pass me. I pass the office of the Department of Motor Vehicles and remind myself again to check when I must renew my driver's license. Riding through Fraternity Court at the university, I remember students in my classes who room there, remember how well they are doing in my courses, and then think of revising one course to be able to reach these students better.

Home Ranges of Animals. Roger A. Powell, Oxford University Press. © Roger A. Powell (2025). DOI: 10.1093/9780191949142.003.0003

Not just the hill but the route of my whole commute that I used for 30 years, with all the permutations for different weather and traffic conditions, is clear in my memory. When I think of that route, I think of it linearly from one direction (home to university) or the other (university to home) but never do I see it from above, like a map drawn on paper (or a GoogleEarth image). I can place my mind's eye in any place along the route and see the apartment buildings and parking lots, houses, woodlots, and shopping centers that line the route. I can envision the travel options projected forward, with various options occurring at distant points along the main route. My mind has an excellent grasp of the spatial relationships among different routes and of each route with its surrounding environment. I know which routes are longer or steeper, which expose me to more traffic, how the different routes cover space, which streets lead to dead ends and how those dead ends relate to the routes that work. When exploring a new route and looking ahead I see a hill, I know from past experience that I need to increase speed to tackle that hill.

My bike route is a major part of my cognitive map of Raleigh, North Carolina. That whole map is an odd amalgam of travel routes, the cityscape perceivable from my bike route, specific experiences on specific days when I learned something new, and, when travelling in a given direction, which places I should encounter first and which I encounter later. The map also has blanks where I do not know areas. It is not a Euclidean map but I can force myself to draw a Euclidian map from my cognitive map. The drawn map would be sketchy.

3.1 Definition of a Home Range and History of the Cognitive Map Concept

Home Range: That part of an animal's cognitive map that the animal keeps up to date; includes places not visited recently but that are held in memory for future, specific use.

The main goal of this chapter is to provide background on the concept of a cognitive map and to show how an animal's cognitive map encapsulates the concept of a home range, leading to my definition of a home range. "Cognitive" refers to cognition[1], which is the mental process of knowing and includes awareness, perception, reasoning, and judgment. Thus, discussion of cognitive maps involves discussion of the brain region called the hippocampus, where the locations of memories and their context are stored, recalled, and updated. The chapter first presents some history of the cognitive map concept and its location in the hippocampus, ending with an overview of what the hippocampus does. After that comes discussion of different ways that animals think, incorporating input from their senses and how those inputs activate the hippocampus. All of the preceding material provides background on how animals navigate using memories and how they plan future movements. With that foundation, I explain why I no longer find Burt's (1943) definition of a home range adequate or accurate. Finally, I repeat my definition of a home range and discuss how to use that definition to gain important insights and understanding about the home ranges and the biology of animals.

If biologists are to understand animals' home ranges, we must crawl into animals' brains to understand their cognition of their environments. From approaches that differ from but are parallel to that of cognitive maps, Real (1991, 1992, 1996) and Kashetsky et al. (2021) documented animal cognition and argued that understanding cognition is critical for understanding animal behavior. Looking at familiarity distributions (those built from animal location data) as maps will not explain why animals do what they do, nor explain relationships between animals and their food, predators, or mates. When analyzing familiarity distributions, a biologist wants to learn something about her study animals, which is learning what, where, when, and especially why they do what they do. The only way to do that is to try to understand what the animals think. All biologists *do* do this, at least to some degree, when they generate familiarity distributions. I suggest that the deeper we go in that direction, the more we can learn. Understanding

[1] Unfortunately, no universally accepted definition of "cognition" exists. Cognition is related to consciousness and sentience, also lacking universally accepted definitions (Buckner 2015, Akagi 2018, Finkel 2023, Gough 2023, Lee 2023, Nussbaum 2023, Velazquez 2024). As Velazquez (2024) noted, "Consciousness? It's all in the mind."

that an animal's cognitive map contains her memory of her world, how well she can use that world to meet her requirements of life is a key to understanding the animal, her behavior, and her resulting ecology (Figure 3.1).

An animal's cognitive map is instrumental for navigation, and it also forms memories of travel routes, places visited, details of those places, and also travel and places planned to be visited in the near future. Importantly, the cognitive map organizes relationships among memories of places and routes, especially the characteristics they have in common, such as closeness of visits in time and location. Regarding memories, a cognitive map resembles an internet search engine, or a card catalog for the library of the brain; the cognitive map knows where to find a memory, knows how to use it, and facilitates using the memory for planning (Eichenbaum 2017, Leviton 2022). An animal's cognitive map can plan multiple navigational strategies, but also organizes and curates spatial and nonspatial memories and other knowledge that emphasize relational organization. His cognitive map organizes tasks abstractly, allowing generalizations and inference (Courellis et al. 2024). An animal's cognitive map has been called its cognitive control center (Tolman 1948).

Tolman (1948) introduced the concept and the term "cognitive map." In describing research on rats in mazes, he conjectured that learning a maze was more than simply a matter of stimulus and response. Instead, he believed that "something like a field map of the environment gets established in the rat's brain" (p. 192) and that the responses of an animal's brain are complicated. Although this was not an entirely new concept (Turner 1907, a little known publication), Tolman envisioned not so much a map as a

> map control room ... the incoming impulses are usually worked over and elaborated in the central control room into a tentative, cognitive-like map of the environment. And it is this tentative map, indicating routes and paths and environmental relationships, which finally determines responses (p. 192).

Tolman also expanded the concept to include "comprehensive maps" (p. 208), which he considered to be maps of broader categories of

Figure 3.1 A descriptive familiarity distribution is not the same as an animal's home range. The former is a map (usually smoothed in some manner) of places that an animal has been documented to have visited during a specified period of time. It quantifies a researcher's perspective of the animal's movements. A home range takes the animal's perspective and is the animal's cognitive map of its environment, including, for example, food patches, when each patch was last visited, how much the patch is expected to have renewed since being foraged, and the quality of the patch. It also includes the animal's landscape of fear and how the landscape of fear affects decisions about travel routes and which foraging patch to visit next, how various places in the home range change seasonally or daily, and so on.

behavior, such as behavioral plans for actions, including behavioral plans for interacting with conspecifics. Here, he included the mental plans, or maps, for a person's interactions with other people, such as figuring out how you will act the next time that you must sit at dinner next to your uncle who is the radical extremist (my example). Tolman did not elaborate on where or how cognitive maps might be housed in animals' brains but his conception of the multiple contents and jobs performed by an animal's cognitive control room has been confirmed. The concept of a cognitive map resembles in some ways von Uexküll's (1934) concept of the *umwelt* but, even though von Uexküll was a biologist, an organism's *umwelt* has been of more interest to philosophers than to biologists.

Unfortunately, the name "cognitive map" can mislead biologists not to understand that a "cognitive map" is way, way more than just a Euclidian map, or a "drawn" map in an animal's mind. A cognitive map may not have any visualization of space at all; echolocating bats use audition in their cognitive maps and procellariiform seabirds develop olfactory cognitive maps that extend to underwater land forms and currents (Nevitt 2008, Nevitt et al. 2008, Depose and Nevitt 2008). In this book, I use "home range" and "cognitive map" somewhat interchangeably. When writing of an animal's perspective, I generally use "home range." When writing about where that perspective is housed in the animal, I generally refer to "cognitive map."

Research on the hippocampus and cognitive maps took off starting in the 1970s (O'Keefe and Dostrovsky 1971, O'Keefe and Nadel 1978), and has been of interest to field biologists studying animals' home ranges almost from that beginning (Peters 1978). Research has been done on diverse mammals, birds, other vertebrates, and even insects. Cognitive maps connect memories of place and time, wayfinding, linear and relationship thinking, and in humans include music, stories, planning how to deal with difficult situations, and my planning how to arrange the words in this sentence.

Over 20 years after Tolman's work, O'Keefe and colleagues (O'Keefe and Dostrovsky 1971, O'Keefe and Nadel 1978) discovered that specific cells in the hippocampus that O'Keefe called "place cells" responded to rats' locations: as a rat moved around in an enclosure, the place cells fired in a matching pattern within the hippocampus. He concluded that the hippocampus must be the location of Tolman's cognitive map. O'Keefe and Nadel asserted that orientation is one of the most important aspects shaping human minds; I would add that the same goes for orientation being important for all organisms. O'Keefe and Nadel argued that orientation is so special that it needed a specialized area of the brain, the hippocampus. Because the hippocampus is an old part of the vertebrate brain and has been conserved by natural selection, all vertebrates must have the foundations for cognitive maps. Strong evidence suggests that even insects have cognitive maps (Gould 1986, Witter et al. 1989, Cheeseman et al. 2014).

Decades before the start of research on cognitive maps, research had shown that the hippocampus was involved in memory, especially the memory of facts and events (declarative memory; reviews: Eichenbaum et al. 2016, Sugar and Moser 2019). Starting with O'Keefe's research, research on the role of the hippocampus and its associated brain regions (what I call the hippocampus complex) in navigation and orientation exploded (Eichenbaum et al. 2016, Sugar and Moser 2019). Most of this research has used laboratory rodents as model animals, due mostly to convenience and to past research protocols for training rodents to decipher mazes (Eichenbaum et al. 2016). Nonetheless, research on birds, especially pigeons but also owls, has demonstrated that the hippocampus of birds is also involved deeply both with memory and with navigation and spatial orientation (Montagnese et al. 1996, Gagliardo et al. 2009, Mora et al. 2012, Blaser et al. 2013, Čapek et al. 2017, Padget et al. 2019, Henrique et al. 2021, Payne et al. 2021, Agarwal et al. 2023). Though birds' brains are small[2], they have similar numbers of neurons

[2] That birds' brains are organized differently than mammals' brains is not surprising, given that birds have experienced selection to have highly functional brains that are light. Most mammals have the luxury of being able to carry heavy brains. Bats, however, have disproportionately smaller and lighter brains than do other mammals of similar weight

as do mammals' brains and the brains of corvids and parrots have neuronal densities similar to primates (Herculano-Houzel 2020, Stacho et al. 2020, Ströckens et al. 2022). Despite the anatomical differences in the hippocampus among vertebrate groups (Montagnese et al. 1996, Rodriguez et al. 2002, Naumann et al. 2015, Striedter 2016, Payne et al. 2021, Rossi and Wright 2021), the circuit mechanisms are similar with cells equivalent to place cells and other dedicated cells (Payne et al. 2021).

3.2 Memories and the Hippocampus

The hippocampus complex contributes not only to explicit remembering but also to an expansive suite of cognitive operations that stretches the concept of memory, including cognitive mapping and navigation across environments and time, exploring actively, guiding choices of behaviors, making decisions and solving problems adaptively, imagining the future, anticipating upcoming stimuli, reasoning inferentially, thinking creatively, learning actively, and understanding and developing language (Cohen 2015, Moser et al. 2017, Tsao et al. 2018, Purandare et al. 2022, Steudler and Ólafsdóttir 2023, Lai et al. 2023, Liu et al. 2023, Chettih et al. 2024, Tacikowski et al. 2024).

The hippocampus complex has place cells, grid cells, border cells, head-direction cells, time cells, and, undoubtedly, other cells with those types of roles that allow an animal to orient herself within her environment, and within time, but also to deal on multiple scales with the features and complexities within that environment ("environment" including dealing with your radical uncle; O'Keefe and Dostrovsky 1971, Fyhn et al. 2004, Sargolini et al. 2006, Leutgeb et al. 2006, Kjelstrup et al. 2008, Pastalkova et al. 2008, Solstad et al. 2008, Ekstrom et al. 2017, Høydal et al. 2019, Nieh et al. 2021). Place cells have developed in pre-weaning rats, who are able to remember early movements around their nests and to learn how to navigate (Bjerknes et al. 2018). Youngsters are, in fact, learning how to learn (Adolph 2019) and the more complex their environments are as they grow and as they learn to become efficient at navigating, the better they will be at navigating in complex environments as adults (Coutrot et al. 2022). Place cells, grid cells, and probably cells of the other types respond to resource levels and rewards and the hippocampus guides memories of those resource levels. The cells do this on different environmental and temporal scales and the cells also update memories with new information (Boccara et al. 2019, Butler et al. 2019, Saleem et al. 2018). They remember important places but not unimportant places (Terada et al. 2022), and have flexible filtering mechanisms that facilitate putting important memories into long-term storage (Chung et al. 2021, Lee et al. 2021, Terada et al. 2022, Yang et al. 2024). A population of place cells can maintain a stable estimate of an animal's location over several days, even while individual cells change their activity patterns (Ziv et al. 2013), maintaining synchrony when an animal sleeps and when the hippocampus is otherwise inactive (Gardner et al. 2019, Almog et al. 2019).

Hippocampus size varies with relative selective pressures on cognitive mapping abilities and on spatial memory (Krebs et al. 1989, Jacobs and Spencer 1994, Conrad et al. 1996, Clayton et al. 1997). The hippocampus forms a cognitive map of physical space, supports navigation and planning of movements, and provides the rest of the brain with a schema of a familiar environment or of a familiar decision-making situation. The schema contains memories of places and concepts connected together according to the times, distances, directions, and organization, including relationships other than positional relationships, of the real world places and concepts, as perceived by an animal in its daily life. This is episodic or associative memory and establishes what an animal knows about the state of the world (Shen et al. 2022, Terceros and Rajasethupathy 2022, Steudler and Ólafsdóttir 2023, Liu et al. 2023, Chettih et al. 2024, Davies et al. 2024). Thus, the hippocampus complex is involved with declarative, episodic, and spatial memory, and the recall of relatively detailed remote memories requires hippocampal involvement.

(Venditti et al. 2024), for which the logical hypothesis is that bats have evolved light brains as an adaption for flight.

The hippocampus is activated when animals encode and retrieve memories, especially memories of facts and events, which in turn involves the relationships among experiences, whether those relationships are spatial, temporal, or due to context. New experiences with a context common to past experiences, such as location or participants or background sounds or time of day, reactivate those past memories. Experiences occurring close in time can link memories of the experiences, and the shorter the time between 2 memories the more likely that they will be recalled together. Recalled memories may be modified to accommodate new information, promoting abstract knowledge of the relationships among distinct experiences. Such integrated memories can generate novel behaviors across a variety of tasks representing inferential reasoning, concept formation, and generalization. Integrated memories also lead to the formation of adaptive memory that anticipates future use of the memory.

Consequently, animals can plan future movements sensitive to where they are within their environments (Kjelstrup et al. 2008, Solstad et al. 2008, Bastos et al. 2024, Evans et al. 2024). Fruit bats, lab rats, and presumably other animals, base the time, direction, and distance for future movements on memories handled by their hippocampi (Ormond and O'Keefe 2022, Sarel et al. 2017). For laboratory rats, ensembles of place cells produce independent forward- and reverse-ordered sequences, apparently part of planning future movements and evaluating past movements, critical for hippocampus-dependent, memory-guided behavior (Epsztein 2022, Wang et al. 2020, Widloski and Foster 2022, Mishchanchuk et al. 2024.). Evaluations of past movements are included within cognitive maps and can lead to spatial partitioning of resources among animals without conflict (Lourie et al. 2021).

Stella, a domestic dog owned by Christina Hunger, a language pathologist, and her husband, was given the freedom to express her wants, desires, and plans using an "augmented and alternative communications device" that allowed her to press buttons (mounted on a large piece of plywood) that then broadcast English words (Hunger 2021). Stella did not simply memorize the names of many objects because she used a vocabulary of at least 4 dozen words that included verbs, adverbs, adjectives, and nouns, which she used correctly. She demonstrated that she planned routes to places that had resources important to her and that her choices of places differed depending, at least, on her needs at a particular time, on her responses to weather, and on the wishes of her people. Stella "spoke" of her plans and negotiated plans with her people. She could ask to go to the dog beach or to the dog park ("outside, beach," "outside, park") or to go outside to be with Jake as he mowed the lawn ("outside, Jake"). Her cognitive map clearly included her surroundings, which she was able to describe in part through her discussions with her owners about potential activities for a day. She knew the details of the routes to the dog beach, to the dog park, and for a neighborhood walk. Stella demonstrated that she thought with forward- and reverse-ordered sequences. While on an outing, and without her talking device, Stella could not express in English words her forward and reverse thoughts. Upon returning home, however, Stella's spoken thoughts were consistent with the research on laboratory rats producing forward- and reverse-ordered sequences. Stella is not the only dog who plans ahead. Indeed, many, many people depend on their guide dogs to think several steps ahead (not footsteps but steps in a plan; Bastos et al. 2024, Holland 2024). Dogs with buttons express their thinking via these buttons (Bastos et al. 2024).

Altogether, the hippocampus complex is, perhaps, a bit like an old library card catalog. Daniel Levitin explained this metaphor: "... memories aren't actually stored in the hippocampus the way you might store a book on a shelf. In the library of the brain, I think the hippocampus is the card catalog that tells you what shelf the memory is on, in which wing of the library" (Leviton 2022: 7). The hippocampus complex is essential to spatial navigation via a cognitive map but the cognitive map includes relational organization and flexibility beyond spatial navigation. Hippocampal networks map navigational strategies and, in addition, other spatial and non-spatial memories and knowledge that are related (Eichenbaum 2017). Thus, the hippocampus complex curates memories old and new, guiding organization of memories to facilitate planning of imminent and future activities.

3.3 Ways of Thinking

Humans think in at least 3 modalities: verbal, visual–object, and visual–spatial (Ferguson 1977, Kozhevnikov et al. 2002, 2005, 2010, Grandin 2022). Humans who are verbal thinkers think in words; those who are visual–object thinkers think in clear images, those who are visual–spatial thinkers think in categories and how subjects are related, though much overlap occurs with visual-object thinking. Grandin (2022) explained these modes of thought using an informal example. She asked people to draw a steeple. Visual–object thinkers draw a specific steeple, often with much detail. Visual–spatial thinkers think of steeples as a category, some with points, others with flat tops, others yet with onion shapes, and so forth and will draw a steeple that may have characteristics of several different kinds of steeples. Verbal thinkers often have trouble sketching a steeple at all and usually sketch two parallel lines with an inverted "V" on top; verbal thinkers might add notes and descriptive words with arrows to parts of the sketch.

Verbal, visual–object, and visual–spatial thinking lie along multi-dimensional continua. People also do some olfactory thinking (when asked to think of food, for example), some touch thinking (patting a dog), auditory thinking (music), and so forth (Grandin 2022). Finding words to describe what I have "olfactorized" (in contrast to having "visualized") can be difficult. I can describe the urines of foxes and possums as both smelling somewhat like dilute anal gland spray of striped skunks, all of which contain mercaptans, but possum is a bit "sweeter" than fox, which is sweeter than skunk (if "sweet" can describe a smell). When I hike in northern conifer forests on a warm summer's day, I anticipate smelling the terpenes from the transpiring conifers and olfactorize that smell. I can not, however, describe the smell because it has no words associated with it and smells like nothing else to which I might compare it. Yet, I love it. Likewise, I olfactorize the autumn smell of forest litter decomposing and the smell of balsam poplars in mid-summer, and seek these smells as I breath while hiking, These smells, too, have no words, so I can not describe to someone else how they smell. Nonetheless, they represent my olfactory thinking, which complements my visual–spatial thinking.

In my life I have handled thousands of wild mammals (and some birds, herptiles, and insects). Healthy animals smell good; they smell healthy. When I bury my nose into the fur of a fisher, or a black bear, or a wolf, or a weasel, or a deer mouse they all smell different but the smells have "healthy" embedded within them. I strongly suspect that olfactory thinking (and auditory thinking and touch thinking) is, somehow, more closely related to visual thinking than to verbal thinking but not necessarily physically related within the space within our brains. If humans could see in the infrared and ultraviolet, how different would our visual thinking be?

Animals are not verbal thinkers (Peper 2020, Grandin 2022); importantly, verbal thinking is not required for complex reasoning (Fedorenko et al. 2024). Given the multiple modes of non-verbal thinking, we must anticipate that animals as diverse as insects, cephalopods, and chordates think in sensory modalities (Reser 2014, Grandin 2022) and that individuals within a species will have different ways of thinking, just as individuals of any species have different morphologies and personalities (Cabrera et al. 2021). Biologists who wish to understand their animals gain from insight into how the animals think. This job is undoubtedly a greater challenge for human verbal thinkers than for visual thinkers of either modality, who do much of their thinking without words. People who are strong auditory thinkers (musicians) and, especially, those who are strong olfactory thinkers probably have head starts toward understanding other mammals, since most other mammals emphasize audition and olfaction (e.g. Gillingham 1986). People who are strong visual thinkers may have head starts toward understanding birds, and maybe insects, though humans have severe limits to their vision; birds and insects can see wavelengths that we can not. Just as humans have 2 modes of visual thinking, I expect that other animals have multiple modes of thinking at least auditorily, olfactorily, chemotactily, and electrically (Allard et al. 2023, Kang et al. 2023). Verbal thinking may be a category of auditory thinking.

When thinking, animals stimulate their hippocampus complexes with different sensory stimuli. Primates stimulate the hippocampus complex through eye movements and, presumably, what they see (Wilming et al. 2018), and rodents and humans stimulate the hippocampus complex with odors (Cohen et al. 2019, Zhou et al. 2021, Yi et al. 2021, Poo et al. 2022). The olfactory bulbs of a dog have a direct neural connection to the hippocampus complex (Andrews et al. 2022); other senses do not have such direct connections. Odors, sounds, and even social partners stimulate place cells, and memories of odors remain in rats' hippocampal time cells (Wood et al. 1999, MacDonald et al. 2013, Aronov et al. 2017, Omer et al. 2018). Weasels forage using audition more than olfaction, and least through vision (Gillingham 1986), stimulating their hippocampi via all 3 senses. Cats follow the trajectories of their out-of-sight owners by triangulating on sounds of their owners' movements (Takagi et al. 2021), suggesting that hearing stimulates their hippocampus complex. When people "visualize" a past event, the memories can be complete with vision, sound, smell, taste, and touch (Liu et al. 2022) and we should expect the same for animals of other species (Cabrera et al. 2021).

The biological null hypothesis has become that all animals with a hippocampus complex stimulate that complex with inputs from all the senses that they use daily for foraging, exploring, travelling, and otherwise living in their environments. Senses provide information on all 3 timescales: past, present, and future. Vision provides information of the present but also of the past, from track and sign that animals, including humans (in a big way), have left. Hearing provides information about the future when an animal hears sounds of another animal, for example, who can not be seen but who will soon become visible. Scents can provide information on all 3 scales. Like vision, scent provides information about animals who were at a place in the past and left scent. Wolves can perceive where humans have passed for at least 3 days (Nelson 1973). Like hearing, scent can provide information on the future when a scent is perceived on the wind of an animal that is out of sight. I documented that fishers could smell at least 200 m when a fisher track I followed turned sharply and led to one of my baited traps, which had not been visible to the fisher (Powell 1977). Search and rescue dogs regularly follow scents of people that they will meet in the future. Bumblebees and domestic chicks can experience an object or place using one sense and can later recognize that object or place using a different sense, the same way that I can reach into my pocket and feel my pocket watch while my brain provides me with a visual image of the watch.

These abilities suggest that the hippocampus can be stimulated by all senses (Solvi et al. 2020, Rossion 2022, Tyree et al. 2023, Versace et al. 2024). A troop of yellow baboons studied by Stuart and Jeanne

Altmann lived in an area with diverse foods including 2 patches of particularly desired fruits that fruited only rarely. One small patch was located within the other food patches used regularly by the troop but a big patch was located across a stretch of desert avoided by the baboons. One year the baboons found abundant fruit in the small patch, which they devoured, and then they headed across the desert to the large patch, which they likewise demolished (Altmann and Altmann 1970, Spencer 1992). Cephalopods have cephalopod-specific chemotactile and electrical receptors that they use to make behavioral decisions (Allard et al. 2023, Kang et al. 2023) and some toothed whales have electro-reception (Hüttner et al. 2023), all of which presumably stimulate the hippocampus.

Clearly, the hippocampus receives incredibly diverse information about its animal's environment, which it organizes, remembers, and uses when planning future activities. When an animal plans a foraging route, for example, the animal does not "visualize" those travel routes and resource patches the way humans do. The animal will use at least olfactory thinking and auditory thinking as well as visual–spatial thinking to understand where she will go.

3.4 Navigation

Different animals learn to navigate emphasizing different sensory stimuli and using different modalities of sensory thinking (Eichenbaum et al. 2016, Wilming et al. 2018, Wehner 2020). Many insects navigate using dead reckoning (continuous re-evaluation of distances travelled in changing directions), getting directions from the sun and using landmarks for small-scale locations (Turner 1907, O'Connor 2019, Wehner 2020, Freire et al. 2023). Dead reckoning, however, is plagued with the problem of accumulating small errors that become big errors. Although much of the original research on cognitive maps appears to have been associated with the concept of dead reckoning, the hippocampus complex can handle myriad types and sources of information (Ekstrom and Rangath 2017, Nieh et al. 2021).

Vertebrates appear to use flexible combinations of a little bit of dead reckoning along with a lot of direction finding and complex landmarks, which include, but are not at all limited to, physical structures (identifiable mountains, lakes, big trees, rocks, plants), slope, wind direction, snow patterns, shadows, odors, and sounds. Animals also navigate by using stars, by detecting the earth's magnetic field, and by hearing infrasound from distant geological features such as mountains (Wehner 2020, Kashetsky et al. 2021, Dacke et al. 2021, Xu et al. 2021, Buehlmann 2023). With 3 navigational tools, 1) a target direction, such as north, 2) time of day (in solar time), and 3) direction of the sun compared to the target direction, an animal (including a human) can calculate a missing tool if he or she knows the other 2. The same is true for using the moon at night, which includes the 4th tool of the phase of the moon (known from the moon's shape).

Members of some human cultures always know compass directions wherever they are. Members of other human cultures navigate hundreds of kilometers across landscapes using landmarks, the stars, and stories and music inherited via oral history (O'Connor 2019). Such navigation is usually linear in the sense of visualized travel routes with relative places along a route associated with relative time (Ingold 2011). Complexity of landscapes experienced during childhood affects human abilities to navigate as adults (Coutrot et al. 2022), which generates the following hypothesis: young animals born (or hatched) during food scarcity that requires complex search navigation are more adept at searching as adults than are animals born into food abundance. I also hypothesize that most mammals olfactorize routes and auditorize routes more than they visualize them.

Laboratory rats, Egyptian fruit bats, and humans track the locations of conspecifics (Danjo et al. 2018, Omer et al. 2018, Stangl et al. 2021), and domestic cats follow their humans (Takagi et al. 2021), suggesting that group-living animals incorporate the locations and movements of group members into their cognitive maps and use that information to navigate.

3.5 Hippocampus and Planning Moves

How the hippocampus interacts with the prefrontal cortex to change routes when an animal is faced with an exploitation–exploration dilemma depends on the animal's cognitive map of alternative routes (Domenech et al. 2020, Steixner-Kumar and Gläscher 2020). Fox squirrels learn complex maneuvers to deal with challenges of moving (walking, running, climbing, leaping) in a complex 3-dimensional environment and use what they have learned to enable future travels (Adolph and Young 2021, Hunt et al. 2021). As conditions change across days and weeks, and presumably years for long-lived animals, the underlying hippocampal codes are stable, except for new information learned about the environment (Liberti et al. 2022).

In addition, an animal's movements should depend on her nutritional state. Although novelty-seeking is apparently not controlled by the hippocampus, the hippocampus is critically involved in novelty-seeking, which is distinct from reward-seeking (Ogasawara et al. 2022), suggesting that animals may be prone to explore when otherwise sated. Ingested non-essential amino acids stimulate exploration and food-seeking; having an excess of nutrients not needed stimulates looking for nutrients needed to balance the diet (Viskaitis et al. 2022). From another perspective, resources with low travel costs or that balance the diet should have added value. I showed that fishers appeared to rank prey by energy return (Powell 1979b), consistent with optimal foraging models (Charnov 1976a, Pyke 1984). Those ranks should not change through time but should lead to predictable changes in diets as prey populations change. Thus, an animal might not update her cognitive map for low-ranked prey when high-ranked prey are abundant.

Now, what does her cognitive map mean to the red squirrel Shortface, whose home range is east of my house? She knows when mushrooms are ripe on the top of the ridge. From past trips across her home range, she plans the route she will take to those mushrooms as she starts climbing a nearby balsam fir tree. She keeps track of changes in the bark of the tree trunk by scanning back across where she has already travelled just as she scans and plans her forward movements as she goes. Reaching a long, horizontal branch, she notes how the bark feels on her feet. She knows exactly how much that branch will sag as she reaches the end and, therefore, exactly where she must start her jump. She also knows how the tension in the branch changes with rainfall, humidity, and temperature. She knows the short leap to the branch on the next tree and she knows how that branch will sag when she lands. If she mistimes the jump slightly, that misjudgment goes into her cognitive map along with the day and the weather and the wind and the feel of the bark and how hungry she was. The smell of the terpenes from the pines tells her that the trees are hot today, which affects branch tension and ease of harvesting cones. The breeze from the northwest means the weather will stay clear. The direction of the breeze tells her the directions from which a broad-winged hawk might come as she climbs down the landing tree. And sounds: if she hears her neighbor squirrel Notch-ear "Chir-r-r-r-r-r-r" to the south, if she hears a raven "Quork," all that and more updates her cognitive map. This is how she thinks of her home range, everywhere, over all parts of it. Her cognition includes the times and places and the resource to harvest next and how the wind and the sounds and the smells make what she knows about her world. She does not just "visualize" her home range but, instead, lives it in hundreds of dimensions because that is how the hippocampus complex works.

The raven that flies over my house does see the world from above and may have a Euclidian map in his head but also knows what routes to take to get to the Burntside River and knows the hillside that is great for wind surfing on windy days. Because his vision sees the world all around him, any visual Euclidian-type maps he has are 3-dimensional. He also knows and maps what the different kinds of wind mean for ease of travel to different places on different days. Humans have lots of words for snow: snow, graupel, sleet, hale, squall, stellar, slush, frozen, drift, fluffy, fine, rime, flurries, blizzard, dump. A raven must have hundreds of concepts of wind and air currents that he remembers (perhaps as feather movements, a sense of touch) both in general and in association with specific places and conditions. He also knows routes through the trees and not just over them. Those are parts of his cognitive map and he uses that information to plan a foraging route.

Cognitive maps are probably limited to 2 physical dimensions for large prairie mammals like pronghorns or bison and for large forest mammals like moose, but 3 physical dimensions for prairie dogs, deer mice, red-backed voles, and black-footed ferrets, with the home ranges going down, and for red and other tree squirrels, bats, and birds, with the home ranges going up, and for weasels going both ways. All home ranges have the additional dimension of time and probably sensory dimensions.

3.6 Home Range Definition

With this background, I no longer find Burt's (1943) definition to be the best way to conceive an animal's home range. First, Burt's definition is limited to areas that an animal visits regularly and does not include areas that the animal knows but visits only rarely, perhaps only in times of critical need. Female desert tortoises when forming eggs will travel far out of the areas used daily to reach a local place with high calcium content in the soil, which they eat (Marlow and Tollestrup 1982). During an autumn with a failed hard mast crop, my colleague Sue Mansfield was following a female black bear with cubs in northern Minnesota. This bear left her home range (known from multiple years of study) to visit an area 25 km away. Sue visited the area and found a rich acorn crop. The family stayed in the oaks for a few days and then headed back home. Travel both to and from the distant site was in nearly straight lines. I had a similar experience following a female with cubs in the Blue Ridge Mountains but could not verify the acorn crop (Powell et al. 1997). From the nearly straight lines of rapid travel of both females, I strongly suspect that they knew of the abilities of those distant oak stands to produce food in years of food scarcity. If my interpretation is correct, then I consider those distant food patches to have been parts of those gals' fundamental home ranges in other years as well because they were familiar with the patches and knew that they could use those patches to obtain food when needed. Both gals took their cubs with them, meaning that those cubs then knew those patches, too.

Bears are not the only animals that can remember over long periods of time. Some bats (who can live to be decades old) can remember for at least 4 years (probably a gross underestimate but the only measurement ever reported; Dixon et al. 2022).

An arctic fox may be familiar with a 100 km^2 area yet use regularly only a small portion with concentrated food during a period of study (Frafjord and Prestrud 1992). The fox can sample distant areas with the potential to have food (by smell, for example) and can thereby determine which areas actually do have food at any particular time. Many mammals can smell and hear over a kilometer and can see a few hundred meters or more in appropriate habitat; birds flying can see much farther and hear at least as well. Animals do not need to be in a specific small area physically to know whether berries or acorns are ripe, or whether competitors have been bedding there regularly, or whether the vole density is high. Thus, an animal's fundamental home range includes critical areas that it may not visit regularly and includes areas sampled remotely to learn if they have important resources. Ospreys nesting in a breeding colony monitor the directions in which other colony members fly when leaving and returning to the colony. They deduce the locations of productive and unproductive lakes by noting which ospreys return to their nests from which directions with fish (Hagan and Walters 1990).

Related to the problem of visiting a site vs keeping tabs on it from a distance, Burt's definition of a home range is not from the animal's perspective. It is *post hoc* and descriptive from a biologist's perspective, like a familiarity distribution built from location data. To understand best why an animal uses his home range as he does, and to understand what he gains from his home range, a biologist needs to understand as well as possible how the animal himself perceives and remembers his home range. Models of animal movement hypothesize that animals in heterogeneous environments benefit from "knowledge" of their environments (Saarenmaa et al. 1988, Folse et al. 1989, Turner et al. 1994, With and Crist 1996, South 1999, Stillman et al. 2000, Matthiopoulos 2003a,b, Moorcroft and Lewis 2006, Dalziel et al. 2008, van Moorter et al. 2009, Spencer 2012, and more). Such knowledge can affect an animal's fitness. Adult or juvenile animals in familiar environments have lower mortality and higher reproductive output than do juveniles dispersing through unfamiliar landscapes (Blanco and Cortes 2007, Gosselink et al. 2007, Soulsbury et al. 2008). For an animal to establish, and to become familiar with, a home range requires time, which leads to site fidelity, and biologists have used site fidelity has been used to define whether or not animals have established home ranges (e.g. Swihart et al. 1988, Spencer et al. 1990).

Consequently, I repeat my definition of a home range presented at the beginning of the chapter:

That part of an animal's cognitive map that the animal keeps up to date; includes places not visited recently but that are held in memory for future, specific use.

Thus, an animal's cognitive map, as her home range[3], allows her to make decisions that affect her fitness, such as where to search next for food, how to reach a hunting site while minimizing chances of becoming someone else's meal, and what parts of her home range overlap with the home range of a potential mate. Her cognitive map includes areas that must be avoided, such as large bodies of water if she is terrestrial or land if she is aquatic. Her cognitive map has a search engine entry, a card catalog card, for every aspect of her conscious life. She keeps her cognitive map up to date via all of her senses.

And her home range is not a map of a place but a concept of the place with all of its characteristics that can be perceived by senses, and be tied together in memory by timing and other relationships, and by physical distributions of resources.

Modelling the perceptual and memory processes that produce cognitive maps and modelling the way that animals use their cognitive maps is the functional approach best able, in my mind, to provide insight into home ranges, into how animals build their home ranges, and into how they maintain them (Powell 2012a, Powell and Mitchell 2012, Spencer 2012, Stephan et al. 2014, Arraut et al. 2017, Gehr et al. 2020, McKeown et al. 2020, Rheault et al. 2021, Nams et al. 2023, Powell et al. 2025). This functional approach will lead biologists away from overemphasizing home range area only (Chapter 12). Thus, this book emphasizes using creativity, curiosity, and imagination to develop models that

[3] Spencer (2012) and Mike Mitchell and I (Powell 2012a, b, Powell and Mitchell 2012) developed similar definitions for a home range. Earlier, I implicitly defined a home range as a cognitive map (Powell 2000, 2004). Peters (1978) understood complexity of cognitive maps with respect to navigation. Some other authors have referred to cognitive maps but defined them as animals' mental maps of where they have travelled, including nothing more than locations.

generate hypotheses pertinent to what animals consider to be their home ranges.

Think of your own cognitive map. It includes all the places you go regularly, the distributions of resources, and your travel routes. It includes sales on milk and the sale on shoes that occurs each spring at the neighboring mall. If a friend tells you of a new grocery store that is being built, that information goes into your cognitive map as do updates on construction, even though you may never go to the site before the store is open. If a new road is built, you may test it as a short cut. Your neighbor's habit of mowing his lawn early each Saturday morning is in your cognitive map; it helps you to plan your Saturday.

Now, as a biologist, you want to gain that type of familiarity with the home ranges, the cognitive maps, of each of your study animals. Your study animals, however, perceive a different world than humans perceive. Using every tool available is required to learn about what animals perceive and understand.

At present, reading cognitive maps of animals in the field is logistically difficult but technology will improve the situation. Rattenborg et al. (2008) attached recorders to the tops of 3 brown-throated three-toed sloths and extended electrodes to the skulls of the sloths. The records generated electroencephalogram and electromyogram data used to monitor sleep and wakeful cycles. A number of research groups have installed electrodes into hippocampi of Egyptian fruit bats that broadcast to recorders mounted externally on the bats (Sarel et al. 2017, Liberti et al. 2022,Forli and Yartsev 2023). Sarel et al. (2017) monitored the firing of 309 hippocampal cells, including 101 place cells and 58 goal-directed cells, as bats followed a trained flight pattern. The goal-directed cells registered angles compared to the ultimate goal and place cells documented the progress of each flight. Such technology opens a door for the future when field biologists might monitor the hippocampus complexes of free-living, wild animals.

For starters, imagine that one can collect perfect and complete location data, including time, for an animal across a biologically consequential time period. Here are some hypotheses that would lead a biologist to collect such data. 1) A forager returns to food patches at the times predicted by resource renewal rates and information loss (such as new berries ripening, or prey reducing wariness after a predator forages in a habitat patch, and competitors truncating such renewal). 2) Resource patches that are not visited can, nonetheless, be perceived (suggesting the importance of building familiarity distributions based on perception of scent, hearing, and other senses). 3) Familiarity distribution boundaries align with physical features on a landscape and with boundaries of neighbors' territories. 4) Territoriality and familiarity distribution overlap change with the productivity of limiting resources. Tests of such hypotheses provide insight into animals' cognitive maps.

Cognitive maps contain myriad information: characteristics of resources, of topography, and of other aspects of a landscape (jumbles of rocks, tangles of coarse woody debris on the ground, etc.), of the distributions and movements of conspecifics and of predators, and more. A biologist does not have to think of target animals' cognitive maps to be able to test hypotheses about most day-to-day activities of study animals, nor to be able to map where the animals spend time. The animals, however, think about those day-to-day activities couched within all the other information that is in their cognitive maps. They *always* think about their locations within the context of that other information.

Human brains are busy and, therefore, use shortcuts when they can (O'Connor 2022). Here is an example. I bought prescription hearing aids some 10 years ago. About 6 months after getting the hearing aids, and enjoying many sounds (especially birds) that I had not heard in years, I realized that I had just then started hearing many new, quiet sounds. After 6 months, my brain was realizing that it could hear again many sounds that had been *very* faint before I started using hearing aids. Years, maybe decades, ago my brain had stopped trying to hear those very faint sounds and took the short cut to ignore them. Now, with hearing aids, my brain could hear those sounds well enough again and, after 6 months, awoke the very old habits of listening to them. Re-awakening my brain took 6 months. Now my brain lets me hear those previously ignored sounds. A human brain habituated to thinking about an animal's locations only, without context, will not consider the extended context unless forced to consider it. Consequently, a biologist who habitually avoids all that messy context, even though he pays lip service to knowing that the context is important, will not think about that context when it is important to his research, which could be often. Consequently, biologists *do* need to think of animals' home ranges as their whole cognitive maps and to keep the messy context in their minds.

The world has beautiful, mind-boggling complexity. I want to think about it.

CHAPTER 4

Movement

After 20 plus years of doing research on black bears in the Blue Ridge Mountains, I could hike anywhere in my 250 km^2 study area without getting lost. From high points, I had seen each mountain and where each ridge rose between parallel creeks and rivers. I could walk a ridge side, then hike downslope to cross a creek and up the other side to reach wherever I needed to go without ever having walked there before. My hiking, my movement through the forest, allowed me to know the distributions of habitats and berry patches and new clearcuts. My cognitive map for those 250 km^2 has an amalgam of views from mountain and ridge tops along with a huge number of specific routes through the forest on the ground, along with distributions of ecological communities. It includes

Home Ranges of Animals. Roger A. Powell, Oxford University Press. © Roger A. Powell (2025). DOI: 10.1093/9780191949142.003.0004

the view from Turkey Spring Gap to Little Pisgah Mountain and how the trail between the two places hugs the ridge side. It includes the plummet down the trail into Thompson Creek Cove. Hiking through my study area, and driving its roads, built my cognitive map. Every move renewed and updated my cognitive map and introduced me to new places and to their ecological communities.

4.1 Movements

Movements constitute the basic construction material for a home range but, as already noted, an amalgam of movements does not constitute a home range. Movement through an environment begins to inform a dispersing animal, who does not have a home range, where to find some resources and what the travel routes between resource patches might be. Only when the dispersing animal finds ample good resource patches to cause him to confine movements within a limited environment can he develop a home range. If he does not find enough, he moves on. An animal's movements across his landscape inform him of resource timing and availability, travel routes, neighbors, predators, and everything needed for his cognitive map. Thus, a home range is an emergent property of movements (Börger et al. 2008, Spencer 2012, Kenkro and Giugglioli 2019). And the more an animal moves through the physical expanse of his home range, the more he learns new things and updates his cognitive map. The animal's cognitive map then allows him to return purposefully to places with specific resources at specific times under specific conditions. Consequently, specific movements emerge from having a home range, forming a symmetry between home ranges and movements: dispersal and subsequent movements generate and update a home range and the home range then generates future movements and more updates.

By studying an animal's movements, a biologist can learn about the animal's realized home range during a period of study and about the animal himself. Horne et al. (2020) differentiated between a collection of movement trajectories, which they called "occurrence distributions," and estimates of how animals distribute time spent across landscapes, which they called "range distributions." When movement data are not complete (that is, have large temporal gaps between locations), the unknown parts of movement trajectories can sometimes be estimated by building probability distributions for potential routes between each pair of consecutive location estimates. Brownian bridge estimators, kriging and other methods are often used for this job (Bullard 1999, Horne et al. 2007, Powell 2012a, Fleming et al. 2014a, 2014b, 2016).

If all the gaps in movement data can be filled, the resulting trajectory still does not represent an animal's home range because it is only the combined movement paths. The occurrence distribution lacks all the additional information contained in the animal's cognitive map, such as memories of the locations of resources, locations of obstacles, travel routes of predators, the animal's understanding of his own present nutritional status, and his plans for future movements (as just a small example). Just as descriptive population modelling describes what populations of animals do but does not provide insight into what each individual animal does, describing movement patterns (populations of movements) provides little to no insight into the drivers behind each movement by an animal. An animal's travel on a given day depends on the animal's concept of his home range. Consequently, the better a biologist understands the cognitive maps of his study animals, the better he can understand individual movements and collections of movements.

Assigning actual times along movement trajectories allows analyses otherwise not possible, such as estimating energy expenditure (Powell 1979b, 1981, Powell and Mitchell 2012, Nickel et al. 2021). By estimating energy expenditure and documenting ingestion, a biologist can estimate nutritional status of an animal, gaining key insights into how nutritional status affects preferences for resources. This can be an entry point for learning more about the animal's cognitive map.

Examination of movements provides understanding of these critical building components of home ranges. From here, this chapter introduces definitions of movement terms and ways to quantify movements. It then presents several options for filling gaps in movement paths and discusses when filling gaps might be necessary, which involves understanding autocorrelation. The chapter ends presenting 2 methods for identifying dispersal movements.

4.2 Terms

The word "move" fills 2½ pages (roughly 8 columns) in the Oxford English Dictionary I have in my study ("Compact" edition, 1971). The words "move" and "movement" differ in that "move" can be a noun or a verb, while "movement" is a noun only. "Move" as a noun is almost always interchangeable with "movement." Sometimes readers are forced to use context to figure out whether "move" is a noun or a verb. Not all definitions of "move" and "movement" apply to animal behavior and ecology, for which disciplines the term "move" refers to an animal's change in position. Granted, wind picking up a piece of dust or a plastic bag can affect organisms but in biology the term "move" generally refers to what organisms do. And, although many plants do move (i.e. change position, either a part of a plant or a whole plant) and some of those movements have been described as intentional (Mao et al. 2023, Baluška et al. 2009, Kaspar et al. 2021, Calvo 2023), I limit discussion in this book to intentional movements of whole, complex animals. The sub-discipline of movement ecology includes unguided movements (Nathan 2008, Nathan et al. 2008). Although originally conceived as a mind-boggling, all-encompassing model of all types of movements[1] (Nathan et al. 2008),

[1] I remember clearly holding the marker and filling the whiteboard when we wrote out the entire model for the first time

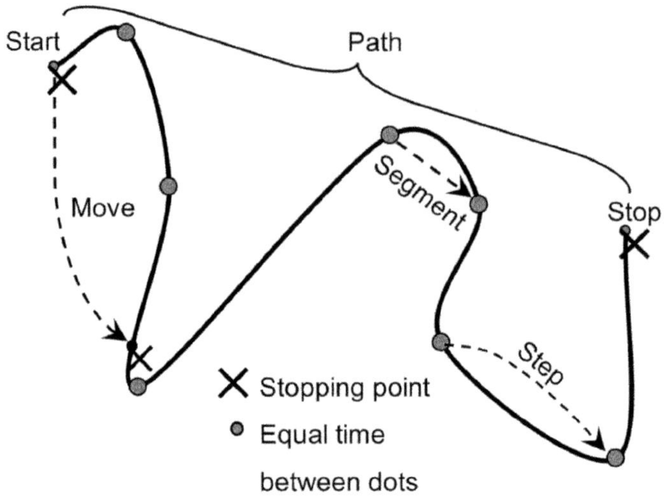

Figure 4.1 The trajectory of a movement (solid, dark curve) appended with the major terms used to describe movements. The entirety of the movement trajectory constitutes the path.

movement ecology has become a name applied to ecological and behavioral research that incorporates animal movements in some way.

Movement is the foundation for much of behavioral ecology (Nathan et al. 2008) and provides important parts of the behavioral aspects of a home range defined as a cognitive map. Animals' movements affect habitat (via herbivory, seed dispersal, downstream effects of predators' movements on prey), community structure (via mutualisms, competition, predation, parasitism), and population distributions.

The topic of movement has its own jargon[2] that includes many terms that are not standardized, are sometimes ambiguous (Favreau 2006), and sometimes do not exactly match the street definitions for the terms. Some movement behaviors have several names and sometimes a single word is used by different authors to mean different behaviors

[2] Etymologically, "jargon" is "nonsense." It derives from Old French, where it meant the unintelligible noises of birds. By the 1600s, it had moved to English and to mean the unintelligible noises of human communication. Now it refers to particular, restricted meanings of words within a certain profession that are unintelligible to people outside that profession. Never, it seems, has jargon ever connoted transparent communication.

Consequently, defining terms in publications, and in conversations, is always a good idea. For this book, here are definitions pertinent for movement (almost all as defined by Turchin 1998).

A **path** is a "complete spatio-temporal record of a followed organism from the beginning to the end point" of an observation. Thus, an "occurrence distribution" is the same as a "path" but is 2 words and 22 letters in contrast to 1 word and 4 letters. From here on, I use "path." A **step** is the change in location over a fixed time. A **move** is a "**segment** of a path between 2 consecutive stopping points." A **segment** has also been defined as the straight line between 2 sampling points, such as 2 telemetry locations. I use "segment" to be measured along a path whether the path has been documented continuously or constructed from locations separated in time (Figure 4.1).

A **bout** is a period during which an animal participates in a single behavior. The behavior is defined by a biologist but the time or distance is defined by the animal. Often bouts are easy to distinguish, such as when the behavior is defined simply as "active" or "resting." Sometimes, however, bouts are not so easy to define. For example, when whales come to the surface to breath, they may exhale and inhale several times, each separated by a short, or long, time below the surface. Deciding how much time between breaths defines separate breathing bouts requires objectivity (Fagen and Young 1978). Fagen and Young used a break in a log survivor function to define a bout. Westcott and Graham (2000) defined a bout as the "period when the pauses between consecutive movements are equal to, or shorter than, the time of the previous movement."

Distance can be defined by the line (a line is straight) between 2 locations or between 2 locations along an animal's actual path of travel. I define distance as measured along a path whether the path has been documented continuously or constructed from locations separated in time.

Dispersal is the outward spreading of an organism from her point of origin. For most vertebrates, the term refers to movement away from the place of hatching or birth that leads a vertebrate to find a place to settle and to establish a home range. **Migration**, in contrast, refers to repeated, periodic movements from one place to another, most often seasonal movements over long distances.

Path shape can be described as linear, directional, zigzag, meandering, sinuous, tortuous, or circuitous. Quantifying the turns in a path allows analyses. **Sinuosity**, **tortuosity**, and **circuitousness** appear to be words used most consistently, though the latter might be confused with a path that must go in circles. The changes in direction between consecutive straight line segments of a path can be quantified in numerous ways (Goss-Custard, 1970, Caro, 1976). If angles to the right (or left) are quantified as degrees positive and to the left (or right) as degrees negative, the mean angle quantifies directionality and the variance or standard deviation quantifies sinuosity. Figure 4.2 shows the path of travel of a male fisher in the northern Sierra Nevada of California, USA. His GPS collar relocated him every 7 minutes during an activity bout that lasted 5 hours and 16 minutes. His mean turning angle was 5.91°, showing that much of his bout was travelling in roughly the same direction; the standard deviation of his turning angles was SD = 78°, showing that his travel had some sinuosity. Fishers tend to travel in reasonably straight lines at a rapid pace between good foraging sites, where they forage by moving slowly and sinuously, searching for prey (Powell 1978), which is what this fisher did. Figures 4.2 and 4.3 show that when he foraged sinuously (turning in sharp angles, the sites marked with stars in Figure 4.2) he moved slowly; when he travelled linearly between foraging sites, he travelled faster.

For continuous data, measuring changing angles of direction is difficult at best.

Sinuosity has been quantified often as $\frac{\textit{Path Length}}{\textit{Line Length}}$, the distance measured along a path divided by the straight-line distance from the start to the stop of the path. I have even used this approach to quantify foraging patterns of fishers and weasels by measuring their trails in snow (Powell 1978).

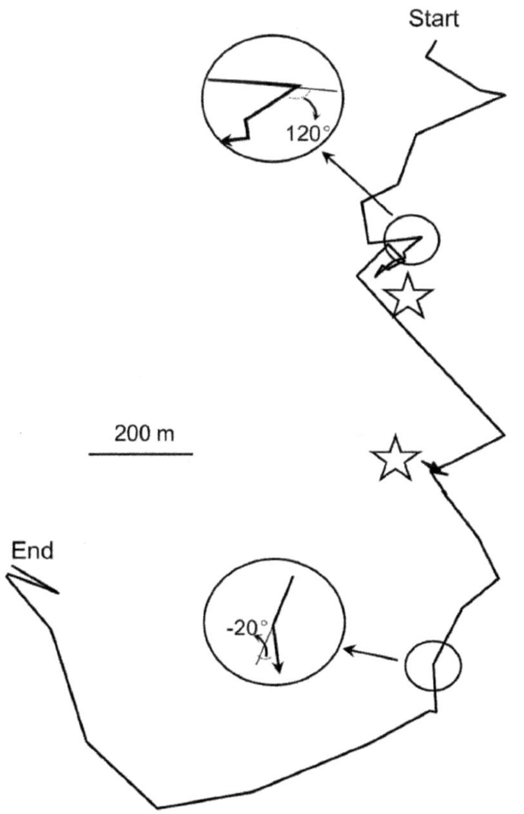

Figure 4.2 Path travelled by male fisher 35734 in the northern Sierra Nevada of California, USA, during a travel bout on 24 October 2016. His GPS collar relocated itself every 7 minutes. His mean turning angle was 5.91° ± 78° (sd). Fishers tend to travel in reasonably straight lines at a rapid pace between good foraging sites, where they forage by moving slowly and sinuously, searching for prey. This path shows 2 foraging sites, marked by stars. The 2 circles show how I measured turning angles.

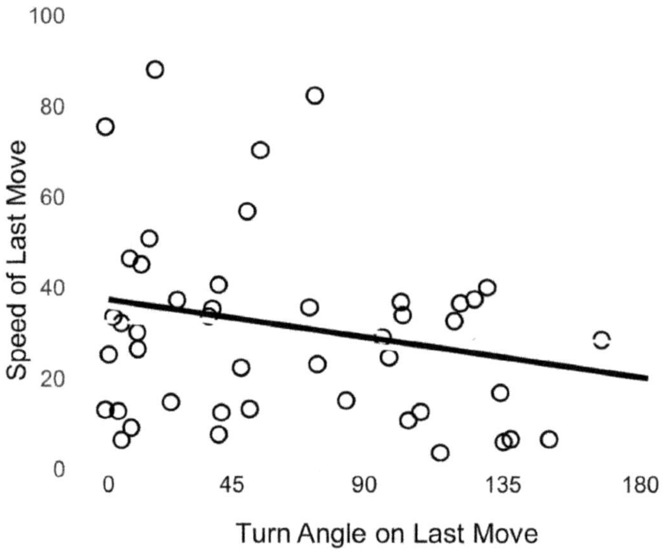

Figure 4.3 Plot of travel speed vs last turning angle for the male fisher whose travel path is shown in Figure 4.2. Sharp turning angles and slow speed are characteristic for foraging movements while obtuse angles and fast speed are characteristic of travel.

A problem with this quantity is that it goes from 1 to ∞; having no upper limit is awkward and can bias analyses. The inverse of this quantity, $\frac{Line\ Length}{Path\ Length}$, is a measure of linearity and goes from 0 to 1, where 1 = linear. If the goal is to measure sinuosity, then a more sinuous path having a smaller value is awkward. Thus, another way to quantify sinuosity is 1 – linearity, or

$$Sinuosity = 1 - \frac{Line\ Length}{Path\ Length},$$

which goes from 0 to 1 and 1 is most sinuous. Using this definition, fisher foraging for snowshoe hares in dense conifer habitat had sinuosity 0.35 (Powell 1978). Fishers frequently changed directions on a meter-by-meter basis, snooping under low conifers and following hare trails (Figures 4.2 and 4.3). In stark contrast, fishers foraged for porcupines by travelling through open hardwood forest in nearly straight-line travel from porcupine den to porcupine den, or from snowshoe hare habitat to porcupine den, with sinuosity 0.03. These long linear moves showed that fishers knew the locations of porcupine dens and planned visits to those dens, demonstrating use of their cognitive maps. Weasels foraging for mice and voles moved with sinuosity 0.32, which was nearly the same as the sinuosity of fishers foraging for hares.

Fractal geometry has been used to quantify sinuosity using dimension D (sometimes denoted as dimension F; Bovet and Benhamou 1988, Loehle 1990, Sugihara and May 1990, Turchin 1991, Gautestad and Mysterud 1993 1995, Bascompte and Vilà 1997, Hagen et al. 2001, Socha and Zemek 2003, Benhamou 2004). D has some practical characteristics. For a linear path (a straight line), $D = 1$. For Brownian (random) motion, $D = 2$. The larger D becomes above 2, the more sinuous the path and the more it folds back on itself. Paths with $D \geq 3$ tend to stay contained around the origin of the path. For diverse mammals, D has averaged <2 (Gautestad and Mysterud 1993 1995, Bascompte and Vilà 1997, Gautestad et al. 1998), showing that movements are neither all linear nor all random.

Scrutinizing location data on smaller and smaller scales leads to clumps of locations within clumps within clumps (Hooten et al. 2017). Gautestad and Mysterud (1993, 1995) argued, therefore, that animals use their home ranges in a multi-scale manner, which makes ultimate sense. A close approximation for D is

$$D = \frac{\log(n)}{\left(\log(n) + \log\left(\frac{2r}{L}\right)\right)},$$

where n is the number of steps in the path, $2r$ is the longest distance between 2 points on the path, and L is the total length tracing the path (Bascompte and Vilà 1997).

The male fisher cited above (Figures 4.2 and 4.3) had $D = 1.33$, documenting that the entire path was predominantly but not completely linear. All the different measures of sinuosity calculated for this fisher have given consistent results.

Movement paths also depend on the status of an animal being studied. Female black bears with cubs have movement paths with higher D in spring, sometimes with $D \approx 3$, than in autumn ($D<2$), presumably because in spring small cubs have limited mobility, while in autumn, when cubs are very mobile, a bear family has high total energy requirements and travels far to find adequate food (Figure 4.4). Female black bears in breeding condition in spring (and without cubs) have low D because they make long movements to find males.

Benhamou (2004) argued that fractal analyses are not appropriate for animal movements because fractal geometry is scale independent, whereas the minima for step lengths for animal movements depend on body size, making step lengths scale dependent. Nonetheless, many aspects of animal movement approximate scale independence. Optimality modelling and empirical data show that both fine-scale and large-scale characteristics of food availability in patchy environments affect

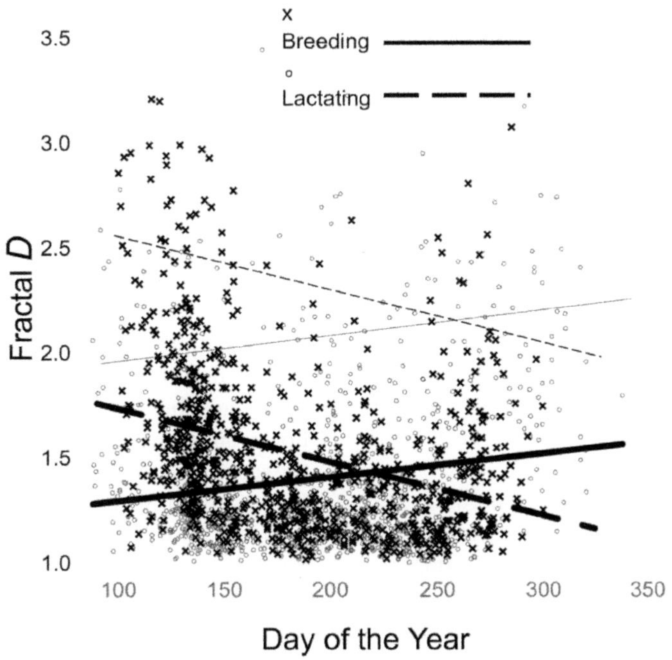

Figure 4.4 Fractal dimension D is an index of travel sinuosity, with values close to 1 associated with straight travel and values >3 with travel that bends back on itself, keeping an animal in a given place. Female black bears with cubs have sinuous travel in spring when travel by cubs is limited. As the year progresses, they have less sinuous and longer movements and they and their cubs meet increasing food requirements and travel to foraging sites. Females in breeding condition have long, straight travel paths (low value for D) in spring, when they seek males for breeding but travel paths become more sinuous as the year progresses. The narrow lines are upper 95% confidence intervals for regressions. These VHF locations were separated by 2 hours during telemetry sessions that lasted from 8 to 48 hours.

movements (Curio 1976, Krebs and Kacelnik 1991). Thus, an animal's decision to remain in or to leave a food patch depends not just on the availability of food within the patch but, also, on the availability of food across its home range and on the locations of the other patches of food.

4.3 Filling Gaps in Movement Paths and Autocorrelation

Often, being able to fill the gaps in movement paths allows calculation of familiarity distributions using all the information available. Otherwise, interpolations over gaps in paths are most useful when research goals include identifying specific travel paths, such as locating where animals cross roads (Horne et al. 2007, Sawyer et al. 2009). Building complete travel paths allowed Kays et al. (in press) to discern highly specific boundaries for territories of kinkajous, some of which showed extremely distinct though sinuous boundaries. New methods of analyzing travel paths without necessarily filling gaps are gathered under the name "path segmentation" (Gurarie et al. 2009, Edelhoff et al. 2016, Mahoney and Young 2017, Getz 2022).

Under ideal conditions, direct observation provides continuous data on an animal's travel path along with associated times and other data, such as foraging, scent marking, and vigilance. For much research, however, logistical problems prevent direct observation. Snow tracking shows continuous movement paths and many associated behaviors but usually lacks precise information on time and is seasonally restricted. High resolution global positioning system (GPS) telemetry can provide sub-meter precision but is limited presently by battery life. Likewise, global positioning system telemetry with video collars can provide almost as much data as direct observation but battery life, characteristics of photovoltaics, and the bulk of the electronics units are limiting. The consequence is that most data on movement paths have gaps; in fact, the data are dominated by gaps. The time attributable to gaps is orders of magnitude greater than the time attributable to locations.

Various approaches can build probability distributions for the path that an animal followed between 2 successive location estimates as long as the 2nd location is limited by the 1st location, called statistical autocorrelation because the second location will be correlated with the first. This happens when an animal is able to move from the first to the second location by travelling at a biologically possible speed but could not reach all places within his home range in the time between the 2 locations. If the second location is not limited by the first location (that is, the animal could have reached anywhere in its home range in the time available) then the locations are not statistically autocorrelated and the gap can not be filled. The approaches for filling gaps are best used for locations within a behavior bout, for which movement characteristics are similar.

Interpolations across data gaps add ridge-shaped kernels, which I call bridges, between consecutive location estimates; the heights, widths, and shapes of the bridges depend on the time and distance between locations and the characteristics of the animal's movement (Bullard 1999, Horne et al. 2007, Noonan et al. 2019a). The ridge-shaped kernels can be calculated from an animal's distribution of travel speeds and the distribution of potential random routes between 2 locations, which can be estimated from location data. Bridges are low and wide at their centers compared to their 2 ends. The bridges become very diffuse as data gaps approach the limit of autocorrelation (Bullard 1999).

Kriging interpolates between location estimates, assuming bivariate normal distributions of spatial variables along a predicted path, basing the predictions on prior covariances (Krige 1951, Fleming et al. 2016). With abundant, uniformly timed and uniformly spaced data, kriging provides the least biased predictions for the movement segment through a gap. Telemetry data and other data for animal movements are, however, never bivariate normal (Fleming et al. 2016). Kriging also needs abundant data that are well spaced without clumped locations, while animals often switch among travel and foraging movements, going from highly spread-out locations to highly circuitous, clumped moves (Magowan et al. 2022). Many alternatives to kriging have been computationally demanding (Diggle et al. 1998).

True Brownian bridges are based on an assumption of Brownian motion and are relatively easy to build (Bullard 1999). If information is available about the causes of data gaps (for example, telemetry locations blocked by dense canopy that can be mapped) then asymmetrical Brownian bridges can be built, increasing their accuracy.

Noonan et al. (2019a) developed a technique to calculate the distribution of possible movement paths between each location in a series using the distributions of time between locations, distances between locations, and location error to simulate possible movement paths. From many simulations, they then calculated summary statistics for potential movement paths, which can be used to build bridges between locations.

Horne et al. (2007) described another way to estimate a good width for bridges. For a plot of the autocorrelated location estimates forming the path of an active animal during an activity bout (no rest stops), a biologist connects every other location estimate with a line and measures the distance from each intermediate location estimate to the line between its neighbors (Figure 4.5). The mean and variance of those measures, plus the distance and time taken to travel, each plotted line segment provides the information needed to plot bridges. A biologist must calculate the variance separately for each activity bout; this is the case for modelling movements in general (Matthiopoulos 2003a, Kranstauber et al. 2012). For example, mixing data for travel and for circuitous foraging (both active behaviors) yields estimates of the variance appropriate for neither activity. Topography also affects the variance to be estimated as well. Black bears walk straight down ridge tops in the Blue Ridge Mountains of North Carolina, USA. I often followed their obvious trails down ridges because those trails provided good paths through dense patches of mountain laurel. Mammals on open prairie do not have their travel paths so restricted.

Given that movements within home ranges have been modelled as modified random walks (Gautestad 1998, Gautestad and Mysterud 1993, 1995), calculating the fractal dimension D for different paths could also improve estimation of variances for bridges. Horne et al. provide the maximum likelihood equation for choosing the variance but, given the potential error for all the variables in the equation and given the assumption that location estimates are normally distributed, I prefer to use biological insight to choose the variance for the bridges. Often, a rough approximation suffices.

Another approach to building bridges is to plot ellipses around the successive pairs of location estimates. For an ellipse, the distance from one focal point to any point on the perimeter and then to the other focal point is a constant. If the standard deviation divided by the mean of the distance measures between locations is small, then the mean plus the standard deviation may yield a decent, representative underestimate of the true speed that the target moved going through the locations.

Yet another approach is to divide each line segment between 2 successive location estimates into short segments, each equal to ½ of the kernel band width h used for the entire data set. Then build kernels at each node between segments, increasing h for the kernels as they approach the center of the segment. For this approach, the gaps must be longer than h (gaps shorter than h will be filled by

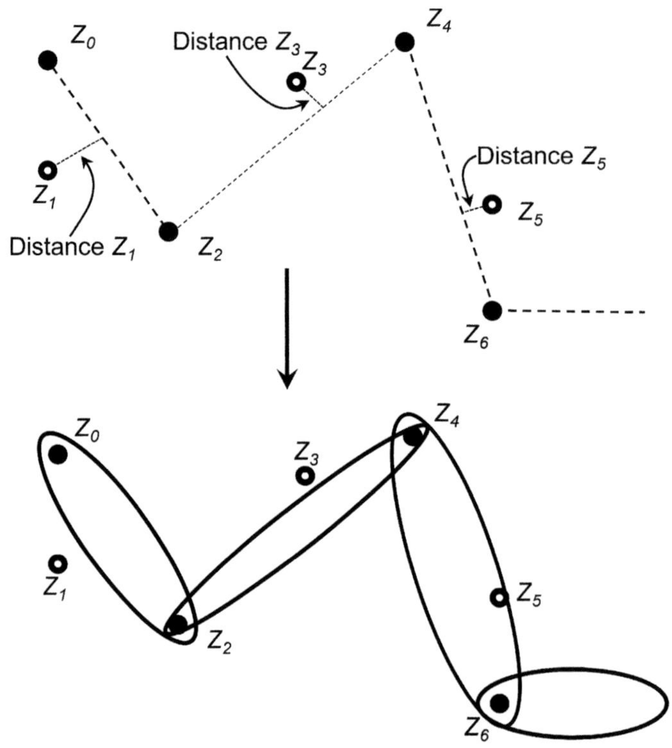

Figure 4.5 Successive locations ($Z_0, Z_1, \ldots Z_6$) of a hypothetical animal's movement path, showing distances from the odd-numbered locations to the lines between successive even-numbered locations. One can use the variance of the distances and travel speed between the even-numbered locations to create Brownian bridges between locations within a behavior bout. Once the relationship is established, bridges can be constructed between all successive locations for which behavior bouts can be identified.

the kernels on the 2 successive location estimates, Figure 4.6).

4.3.1 Quantifying Autocorrelation

Because statistical autocorrelation affects the reconstruction of movement paths, and can affect some analyses, knowing when it exists is important. Unfortunately, measuring statistical autocorrelation is not always straightforward. Swihart and Slade (1985) suggested using Schoener's (1981) measure of independence to quantify autocorrelation. Schoener derived the statistic t^2/r^2 where t^2 is the mean net squared distance between successive animal observations (also called the net squared displacement),

$$t^2 = \frac{1}{n-1}\left(\sum_{i=1}^{n-1}(x_{i+1} - x_i)^2 + \sum_{i=1}^{n-1}(y_{i+1} - y_i)^2\right),$$

while r^2 is the mean squared distance from the center of activity,

$$r^2 = \frac{1}{n-1}\left(\sum_{i=1}^{n}(x_i - \bar{x})^2 + \sum_{i=1}^{n}\left(y_i - \bar{y}\right)^2\right),$$

where n is the number of observations, x_i and y_i are the values for x and y for the ith observation, and $\bar{x}$ and $\bar{y}$ are the means of all values of x and y. Although the expected value of t^2/r^2 is theoretically 2 for independent locations (Schoener 1981), Swihart and Slade (1985, 1986) showed that this was not always the case for empirical data, though high values did suggest independence. My experience has been that, with triangulated, very high frequency (VHF) telemetry data, the values of t^2/r^2 for locations close together in time (short time lag) are usually low, but not always. As the time lag increases, the values sometimes exceed 1.75, and usually show significant dips at long time

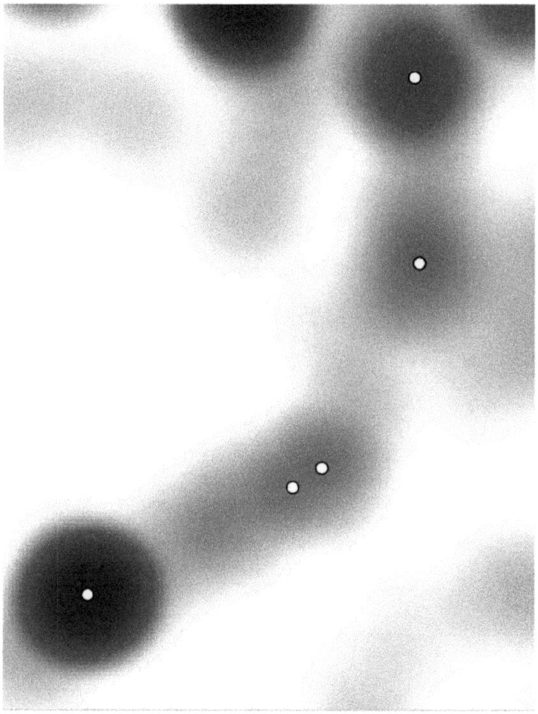

Figure 4.6 These bridges were built by inserting kernels between autocorrelated location estimates for female black bear Dot, followed in northern Minnesota, USA, in 2013. The white dots show GPS locations along one travel path and the bridges built between them. The central bridge is diffuse because the path skipped a location, making the time between the neighboring locations long. The dark circles are where the bear was located a few times, apparently resting. Dot's GPS collar relocated itself every 10 minutes.

lags (Figure 4.7). Too many long time lags can preclude calculating time to independence altogether (Table 4.1). Any collection of graphs of t^2/r^2 *vs* time lags is likely to exhibit some graphs with frequent dips at large time lags (Figure 4.7) because animals return to resource patches when they renew (such as new berries ripening after a patch has been foraged). Table 4.1 shows summary data for 6 populations of 4 species for mean time to independence ($t^2/r^2 \geq 1.75$), mean time to the maximum value of t^2/r^2 for graphs that do not suggest independence, and mean time until a graph plateaus at a relatively constant value whether independence is demonstrated or not. Rooney et al. (1998) presented similar results for mountain hares and bank voles.

Fleming et al. (2014a) introduced semivariance functions as a tool to identify autocorrelation; they provide information on the variation in types of movement used by an animal who generated a location data set. This approach starts by calculating the time lags for all pairs of locations within a location data set (as does Schoener's t^2/r^2) and then calculates the semivariance value, $\hat{\gamma}$, for each value of the time lags. For data at equal time intervals, such as GPS data, the estimated semivariance (Fleming et al. 2014a) is

$$\hat{\gamma}(\tau) = \frac{1}{2n(\tau)} \sum_i \left| x\left(\frac{\hat{t}+\tau}{2}\right) - x\left(\frac{\hat{t}-\tau}{2}\right) \right|^2,$$

where $n(\tau)$ is the number of location pairs with time lag τ, $\hat{t}$ is the mean time for the given pair of locations, and $x(-)$ is the location at the time within the parentheses. The above equation does not work well for data with highly variable lags, as is the case for many very high frequency telemetry data. For such data, the estimated semivariance (Fleming et al. 2014a) is

$$\hat{\gamma}(\tau) = \frac{\sum_\tau w_{ij} \left| x(t_i) - x(t_j) \right|^2}{2 \sum_\tau w_{ij}},$$

where the summations are over all time lags τ, i and j are the times for 2 location estimates, and the weight for each pair of locations, w_{ij}, is

$$w_{ij}(\tau) = w_i \left(\frac{\theta\left(\frac{t_i - t_{i-1}}{\tau}\right) + \theta\left(\frac{t_{i+1} - t_i}{\tau}\right)}{2} \right) w_j \left(\frac{\theta\left(\frac{t_j - t_{j-1}}{\tau}\right) + \theta\left(\frac{t_{j+1} - t_j}{\tau}\right)}{2} \right),$$

where w_i and w_j are weighting functions evaluated for a given $\tau = \left| (t_i - t_j) \right|$, and $\theta(\tau) \equiv \tau$ if $\tau \leq 1, \equiv 1$ if $\tau \geq 1$.

Fleming et al. explained that $\hat{\gamma}(\tau)$ is the average squared distance between locations observed at two different times. The mean semivariance for multiple animals can be calculated to obtain the mean time limit (the mean lag) for autocorrelation.

The values of $\hat{\gamma}(\tau)$ plotted against τ yields a semivariogram. For an animal exhibiting philopatry and who has established a home range, the

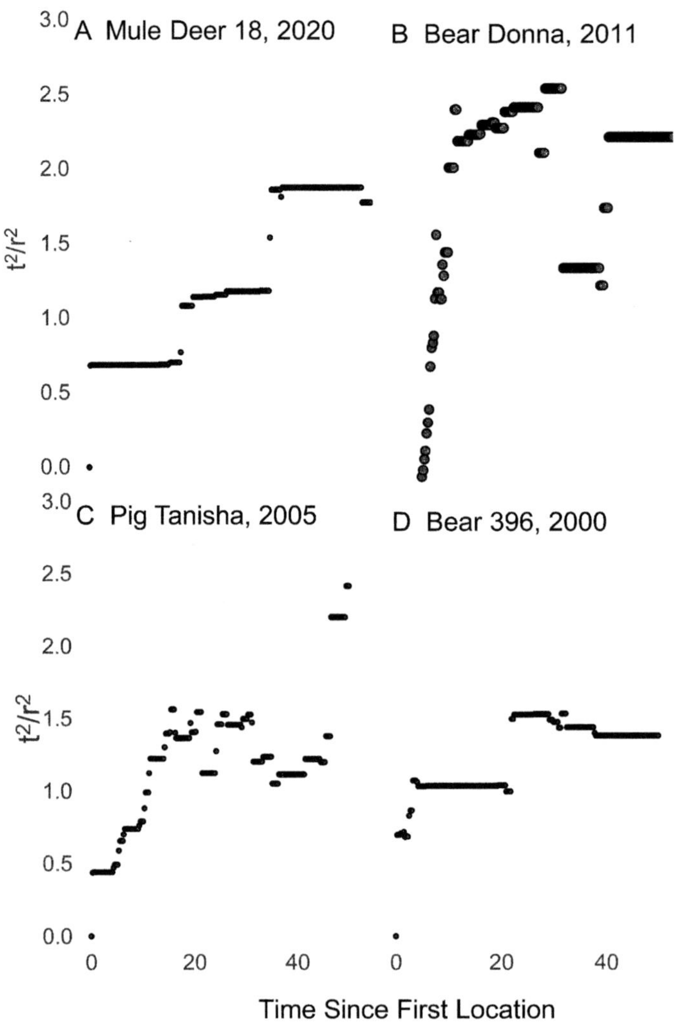

Figure 4.7 Examples of graphs showing values of t^2/r^2 against lag time (hours between success locations) vs autocorrelation. A. Data for mule deer 18 in New Mexico, USA, in 2020. Independence appears to come at around 35 hours. B. Data for black bear Donna in northern Minnesota, USA, 2011. Independence appears to come at around 15 hours. C. Data for feral pig Tanisha in Georgia, USA, 2005. Independence appears to be shy of 20 hours. D. Data for black bear 396 in the North Carolina mountains, USA, 2000. Independence appears to come just after 20 hours. Theoretically, $t^2/r^2 = 2$ at independence but empirical data often plateau at values a bit below 2 (for example, panels C and D). Note that panels B and C exhibit distinct drops in the value of t^2/r^2 after reaching a plateau, suggesting that the animals returned to sites they had visited in the past.

semivariogram rises initially as τ increases levelling off at the time limit for autocorrelation. Fleming et al. illustrate several delightful semivariograms that do, indeed, rise smoothly then levelling to a slope smoothly approaching 0.

Figure 4.8 shows the semivariogram for female black bear 61 in 1984 on my study site in the Blue Ridge Mountains of western North Carolina, USA. This semivariogram was drawn by the online website for the Continuous Time Movement Model[3] (ctmm). Note that the smooth curve fit to the semivariance values plateaus near 30 hours but the actual data do not plateau until about 60 hours.

[3] https://ctmm-initiative.github.io/ctmmwebdoc/. Chapter 11 has more information on ctmm.

Table 4.1 For samples of black bears, fishers, feral pigs, and mule deer, 1) mean time (hours) needed for 2 successive locations to be independent (not autocorrelated), 2) if independence not reached, mean time to reach the maximum value of t^2/r^2 ($t^2/r^2 \geq 1.75$) $t^2/r^2(\pm sd)$ t^2/r^2 plateau D 1 to ∞ $D = 1$ $D \geq 3 < 2$ $D = 1.33$ $D \approx 3$ ($D < 2$), 3) mean time needed for t^2/r^2 to plateau in a graph of t^2/r^2 vs time since the 1st of the 2 locations (time lag). Large time gaps prevented estimation of times and t^2/r^2. Black bears in North Carolina were the only animals studied using VHF telemetry (not GPS telemetry), explaining why so many of those bears lack estimates. N = animal years.

Species and Site	Test	Independence reached ($t^2/r^2 \geq 1.75$)	Mean hours ($\pm$sd) to independence (hr), to maximum t^2/r^2 (if no independence), or to plateau	Maximum $t^2/r^2(\pm sd)$	N
Black bear, North Carolina	$t^2/r^2 \geq 1.75$	Yes	20.3 ± 15.0	4.33 ± 4.27	87
		No	26.3 ± 17.0	0.77 ± 0.60	57
		No	Unable to estimate		51
	t^2/r^2 plateau	Yes	33.3 ± 12.1	3.09 ± 1.80	79
		No	27.9 ± 15.5	0.81 ± 0.59	64
Black bear, Minnesota	$t^2/r^2 \geq 1.75$	Yes	13.6 ± 11.1	3.93 ± 1.20	40
		No	22.5 ± 16.8	1.54 ± 0.18	6
	t^2/r^2 plateau	Yes	37.7 ± 11.3	3.65 ± 1.21	38
		No	24.6 ± 15.7	1.52 ± 0.18	6
		No	Unable to estimate		2
Fisher, California	$t^2/r^2 \geq 1.75$	Yes	2.28 ± 2.35	4.29 ± 0.53	8
		No	0.25		1
	t^2/r^2 plateau	Yes	7.5 ± 1.8	3.34 ± 1.80	3
		No	4.5	1.39	1
		No	Unable to estimate		5
Fisher, New York	$t^2/r^2 \geq 1.75$	Yes	6.25 ± 6.00	3.76 ± 0.75	9
	t^2/r^2 plateau	Yes	13.2 ± 4.2	2.95 ± 1.04	9
Feral pig, Georgia	$t^2/r^2 \geq 1.75$	Yes	12.2 ± 15.2	3.58 ± 0.84	28
		No	26.2	1.71	1
	t^2/r^2 plateau	Yes	27.3 ± 14.7	3.12 ± 0.96	23
		No	23.8 ± 14.7	1.64 ± 0.06	3
		No	Unable to estimate		3
Mule Deer, New Mexico	$t^2/r^2 \geq 1.75$	Yes	11.3 ± 9.7	3.85 ± 2.10	84
		No	7.1 ± 7.5	1.09 ± 0.52	15
	t^2/r^2 plateau	Yes	23.9 ± 11.2	3.14 ± 1.63	31
		Yes	Unable to estimate		44
		No	9.5 ± 8.2	0.96 ± 0.56	18
		No	Unable to estimate		6

Similar to the function proposed by Swihart and Slade (1985; Figure 4.7), semivariograms do not always provide clear cut estimates for the time limit for autocorrelation. Figures 4.7 and 4.8 show the reality of results for trying to calculate the time to independence.

4.3.2 Dead Reckoning

Although global positioning system telemetry data often record locations of animals on short time intervals, location estimates only a few minutes apart will still have gaps. Gunner et al. (2020, 2021) used accelerometer and magnetometer data to calculate angular velocities in 3 dimensions and, thereby, reconstructed movements of African lions, a loggerhead sea turtle, a Magellanic penguin, and an imperial cormorant between successive global positioning system locations. The accelerometer and magnetometer data were recorded at intervals shorter than the location data. Gunner called this approach Dead Reckoning because it reconstructs a route from limited information about short-term changes of body orientation and movement directions. Magowan et al. (2022) used this approach to reconstruct both travel routes and, importantly, to reconstruct complex, local foraging moves of European badgers (Figure 4.9). The foraging moves

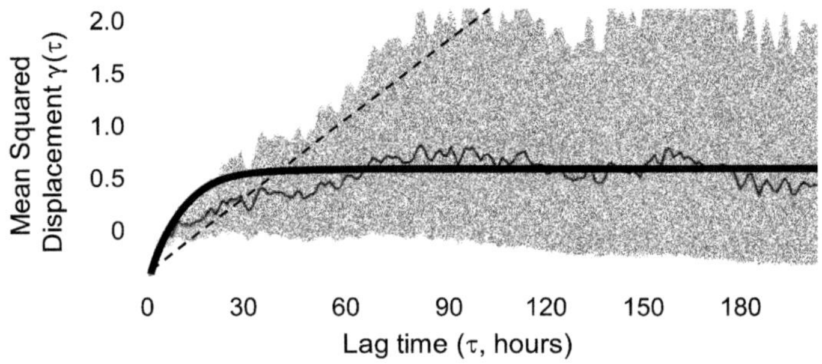

Figure 4.8 The semivariogram for female black bear 61 in 1984. The curve fit to the data for this semivariogram plateaus at about 30 hours, while 32 hours was the autocorrelation limit time calculated for this bear using the approach of Swihart and Slade (1985, Powell et al. 1997). The actual semivariance values do not plateau until about 60 hours. This semivariogram was drawn by the online website for ctmm (https://ctmm-initiative.github.io/ctmmwebdoc/). The dashed line illustrates the semivariogram for a random walk.

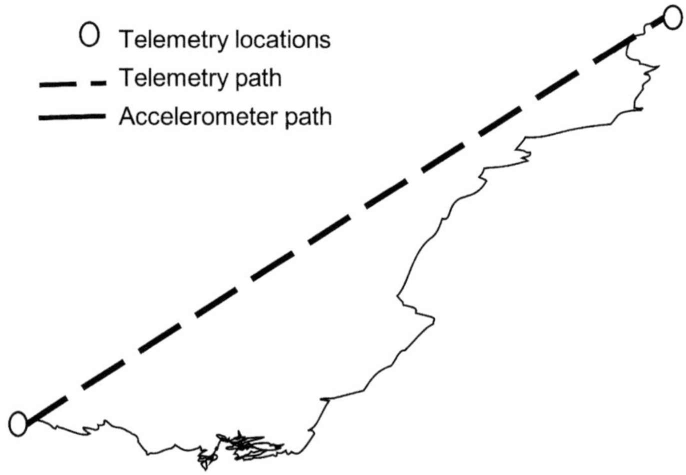

Figure 4.9 A travel path for a European badger reconstructed by Magowan et al. (2022) using Gunner's (2020, 2021) "dead reckoning" approach. Such reconstruction requires accelerometer and magnetometer data accompanying location data. (Figure modified from Magowan et al. 2022.).

showed themselves in the accelerometer data as tight turnings in the middle of what was otherwise a long, fairly straight travel route. Global positioning system locations alone could not detect such foraging bouts. Such Dead Reckoning has the potential to fill telemetry data gaps with better information than available from statistical interpolations, though many global positioning system packages do not provide the detailed accelerometer and magnetometer data, and the accelerometer data must be collected at a high enough frequency to be able to differentiate behaviors (Yu et al. 2023). As technology improves, more details of animal movements will become available (Kays et al. 2015), showing how animals understand their cognitive maps and plan foraging paths.

Estimating travel across gaps in movement paths provides biologists with more information. A biologist must ask whether the added information will allow testing of hypotheses not otherwise possible. Building bridges across gaps in movement paths may be critically important, as highlighted by the European badgers studied by Magowan et al. (2022), or not of importance at all, as illustrated in

Chapter 11, showing that familiarity distributions including bridges usually differ little from simple kernel density familiarity distributions.

4.4 Dispersal

Dispersal is defined as an animal's exploratory movements whose goal is to establish a new home range or territory where the exploratory movements end. During dispersal, an animal moves through previously unknown areas, learns new surroundings, becomes aware of new potential resources, and looks for a place to call home that can support a new resident. This information is encoded into the animal's expanding cognitive map.

Usually, dispersal is obvious, such as when a young adult mammal makes a series of long moves that lead him to be far from his mother's home range. Sometimes, however, dispersal is not so obvious, such as when a daughter settles into a home range next to her mother's. Objective ways to quantify dispersal allow biologists to quantify movement metrics for dispersal trajectories among members of a population, to quantify individual differences, and to test hypotheses related to dispersal (Börger and Fryxell 2012). Mean squared displacement arises repeatedly in studies of random movements and studies of real animal movements (Uhlenbeck and Ornstein 1930, Nouvellet et al. 2009, Fleming et al. 2014b). The central role of mean squared displacement arises through its correlations with the time taken for the displacement and directional changes during displacement (Nouvellet et al. 2009).

Semivariograms, introduced in the previous section, can be used to document philopatry and establishment of a home range. The semivariogram for an animal who shows philopatry and who has an established home range rises and then stabilizes (Figure 4.8). In contrast, the semi-variogram for a random walk increases linearly. Dispersal is not a random walk, since a dispersing animal lingers in places that appear potentially favorable and may then move on.

Börger and Fryxell (2012) introduced the use of net squared displacement across time to identify dispersal, migration and seasonal home ranges, home range shifts, and resident home ranges. For their models, net squared displacement is the squared distance from a starting point to each sequential location in the data set. The starting location and time may be the beginning of the data set or another start point, (such as the center of an animal's natal home range when dispersal movements began; Börger and Fryxell 2012, Spitz et al. 2017).

For the approach of Börger and Fryxell, the model for the expected net squared distance of a dispersing individual, $D_{NS-disperse}$, follows a sigmoid curve that asymptotes at some measure of central tendency associated with the new home range:

$$D_{NS-disperse} = \frac{\delta}{1 + \exp((\theta - t)/\phi)},$$

where t is time since the designated start time; where δ is the distance from an animal's starting point to her new range; where θ is the time to the mid-point of dispersal; and where ϕ is the time required for her to go from ½ of the distance to ¾ of the distance to the new home range.

For a resident animal, the model for net squared distance, $D_{NS-resident}$, rises asymptotically to its mean:

$$D_{NS-resident} = \gamma(1 - \exp(\kappa t)),$$

where γ is the mean net squared displacement and κ is the natural logarithm of the rate constant that quantifies the initial period of increase required for $D_{NS-resident}$ to reach γ.

For an animal that is nomadic, the model for net squared distance, $D_{NS-Nomad}$, rises linearly, representing random movement:

$$D_{NS-Nomad} = \beta \cdot t,$$

where β is a constant.

Börger and Fryxell also provided equations for animals that migrate and for those that have seasonal home ranges and provided equations for models incorporating elevation. One can use information theoretic approaches (e.g. Akaice Information Criteria, AIC) or summed squared error to rank how well the different behavior models describe animals' movement trajectories and, thereby, assign movement bouts to categories.

Smereka et al. (2021) used this approach to differentiate movement categories of pumas. The timing of their initial captures of pumas and their subsequent data appeared to have been very well designed for Börger's and Fryxell's method.

Although a biologist can establish objective criteria for calculating values for most of the variables and parameters in Börger and Fryxell's behavior models, when a data set does not show dispersal vs residency clearly, calculating δ, θ, and ϕ can be difficult. For example, if net squared displacement appears to plateau, identifying when the plateau starts is not always clear, making the

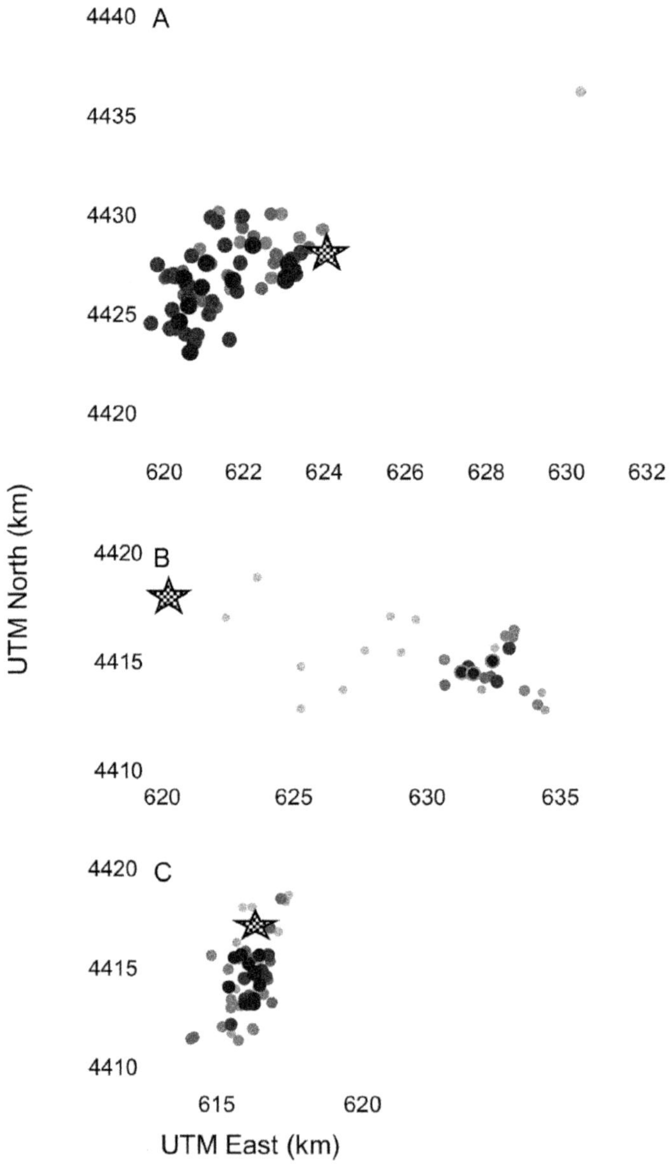

Figure 4.10 Plots of location estimates for 3 fishers. Stars denote where fishers were released. Locations shortly after release are small and grey, grading to large and black at the end of the period of tracking. A. Founding female fisher 17582 established her home range where she was released. B. Founding female fisher 1F111 dispersed over 15 km from where she was released to establish her home range. C. Adult, resident female fisher 20950 maintained her home range in the same area for over 6 years. This fisher's capture site in year 6 and subsequent locations are shown. Note that her capture site was not centrally located.

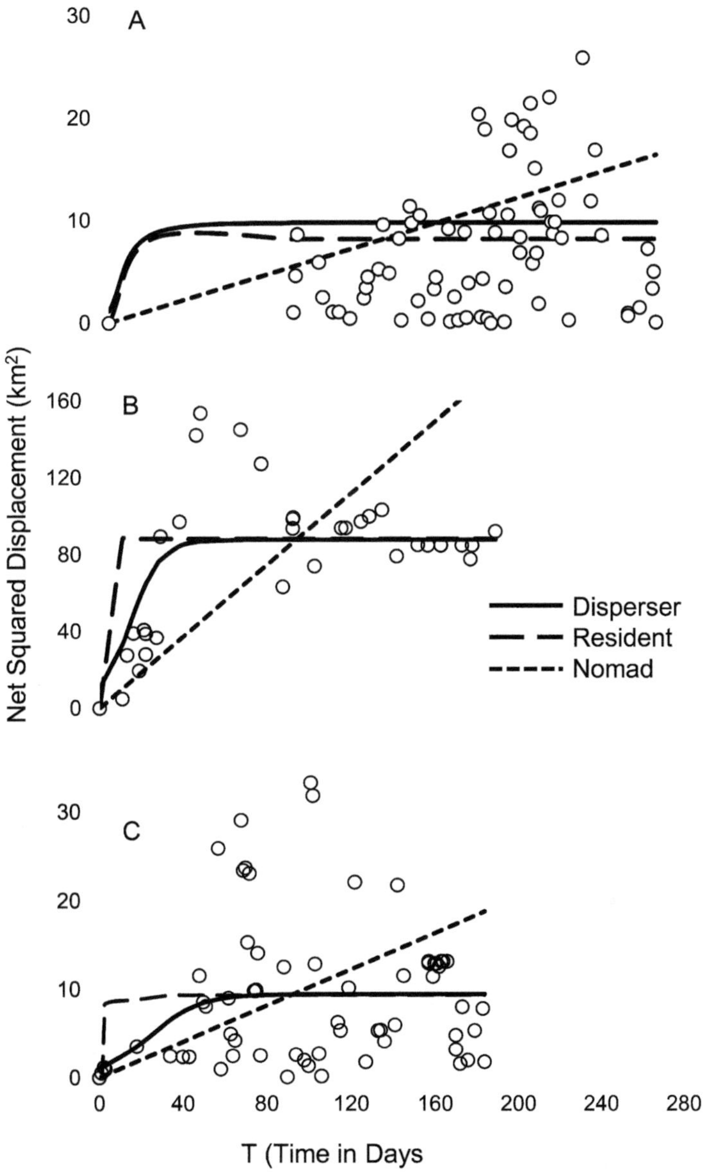

Figure 4.11 Plots of net squared displacement for the female fishers whose locations are shown in Figure 4.10 A, B and C. Dots show the data and the 3 curves show the predicted data patterns for a Disperser, Resident, and Nomad derived from the models of Börger and Fryxell (2012). A. The predicted pattern for a Resident matches the data best and, indeed, this founding fisher did not disperse but established her home range where she was released. B. The predicted pattern for a Disperser matches the data best and, indeed, this founding fisher did disperse and established her home range some 15 km from where she was released. C. The predicted pattern for a Disperser matches the data best according to both AICc and summed squared error analyses. Nonetheless, the fisher had been a resident for 6 years and maintained her home range in the same area all 6 years. Thus, the best fitting model did not match the true status of this fisher. AICc chose the wrong model for 75% of the fisher-years in the data set and summed squared error chose the wrong model about 40% of the time.

mid-point for dispersal and ½ to ¾ of the move also unclear.

Using data from a fisher reintroduction (Facka et al. 2016, Facka and Powell 2021), I ranked how well the behavior models discriminated between movements of known residents vs dispersers. I calculated from the data all model parameters using a priori, objective definitions without relying on data regarding residency. I chose β such that $D_{NS-Nomad}$ crossed the mean net squared distance half way through the period of data collection. I figured that if I used information on residency to affect how I calculated parameter values, that would not have been an independent test of how well the models discriminate behaviors.

My research team released 40 founding fishers in late autumn to early winter over 3 years beginning in 2009. Beginning with year 2 of the releases and continuing to year 8 of the research, we had a concentrated live-trapping campaign during late autumn into early winter each year. We followed almost all fishers using very high frequency telemetry (global positioning system telemetry for a very few). During our trapping campaigns, fishers fell into 5 non-overlapping categories: 1) founding fishers that settled where released (Figure 4.10A); 2) founding fishers that dispersed immediately from their release sites and established home ranges elsewhere (Figure 4.10B); 3) young-of-the-year that had been born into the new population and that had already dispersed and settled on their new home ranges before we trapped them for the first time; 4) young-of-the-year that dispersed after capture, handling and release; and 5) adult residents living on established home ranges (Figure 4.10C). For the 5 categories, and having no prior information about the fishers, initial moves could have represented either the initial period during which net squared distance approached the mean net squared distance for a resident's home range or could have represented a period of dispersal, during which time net squared distance approached the mean net squared distance after dispersal.

Figure 4.11 shows plots of net squared displacement for the 3 plots of locations shown in Figure 4.10. Figures 4.10A, B and 4.11A, B show data that the Börger and Fryxell method identified correctly. Figures 4.10C and 4.11C show data for a fisher that had been resident for 6 years yet was identified by the Börger and Fryxell method to be a dispersing fisher. In all, neither AICc nor summed squared error could discriminate consistently between the models for dispersers, residents, and nomads. The AICc choice was equivocal 17 times (ΔAICc $\leq$ 1), wrong 50 times, and correct only 22 times. Summed squared error did a better job but still chose correctly for only 54 of the 87 follows and was wrong for 33. The trapping and release sites for some juvenile fishers were 1–2 km from the rest of those fishers' locations during the following year, a distance easily travelled by a fisher during day-to-day activities. Bait used for live-trapping could have drawn juvenile fishers with newly established home ranges out of their home ranges, thereby causing the models to confuse resident juveniles with dispersers. Those release sites, however, fell within the ranges of locations for those fishers during future years. Adults with established home ranges were not drawn to traps in adjacent home ranges. Figure 4.10C shows that Fisher 20950 was live-trapped and then released near the edge of her plotted locations for that year but well within the area where she was expected to move.

The method of Börger and Fryxell (2012) can benefit from further testing to find what characteristics of data sets cause the models trouble, thereby guiding the use of the models to situations where they work well. Good tools are needed to identify dispersal moves.

CHAPTER 5

Information, Movements, and Home Ranges

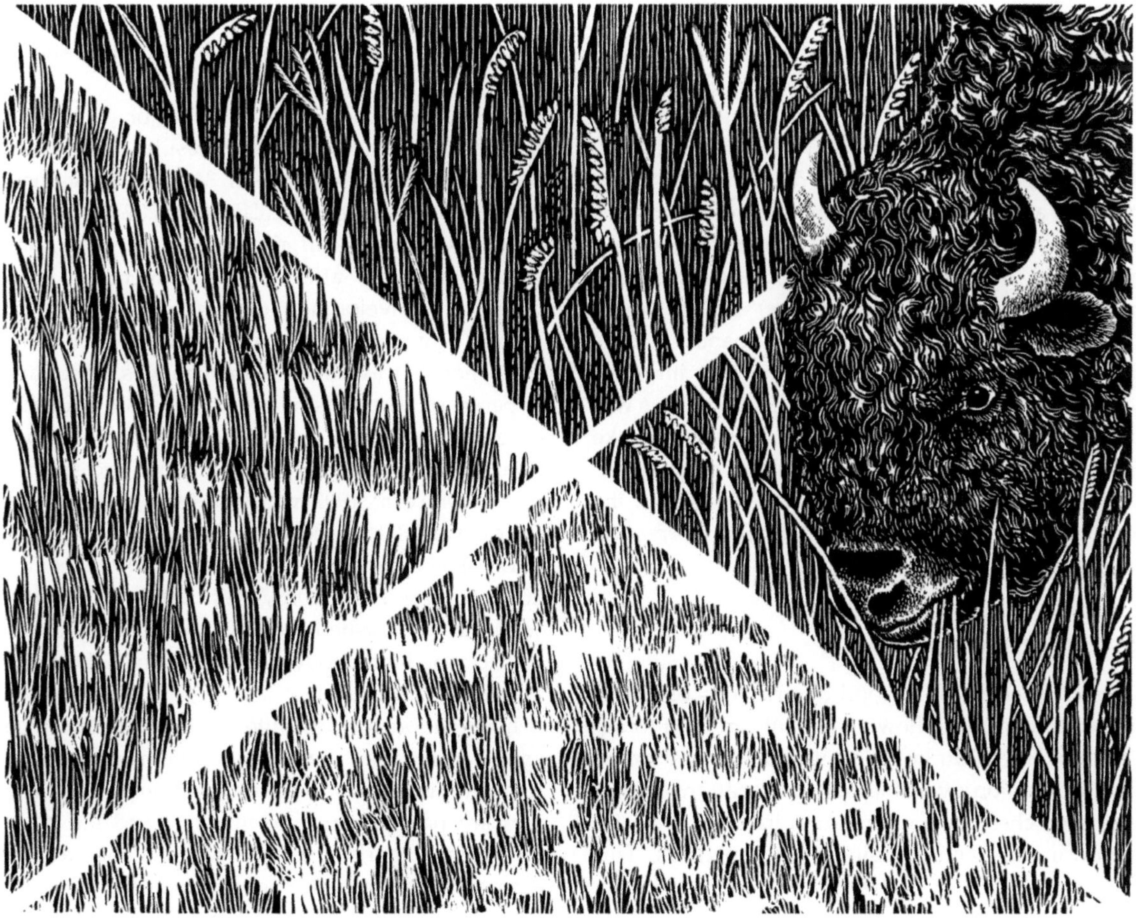

I snowshoed along a fisher track in the deep winter snow of Upper Peninsular Michigan and the track had been going nearly in a straight line for well over ½ km. From following a fisher track in this same area of forest in the previous year, I had a good hunch that this fisher was headed toward a porcupine's winter den that was not far away. The den was a large hole at the base of a big yellow birch tree.

And then, there was the den tree. In contrast to the previous year, however, no porcupine trails led

Home Ranges of Animals. Roger A. Powell, Oxford University Press. © Roger A. Powell (2025). DOI: 10.1093/9780191949142.003.0005

from the den to foraging trees nearby. The den was not occupied this winter. The den entrance at the base of the tree was covered with snow, completely disguising it. The fisher track led right up to the den tree and the fisher left a muzzle print in the snow exactly at the den entrance. Then, without even an apparent pause, the track led on.

The fisher whose track I followed clearly knew the location of that porcupine den and had sought it, hoping for a chance to kill a porcupine. Porcupines are important prey for fishers. Fishers seldom find porcupines under conditions that allow an attack and, even then, killing a porcupine is difficult. A porcupine, though, provides a fisher with possibly weeks of food in winter. Fishers remember where porcupines have their winter dens and check active dens regularly in winter to learn whether the resident porcupine is in a position allowing an attack. Porcupines usually leave their dens in the evening to reach a feeding tree to climb; they return to dens at early dawn. The fisher I followed probably timed his return to this den for evening or dawn. Spatial memory (porcupine den locations) decays slowly but attribute information (den occupancy) decays faster. This fisher had just updated his information about this porcupine den. That information will have affected his future choices for when to make a return visit to this den. The following winter, the fisher would still remember the location of the den but he would need to update his information regarding occupancy.

5.1 Information

This chapter builds beyond descriptions of movements to delve into how an animal can plan foraging movements to be efficient and effective based on information available beyond locations of foraging patches. Such information includes resource renewal rates and foraging by competitors. This information is part of the animal's cognitive map. The chapter begins with a brief overview of how memory and information affect foraging decisions and movements, leading to the emergence and updating of home ranges. The chapter introduces a model for how information affects animals' exploration of new environments and, thereby, home range establishment, and how information decay affects foraging choices after home range establishment. The model also shows how an animal's knowledge of how information decays can lead to maintaining a territory through avoidance of areas where information decays rapidly or unpredictably. The chapter then introduces another model of animals using information to choose patches to include in territories. Next, the chapter introduces my own simulations of foraging animals' movements, which show that return intervals to resource patches change with resource renewal rates and with information loss. Analyses of resource renewal rates and return intervals for foraging black bears and fishers follow. Lastly, the chapter explores briefly how changing landscapes can affect how animals should respond to information loss.

An animal's memories are a major component of her cognitive map, which resides in her hippocampus complex (Chapter 3). Direct tests have established (Ranc et al. 2020, 2021, 2022) what has been assumed for all previous studies of home ranges: Memory affects establishment of home ranges and development of cognitive maps, affects animal movements, and affects choices for when and where to travel within home ranges. That travel depends on locations of and qualities of resources, on encounters with aggressive neighbors, on locations of scent marks from neighbors, and so forth (e.g. Powell et al. 1997, van Moorter et al. 2009, 2016, Burns et al. 2011, Merkle et al. 2014, 2019, Wirth et al. 2017, Menezes et al. 2019, Wehner 2020, Collet et al. 2022, Bridges et al. 2023, Muth 2023). Theoretical work has established the obvious: Memory can be incorporated into models of home ranges (Potts and Lewis 2016).

Information, or attribute memory, differs from spatial memories. Resource abundances, encounters with neighbors, even information about predators are part of attribute memory and are based on recent experiences. The accuracy of attribute memory, of information, decays relatively quickly. Resource locations and environmental features are part of spatial memory, however, which decays slowly (Ranc et al. 2021). How and why an animal returns to a site affects and is affected by the animal's nutritional needs, habitat preferences, and landscape topography. These are central topics in ecology and behavior (Morris 2003, Nathan et al. 2008) and are central components of cognitive maps. Having information about resources can lead animals to value site fidelity over energy maximization (Merkle et al. 2015). Nonetheless, early models of optimal foraging predicted at what reduced value a forager ought to leave a resource patch (e.g. Charnov 1976b) but ignored calculating the optimal time to revisit that patch, assuming for simplicity that an environment can not be depleted and that foraging animals know the locations and qualities of all patches (Charnov 1976b, Emlen 1966, MacArthur and Pianka 1966). Once the concept of revisiting a resource patch was incorporated into models, the basic assumption became that an animal should not return to a patch already foraged until the patch has completely renewed (Kamil 1978, Davies and Houston 1981, Possingham and Houston 1990). This assumption led to the study of traplining patches[1] (treating patches like a fur trapper's trap line, where traps are checked in some order; Ackerman et al. 1982, Anderson 1983, Thomson et al. 1997).

Revisiting patches depends on more than renewal rates, however. Paton and Carpenter (1984) showed that exploitative competition for particular flowers led traplining hummingbirds to return to those flowers sooner than they returned to flowers to which they had exclusive use, thereby beating competitors to the flowers. Menezes et al. (2019) showed that exploitative competition led gerbils to increase visitation rates to resource patches. Thus, the information that an animal has about a resource patch also affects when the patch will be revisited.

Most models of animal movements assume that individuals establish and maintain home ranges based on distributions of resources, competitors, and predators, and that individuals move through their home ranges so as to increase fitness by using the environment efficiently (some baseline examples are by Fretwell 1972, Schoener 1981, Stephens and Krebs 1986, Rosenzweig 1991, Powell 2004, Mitchell and Powell 2004 2008, Stephens et al. 2007). Most models, apart from recent optimal foraging models, are based on assumptions that animals know nothing about the distribution of resources (Emlen 1966, Cody 1971, 1974, Pulliam 1974, Charnov 1976b, Pyke et al. 1977, Pyke 1978, 1981, 1984, Benhamou and Bovet 1989, Bovet and Benhamou 1991, Turchin 1991) or that they know everything about the distribution of resources (Fretwell 1972, Anderson 1983, Ford 1983), neither of which is ever true. Even dispersing animals, who might be assumed to know nothing about the environments they explore, use their senses to learn about the resource patches on new, unknown landscapes before reaching individual patches. Long-term residents often know when competitors have reduced the value of a resource patch. The more a biologist knows about her study animals' cognitive maps, the better to predict the animals' movements and return intervals to foraging patches.

Spencer (1992, 2012) developed a theoretical model for an animal foraging and for the emergence of home ranges on a dynamic and only partially predictable landscape.

5.2 Information, Foraging, and Home Ranges

Spencer (1992, 2012) started with realistic assumptions about cognitive processes that animals use to create, to update, and to maintain their cognitive maps. Their cognitive maps include not just the locations and status of resources but also the other animals of the same and different species who share and change the environment. Animals use all this information to plan future movements, which then contribute to changes in their cognitive maps as they learn about how resources have changed in ways not expected. The process is a continuous feedback loop.

Spencer, therefore, started with the assumption that an animal has "imperfect *expectancies*" of resource availabilities at patches already incorporated into his cognitive map. Those expectancies

[1] Finding the optimal order to visit patches (or a trapper's trap sites) when traplining is similar to the Travelling Salesman Problem, which is to find the shortest route among the salesman's assigned cities when starting and ending at his home. The Travelling Salesman Problem is one of a class of problems called NP-complete that are so difficult that no analytical proof has been derived (a computer proof has been derived for the Travelling Salesman Problem) and no animals (including humans) are able to calculate on the hoof the optimal route to take among the sites. Anderson (1983) was the first to show that a number of easy rules for traplining are "close enough," such as "Go to the patch with the highest value discounted by travel cost."

decay over time to some known average, or background, expectancy as competitors, movements of predators, weather, and unknown events affect the potential of a resource patch to contribute to the animal's fitness. The animal's choices to visit, or not to visit, known patches are goal-driven based on his expectancies and how those expectancies will change over time. Because the expected reward from a patch decreases over time due to not knowing whether some competitor visits the patch, an animal should visit a given patch while its information about that patch is most complete. In contrast, though, many resources renew slowly after being exploited or depressed (Charnov et al. 1976) and an animal should delay visiting those patches until they have renewed significantly. The result is that an optimal time exists for returning to a patch, a time that balances the potential renewal of the patch with the loss of information about that patch over time (Spencer 1992, 2012).

Although Spencer modelled animals dealing with food resources, information about predator avoidance and other resources also fit his model. A site where a predator detects a target animal, for example, has become dangerous for the target animal but the safety of that site renews the longer the animal is away from the site, ultimately reaching the average safety value for the home range (or an average value for the habitat). At the same time, the animal's information about that site decays, just as does the information for food resource at the site, though information decays at different rates for food information and safety information.

With this background, Spencer built the following model. Assume that the expected value, V, for the resources in a patch is a function of resource availability, R, and the proportion of that availability, Δ $(0 \leq \Delta \leq 1)$, that an animal can remove during a visit to the patch. Assume a Type I functional response for foragers (Holling 1965). Thus, at the beginning of foraging in a patch,

$$V = \Delta R.$$

While foraging, the animal reduces V and R to final values of V_0 and R_0, which are the initial values that set the stage for resource renewal and information decay.

Neither resource availability, R, nor the proportion of R that an animal can harvest, Δ, will be the same for all patches of the same resource. Some patches will have a conformation that makes availability of the resource easier or harder to harvest. The safety of patches differs, affecting the time an animal will spend in a patch and, therefore, the size of Δ. Two patches of the same resource with the same danger of predation but with different values of R will likely have a similar giving up density of the resource (Joel Brown's "GUD," Kotler et al.1991) and, therefore, a different Δ.

Real animals learn the landscapes on which they have established their home ranges. Their cognitive maps include specific knowledge of how to move through a berry patch most efficiently, where vole tunnels are most dense, how best to approach each reef from the air so as not to spook fish near the surface, or which flowers in a patch harbor the most nectar. This information, I, has value to a forager and increases V:

$$V = \Delta R(1 + I).$$

Such information, however, is dynamic and can decrease in quality as time passes. At some time after a black bear leaves a berry patch, maybe soon, maybe much later, or after a hummingbird leaves a patch of wild bergamot, a competitor (a conspecific or a member of another species) can visit the patch and reduce the values of V and R back down toward V_0 and R_0 (their values when the forager last finished in the patch), or even below (depending on the competitor), while the forager anticipates that the value of the patch will have continued to renew. For example, rodents' seed caches sometimes get pilfered and, thereby, decrease in size at decelerating rates as time passes (Cao et al. 2018, Dimitri and Longland 2022, Vander Wall et al. 2006). Similarly, a predator in a patch of prey habitat depresses prey availability by making prey cautious, whether or not the predator captures any prey (Jędrzejewska and Jędrzejewski 1990, Ylönen 1989 1994, Ylönen and Ronkainen 1994). Such relationships between experience and performance can be modelled with negative exponential functions (Anderson 1995):

$$I_t = I_0 e^{-dt},$$

where I_t is the degraded quality of information about a patch expected by an experienced forager at time t since leaving the patch; I_0 was the quality of the information to the forager immediately upon leaving the patch and that information decays at rate d. A negative exponential function has no special magic and other functions that decrease at a decelerating rate could be used. At present, we have no biological reason to use any particular function in Spencer's model. Holling (1959) modelled the Type I functional response using a different decelerating function that did have biological logic but that does not apply here because it relates to searching for hidden resources. Luckily, most decelerating functions give the same basic results. Finally, assume that patches renew at a decreasing rate, also modelled as a negative exponential function with intrinsic renewal rate of r, making the expected value of a patch at time t since last leaving the patch to be

$$V_t = \Delta\left[R_0 + (R_{max} - R_0)\left(1 - e^{-rt}\right)\right]\left(1 + I_0 e^{-dt}\right), \quad \text{Eq 5.1}$$

where R_{max} is the maximum resource value for the patch, which most animals can assess, at least approximately. As time passes since a forager has left a resource patch, the value of that patch, V, increases from V_0, rising past the background expected value for patches in the environment, then peaking because the quality of the information about the patch falls (Figure 5.1A). The value of the patch finally asymptotes at the background expected value for patches of its resource type. If the rate of resource renewal increases or decreases, the timing of the peak comes earlier or later, which makes sense (Figure 5.1B). Similarly, if the quality of information decreases faster or more slowly, the timing of the peak comes earlier or later (Figure 5.1C). Using different equations for the curves can change their specific shapes but, nonetheless, all lead to a humped curve and to similar predictions for the relative timing for when a forager should return to a patch.

As long as information has value, an informed forager should return to patches before they renew completely because the hump in the curve of patch value raises patch value above the value otherwise expected (Spencer 2012). Returning to patches leads to the continued updating of a forager's cognitive map, which includes expected resource renewal rate and expected information decay.

Although nearly always true, returning to patches has one exception. When the combined productivities of all the patches that a forager can track is not able to meet the forager's requirements, the forager should become at least partially nomadic until it can find a landscape that can meet its requirements (Chapter 7; Powell et al. 1997). The forager will make the decision to become nomadic based on the information she has gleaned from the landscape, meaning that information does have value for the forager, just not the information needed to establish or to maintain a home range.

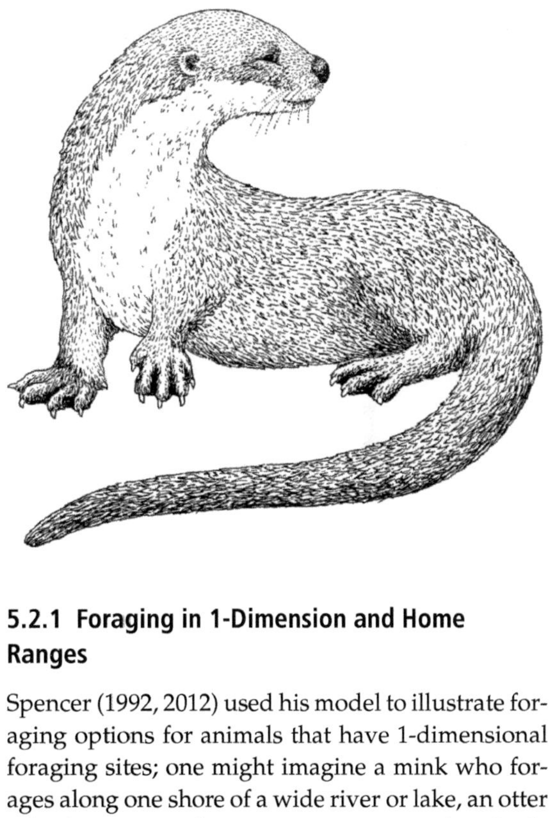

5.2.1 Foraging in 1-Dimension and Home Ranges

Spencer (1992, 2012) used his model to illustrate foraging options for animals that have 1-dimensional foraging sites; one might imagine a mink who forages along one shore of a wide river or lake, an otter who forages in a long, narrow river, or shorebirds foraging along waterways or shorelines (Botton et al. 1994, Davies and Houston 1981, Gerell 1967, Harrington et al. 2009, Karpanty et al. 2006, Sargeant et al. 1973). As such an animal forages along the foraging dimension, she depletes or depresses resources as she moves forward. One might assume

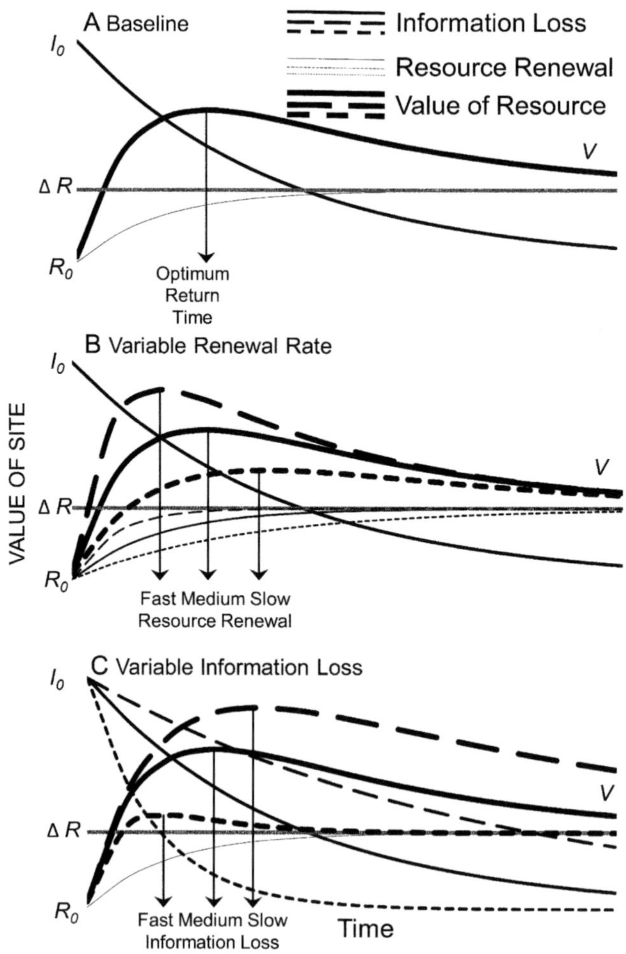

Figure 5.1 Optimum time to return to a patch. I_0 is the initial information available to a forager when he finishes foraging in a patch and leaves. ΔR is amount of resource that a forager can obtain from a patch. V_0 is the initial value of the patch to a forager when he leaves a patch. R_0 is the initial resource availability in the patch when a forager leaves the patch. A. When a forager leaves a resource patch, information available about the patch begins to decay. At the same time, the resources within the patch begin to renew. The optimum time to return to the patch is that time when the value to the forager is maximized, which is when information per resource is maximized. B. If resource renewal occurs faster (long dashes) or slower (short dashes), the optimal return time comes sooner or later. C. If information loss occurs faster (short dashes) or slower (long dashes) the optimal return time comes sooner or later.

that the forager moves from 1 position to the next adjacent position at a constant rate and that the forager is naïve initially about the resource distribution along her foraging path but has an expectation of average resource availability developed from past experiences elsewhere or at this site in times past.

At any time, the forager can decide to continue foraging forward, to stop, or to reverse direction and to forage along the route that she has just travelled. Stopping is a poor option because the value, V, of the immediate position in the foraging site will simply diminish as the forager continues to forage there. The foraging prospects ahead are the expected average availability of resources at not-foraged positions on the landscape, such as what the forager has just been experiencing. The values for foraging positions close behind the forager will be lower than the values for positions immediately

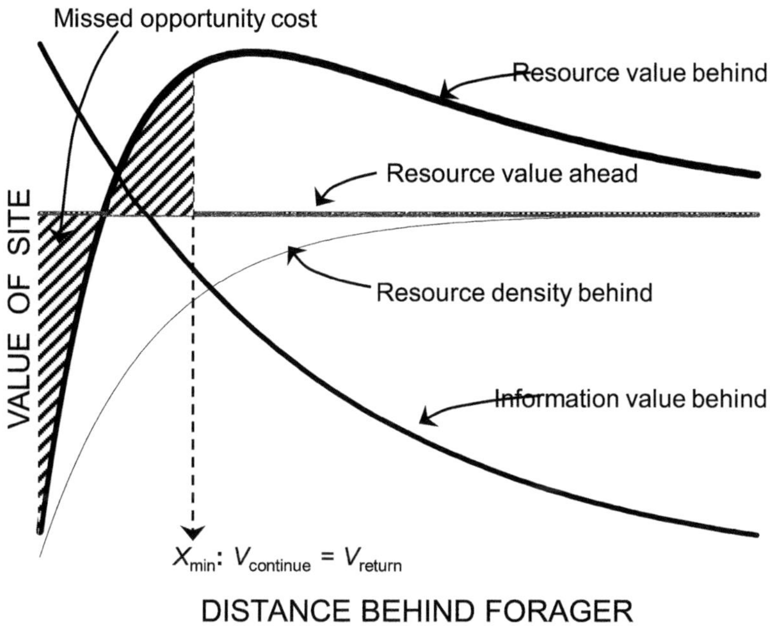

Figure 5.2 At some point, a forager can increase his resource gain by reversing direction and taking advantage of information learned about the resources behind him that he has visited. Immediately upon reversing direction, however, the forager faces resources lowered below the anticipated value of having continued straight ahead (missed opportunity cost). At a certain time and distance of reversed travel, due to information and resource renewal, resource values reach and then exceed the anticipated value for unvisited resources. And at some further time and position, t_{min} and x_{min}, resource values are high enough to compensate for the initial low returns after reversing direction. Beyond that distance and time, foraging at previously visited sites reaps rewards in excess of not having reversed direction.

in front of the forager. Positions just foraged do renew, however, as the forager moves further and further in the reverse direction. And, as the forager has foraged in each position along her path, she has gained information about the resources available in that position, information that will help her improve her rate of resource gain should she return to that position.

Depending on the rate of resource renewal and the value of the information related to improved foraging success at previously visited positions of the site, a time may arrive when reversing direction pays a dividend. The information learned about visited sites diminishes as time passes and, therefore, provides diminishing benefits. The forager must travel a ways back to reach the point where resource renewal and still relevant information compensate for the immediate low returns from reversing direction (Figure 5.2); reaching that point must occur within the time still available for the present foraging bout, T (Spencer 1992, 2012). The value, $V_{x,t}$, at position x and time t since turning around, formerly behind but now in front of the forager, is

$$V_{x,t,} = \left(V_{x,0} + [1 - V_{x,0}]\left[1 - e^{-2rt}\right]\right)\left(1 + I_{x,0}e^{-2dt}\right).$$

When the forager reverses direction, at some point going back the accumulated resource gain from the combination of information and resource renewal may exceed the missed opportunity cost (the anticipated resource gain from not reversing direction). The foraging breaks even at time t_{min} or distance x_{min} from turning around. If the time left in the foraging bout upon reversing direction is $>t_{min}$, then reversing direction will yield more resources than continuing ahead, where the forager has no position-specific information. If the time left in the foraging bout exceeds t_{min}, then the forager has options for reversing and re-reversing its foraging direction depending on how quickly resources renew, how quickly information decays, and whether the forager reaches the starting position of its foraging bout.

One might consider this 1-dimensional exercise restricted to too few realistic situations. Even so, animals foraging in 2 dimensions do sometimes travel through recently foraged resource patches and are faced with situations of choosing between foraging in patches with limited information and average expectations vs traversing patches just foraged to reach other patches for which the forager has important information. Consequently, this exercise does introduce realistic challenges that face 2- and 3-dimensional foragers.

Spencer (1992, 2012) provided a set of characteristic combinations of resource renewal and information loss and the patterns produced for foraging in 1 dimension. In the most realistic combinations, a forager moves back and forth across expanding lengths of habitat until it establishes a dynamic equilibrium between resource depletion or depression with resource renewal. Factors that favor early return, such as high information value coupled with rapid information decay (e.g. as resource distributions change) leads an animal to concentrate activity into a smaller area than would be predicted by resource availability alone, and potentially smaller than required to supply the individual's resource needs in the long term. In such cases, movements resemble a "shifting core" or "shifting foraging area" pattern of home range use, as the forager concentrates his movements in the short term to take advantage of recent information, but shifts to other areas over the longer term because the initial area cannot support his needs indefinitely. This pattern has been described in numerous mammals (e.g., Altmann and Altmann 1970, Gerell 1970, Waser 1976, Milton 1980, Terborgh 1983, Rabinowitz 1986, Jones 1989, Powell et al. 1997, Fryxell et al. 2008). In cases of distributions changing unpredictably (e.g. high information value, but with rapid decay), entire home ranges may shift over time. Under these conditions, fidelity to a localized area may be beneficial in the short term but gross changes in resource distributions favor shifting to new areas as they gain value. For example, the territories of red foxes in urban Oxford, England, "drifted" at a rate equivalent to complete displacement every 13 months (Doncaster and Macdonald 1991; Moorhouse and Macdonald 2005), which the authors attributed to unpredictable food availability, high mortality, and social instability in urban environments.

When resources and their renewal rates are known, information and its decay rate are known, and the time for foraging is long enough, a 1-dimensional forager can set up a "home range," passing back and forth across a stretch of her linear habitat. The forager may cover her entire home range with each reversal or she may reverse a few times in succession in a small area before progressing across a longer area, then doing some quick reverses again (Figure 5.3). Different combinations of resource harvest, resource renewal, and information loss can lead foragers to compartmentalize foraging into one confined area and then into another. Both optimal home range length and total harvest value decrease with increasing rates of information decay and decreasing rates of resource renewal; both changes lower and move the hump in the resource value curve (Figure 5.3).

Different rules for when to reverse direction may yield more resources, such as, for example, not reversing when reversing first proves more productive than continuing forward but waiting to reverse and allowing the values of resources to increase more. Maximizing resources, however, may not be important to an animal. Animals may seek to acquire required resources in the minimum time necessary or within a minimal area (Charnov 1976b, Mitchell and Powell 2004).

Spencer (1992, 2012) simulated landscapes with only 1 forager, whereas real foragers exist within communities with multiple conspecific and heterospecific competitors. A forager who encounters a patch recently depleted or depressed by another forager will gain little resource but gains new information from that patch. Such a forager will recalibrate the expected resources ahead and may benefit from reversing foraging direction. Under these conditions, multiple foragers will tend to space themselves out over time, leading to minimal overlap of home ranges, especially when information has high value. The result will be territoriality via mutual avoidance. This situation matches closely a model proposed by Stamps and Krishnan (1999).

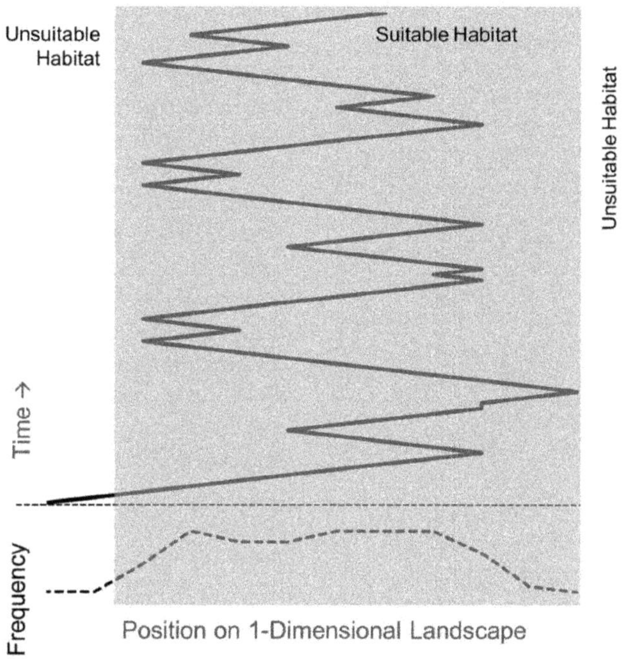

Figure 5.3 Top Curve. A forager's track through time (solid, going upward from the fine line across the graph) on a 1-dimensional landscape, showing reversals of travel direction. The forager moves 1 position unit with each time unit and reverses direction following Spencer's model. Bottom curve (dashed, below the fine line) shows the forager's familiarity distribution along the single dimension.

5.3 Information and 2-Dimensional Home Ranges

Stamps and Krishnan (1999) modelled gain and loss of information somewhat similarly to the way that Spencer (1992, 2012) did. Assume that when a naïve animal explores a new landscape that may or may not already be occupied by conspecifics, he starts building a cognitive map of the area, including what resources are available, where the resources are available, and how the resources can best be harvested or otherwise used. Before the animal approaches a patch, the value of the patch, V_u, is unknown but, having reached the patch at time t, the animal is able to perceive some information about the patch's value, V_t (using scent, sight, hearing, and other senses). The probability that the animal enters the patch depends on V_t compared to the values of other patches that he has already visited.

Revisiting a patch leads to increasingly better understanding of the resources in the patch and how to harvest them. This understanding increases with repeated visits in a decelerating manner toward some maximum amount of information or value for the patch, V_{max}. If an animal encounters a conspecific in a patch, an agonistic encounter ensues. Such an encounter decreases the value of that patch toward a maximum loss of value, F_{max}, in a decelerating manner that depends on the numbers of encounters in the patch. The model for patches that have been visited becomes

$$V_t = V_{max}\left(1 - e^{-Npt/Rp}\right) - F_{max}\left(1 - e^{-Nft/Rf}\right), \quad \text{Eq. 5.2}$$

where

N_{pt} = the number of positive visits that the animal has had to the patch by time t,

R_p = the extent to which each positive visit to the patch enhances the patch's value,

N_{ft} = the number of agonistic encounters that the animal has had in the patch by time t,

R_f = the extent to which each agonistic encounter in the patch reduces the patch's value.

Assume that an animal visits patches and adds a new patch to her territory based in some way on V_t compared to the values for other patches, until she has assembled a territory with enough of the limiting resources so that she can survive and reproduce. Stamps and Krishnan assumed that animals prefer known patches to unknown patches, which can contradict their assumption that animals add patches based on patch value. For example, if an animal encounters a new patch and her abilities to assess the patch from the patch's edge indicate that this new patch is truly superior, she should not prefer patches already visited simply because they have been visited.

This model implies that territory owners have an advantage in agonistic encounters because a contested patch has a higher value to the owner who knows the extent of the resources in the patch and how best to harvest those resources. Although an intruder who has no territory might value the patch highly, the potential cost of injury during an agonistic encounter is also higher for the intruder.

Stamps and Krishnan ran 2-dimensional computer simulations using different sets of parameter values with 10 repetitions per set. Simulated animals established territories with appropriate numbers of resource patches for the various parameter values. Residents had strong advantages over invaders.

One could redefine the $F_{max}(1 - e^{-Nft/Rf})$ in equation 5.2 to be loss of information about a patch because it was visited by a competitor. Likewise, one could redefine $(1 + I_0e^{-dt})$ in equation 5.1 to be loss of patch value due to squabbles with neighbors and exploitation competition. Equations 5.1 and 5.2, of Spencer and of Stamps and Krishnan respectively, are not equivalent but they are similar, each showing a gain in information about resources while foraging in a patch and a reason to avoid patches that are used by competitors. Had Spencer adopted whatever rules Stamps and Krishnan adopted for choosing which patch to visit next, he could have simulated 2-dimensional home ranges as they did. The models of Spencer and Stamps and Krishnan do, however, differ in an important way. In Spencer's model, loss of value for a resource patch stems from loss of information and that loss increases as time passes since the last visit to the patch. In the model of Stamps and Krishnan, loss of value for a resource patch is due to competitors but information about resources does not decrease: an animal has constant knowledge about resource values of patches as time passes.

Morrell and Kokko (2005) used a game theoretic approach to the situation modelled by Stamps and Krishnan (1999) with very similar results. And, as with the model of Stamps and Krishnan, one could redefine the formulas used by Morrell and Kokko to result in contestants avoiding sites where information about resources can not be trusted rather than visiting such sites and incurring the cost of having gone there and found no resource available. As with Spencer's model, this result is territoriality via mutual avoidance.

5.4 Calculating the Value of Information

Animals with site-specific information may search more efficiently than naïve animals, may increase their probability of capturing elusive prey, or may return to productive sites before they are exploited by competitors (e.g. Croze 1970; Gill and Wolff 1975; Schaller 1972; Vander Wall et al. 2006). If P_n is the probability of a naïve forager obtaining a particular resource item and P_i is the probability for an informed forager, then the value of information, I, is indexed as

$$\mathrm{I} = P_i/P_n - 1.$$

Experiments have measured these probabilities and can be used to calculate I for diverse foragers. For example, territorial sunbirds (informed foragers) have higher foraging success by ≈25% ($I \approx 0.25$) compared to intruders (naïve foragers) due to avoidance of recently depleted flowers (Gill and Wolff 1975). Decay in memory for birds and mammals that scatter-hoard food items is very low, with I = 31.7 to 41.5 for Clark's nutcrackers (Vanderwall 1982) and I = 42.3 for a captive Eurasian nutcracker (Balda 1980). Spencer (1992) calculated I for eastern grey squirrels = 1.2 two days after caching, 1.3 after 4 days, and 0.8 after 12 days. Vander Wall et al. (2006) found that yellow-pine chipmunks recovered their own seed caches 3.4 to 6.5 times faster than paired, identical artificial caches or than pilfering from (naïve) squirrels, by using spatial memory, yielding $I = 2.4$ to 5.5.

5.5 Memory, Conflict, and Territory Establishment

Potts and Lewis (2016) developed a model for the generation of territories for animals that have direct conflict (fights) in conflict zones between home ranges and whose memories of the conflicts decay. The safety that an animal experiences in a conflict zone increases when it enters the zone without conflict. Territories are generated when animals flee conflicts quickly and conflicts do not always occur when animals meet. Real animals do remember conflicts and where they occur. The authors modelled information loss to be linear, though memory or information loss is not critical to the biological situation they have constructed. Potts and Lewis claimed that direct conflicts being involved in establishment of territories had not received serious (modelling) consideration but they appeared not to have been aware of the work of Stamps and Krishnan (1999).

5.6 2- and 3-Dimensional Home Ranges

Real foragers, of course, even those that forage at times in 1-dimensional habitats, live on 2- and 3-dimensional landscapes. I developed individual-based models for animals foraging on 2-dimensional landscapes. The models follow multiple individuals for a year at 100 time steps per day as the individuals forage for 10 resources with different annual phenologies, each in 250 patches in each of 10 densities distributed randomly, clumped, or evenly across a 2-dimensional landscape. Each forager seeks the patches with the highest return discounted by travel cost, learns how best to forage in each patch visited, remembers renewal rates, and deals with competitors.

As renewal rates for patches vary from low to high, foragers return to patches sooner, and they return sooner when information decays quickly compared to slow decay (Figure 5.4). And, with appropriate parameter values for resource renewal, travel speeds, and food requirements, simulated foragers establish home ranges with little overlap by avoiding patches that are regularly foraged by competitors. Such home ranges can be considered territories. Thus, this 2-dimensional model mimics the major results of Spencer's (1992, 2012) 1-dimensional model and the results of Stamps and Krishnan (1999).

On my black bear study area in the Blue Ridge Mountains of North Carolina, USA, my research team documented total berry production and patch renewal rates during 12 years. Patch renewal rates correlated with patch productivities. Using berry productivity as an index of renewal rate, I analyzed return intervals in years with different berry and hard mast productivities for the black bears. I also analyzed return intervals for black bears in

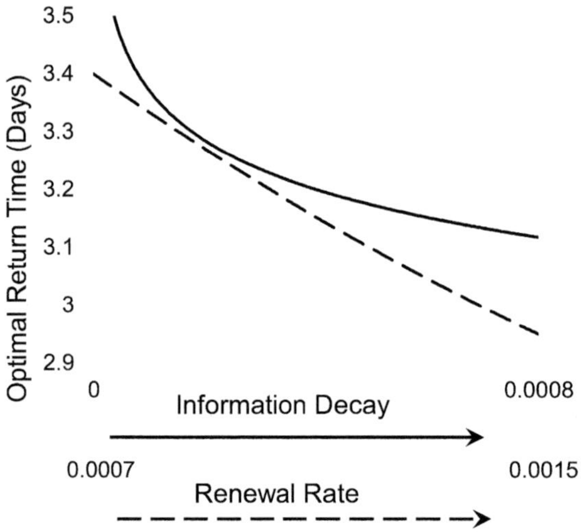

Figure 5.4 Optimal return times for model animals foraging on a simulated landscape with 250 patches for each of 10 resource types with 10 densities distributed randomly. Increasing decay rates for information and renewal rates for resources lead to shorter optimal return times.

northern Minnesota, USA (Powell et al. 2022) and assumed that high food productivity led to high renewal rates at this site as well. For the fishers that I studied in northern California, USA, I assumed that when the major prey population was high, some individual prey would become active again soon after behavioral depression caused by a predator, but, when the population is low, few individuals will become active again soon. I analyzed return intervals in years with different prey densities.

For each black bear and fisher, I looked at every active telemetry location (ignoring resting locations) and calculated the time until the individual returned to that location, if it did return. I used 125 m as the threshold for returning for black bears in North Carolina (a bear that was located $\leq$ 125 m from a site was considered to have returned to the site), 50 m for black bears in Minnesota, and 100 m for fishers. I set 2 days as the minimum return interval, to avoid confusion with an individual resting at a foraging site for a long time. For bears, I analyzed return intervals for summers, when berries were major foods, and autumns, when hard mast was the major food (Powell et al. 1997, 2022). For fishers, I analyzed return intervals with respect to population fluctuations of western grey squirrels and Douglas squirrels separately, together, and then together with California ground squirrels as well, the major prey for the fishers on my study site (Facka 2017, Powell et al. 2018).

For black bears in North Carolina and for fishers, the responses of individual animals differed and, therefore, I blocked the analyses by the individuals to test for the effects of food variation. For black bears in both summer and autumn, return intervals decreased with increasing resource renewal rates (Summer, $F = 1.55$, $p = 0.03$; Autumn, $F = 5.27$, $p < 0.001$; Figure 5.5). Similarly, for fishers for all prey groupings, return intervals decreased with increasing prey renewal rates, though the relationship was most clear for western grey squirrels, which were the fishers' most important prey (Facka 2017; western grey squirrels, $F = 2.32$, $p < 0.001$; Douglas squirrels, $F = 2.33$, $p < 0.001$; both tree squirrels together $F = 2.32$, $p < 0.001$; tree plus ground squirrels, $F = 2.33$, $p < 0.001$; Figure 5.6). Thus, black bears and fishers appear to use information about renewal rates as predicted by Spencer's model.

Choosing the criteria for what constitutes a return to a site must make biological sense and must be consistent with the technology used to collect data. Forest stands, as defined by the foresters who managed my fisher study site, were several hectares in size and tended toward uniformity in tree age. Although topography and physical conditions varied within each stand, they differed much less within stands than among stands. Squirrel populations varied with stand ages and with the dominant tree species, which were managed to be fairly consistent within stands. Telemetry error for handheld, very high frequency (VHF) telemetry averaged 100 m (to the nearest 10 m). Fisher locations averaged > 1 day apart and, therefore, previous locations did not provide much help for project personnel trying to locate fishers on any given day. Thus, the 100 m criterion for a fisher returning to a site makes biological and technological sense.

For black bears, some berry patches were < 25 m in diameter, though most good ones exceeded 100 m, suggesting that 50–75 m would make more biological sense for defining return to a patch than the 125 m distance chosen for North Carolina bears. The mean telemetry error measured for randomly placed collars, however, was 250 m (Zimmerman

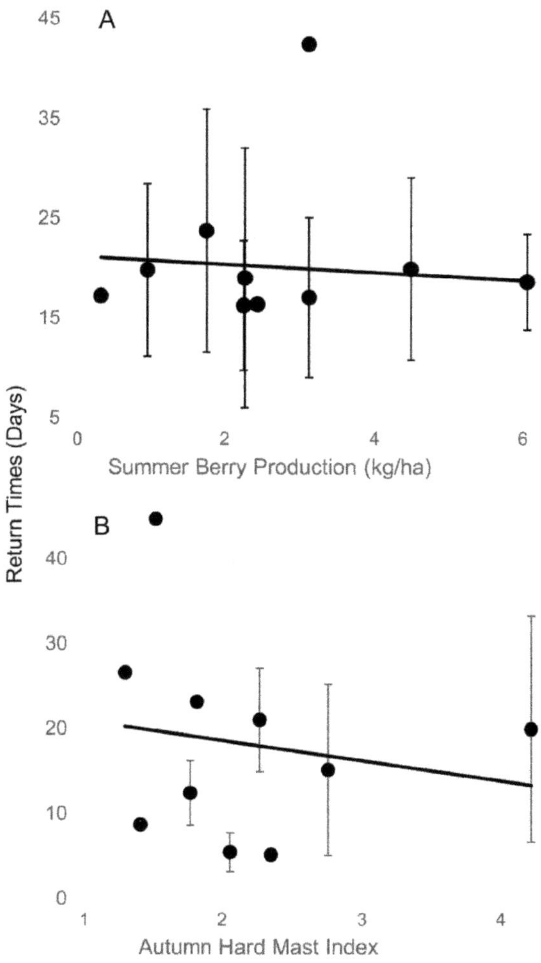

Figure 5.5 Black bears in the Blue Ridge Mountains, North Carolina, USA, returned to food patches more slowly when food renewal times for food were slow. A. Summer return times in response to berry production. B. Autumn return times in response to hard mast production.

and Powell 1995). Bears were located at 2-hour intervals grouped into sessions that lasted for 8, 12, or 24 hours. The result was that previous locations did assist in finding subsequent locations, reducing error somewhat for bear data compared to data from randomly placed collars. I compromised by using the 125 m criterion for bears returning to a site. Here, technological limitations trumped biology. Being forced to use a distance criterion much larger than 125 m would have prevented me from using my bear data to test Spencer's predictions.

The black bears in Minnesota wore global positioning system (GPS) collars that located themselves every 10 minutes. Telemetry error averaged < 35 m. For these bears, I used a 50 m criterion for returning to a site because doing so made both biological and technological sense.

The conditions under which some populations live may preclude behavior consistent with models of behavior. The bears in northern Minnesota were supplemented with food in all years (Powell et al. 2022). Return intervals to supplemental feeding sites increased with increasing production of both summer and autumn wild foods (glm; summer foods, $F = 11.8, p < 0.001$; autumn foods, $F = 6.7$, $p < 0.001$), showing the bears' preferences for wild foods but willingness to eat supplemental foods when wild foods were scarce. The result was that return intervals to wild foraging sites also increased with increasing production of wild foods (glm; summer foods, $F = 7.6$, $p < 0.001$; autumn foods, $F = 36.2$, $p < 0.001$), the exact opposite of Spencer's prediction.

Bears appear to plan return intervals to food patches based on resource renewal and probably on loss of information as well. Other animals surely do the same. Ground squirrels, tree squirrels, ungulates (Brown et al. 1992, van der Merwe and Brown 2008, Abu Baker and Brown 2014), and many, many other animals maintain information on predators in their cognitive maps and change foraging and other behaviors to avoid predation. When a forager leaves a site where he has perceived a predator, the forager's information about that site decays while the chance of a predator still being at the site also decreases, leading to renewed safety the longer since the forager left the site. Consequently, the optimal time to return to a site relative to predation likely differs from the optimal time for food acquisition. Animals do not treat each side of this tradeoff equally because it is a death–dinner trade-off. Nonetheless, animals that face hunger are willing to face risks that sated animals avoid (Real 1981b, Lima and Dill 1990).

5.7 The Down Side of Information

Merkle et al. (2022) argued that for animals to base foraging decisions, and hence home ranges,

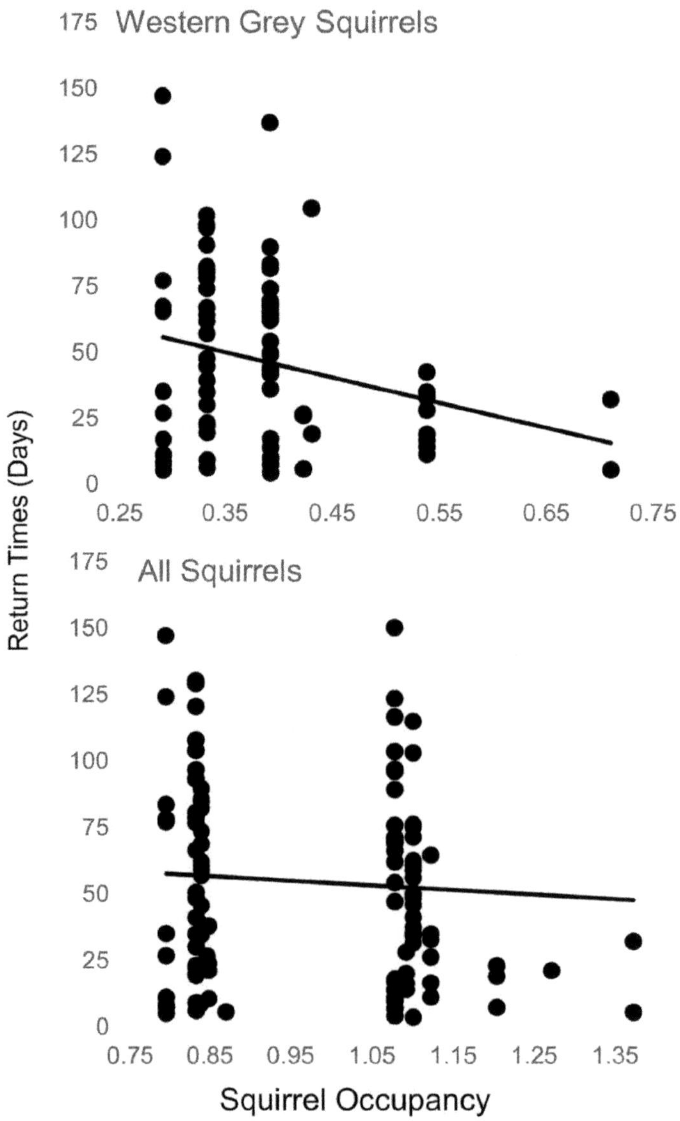

Figure 5.6 Fishers in the northern Sierra Nevada of California, USA, returned to food patches more slowly when prey occupancy was low and prey renewal times were assumed to be slow. Changes in return times were most pronounced for western grey squirrels, the most important single prey for fishers at the study site.

on memory and maintaining site fidelity may be maladaptive in today's world. The world is, after all, changing in many places faster than the pace of evolution for long-lived organisms with low reproductive rates. Sites that change fast, that is, have more rapid decay of information and have slower resource renewal conditions than are natural, lead animals to return to foraging sites quickly even when resource availability or quality is still low. If the models of Spencer (1992, 2012) and Stamps and Krishnan (1999) do, indeed, mimic real biological conditions, animals living on changing landscapes will return to habitat patches quickly, will be forced to expand their home ranges, if possible, or will abandon their home ranges to try to find new ones. New sites, however, may also suffer from rapid decay of information. The result could be habitat loss more extensive than that calculated from changes in vegetation or other land cover alone.

The hypothesis deserves to be tested that return intervals to resource patches have shortened due to increased rates of information loss. The long-term effects of changing resource renewal rates and rates of information loss on animals' home ranges, on their use of environments, and on populations need to be learned.

CHAPTER 6

Mechanistic Familiarity Distributions

When I was conditioning for the Marine Corps Marathon some years ago, I ran a 20-mile workout loop one day that ended with running past a hamburger joint. Despite following a diet that avoids, as much as possible, products of industrial agriculture, the hamburgers smelled ***de**-licious* (overcoming in my mind the stench of grease). When I got home, I told Consie, my wife, about the burger smell and that hamburgers were what my body craved for supper. The state of my body directed the options that we had for supper that evening: we could fix something at home with red meat or go out to eat at a restaurant that we knew served good hamburgers. My nutritional state affected where we would go that evening. Situations such as this affect how animals make decisions about when and where to forage. Animals' knowledge of their home ranges, the locations of particular resources, and how that knowledge affects decisions are parts of their cognitive maps. Mechanistic utilization distributions provide a productive approach for testing hypotheses about why animals make specific decisions, like decisions about where to forage for supper.

6.1 Mechanistic Utilization Distributions for Testing Hypotheses

A mechanism is the process by which something is done. Hence, mechanistic home range models emphasize the processes animals use to build their

Home Ranges of Animals. Roger A. Powell, Oxford University Press. © Roger A. Powell (2025). DOI: 10.1093/9780191949142.003.0006

home ranges. Biologists can use those processes to build testable utilization distributions. The home range models identified as mechanistic fall into 2 broad categories. The first includes models that use some mechanism for simulating familiarity distributions, which are then used to explore questions not related to the mechanism (Börger et al. 2008, Nabe-Nielsen et al. 2013, Tao et al. 2016, Ellison et al. 2020, White et al. 2020, Rohwäder and Jeltsch 2022). Models in the second category start with rules of animal behavior, interactions, and movements hypothesized to be consequential for animals, and the rules are then used to build utilization distributions. Those resultant mechanistic utilization distributions, or mechanistic distributions, can be tested against location data because they were made independently of the location data. Testing them can show which rules explain the data best. These are the mechanistic models and distributions covered in this chapter. Mechanistic distributions lead to new knowledge through the testing of hypotheses related to the biology of study animals used to build the mechanistic distributions (Adams 2001, Moorcroft and Lewis 2006). Such tests provide key insights into animals' cognitive maps.

Familiarity distributions, in contrast, are built from data sets and are, therefore, thoroughly descriptive; they have no theoretical or predictive value (Chapters 1, 11). They can be tested to learn what parameter values for estimators describe a data set best but such testing contributes little to knowledge about why animals behave as they do. Such tests of familiarity distributions can, however, lead to hypotheses related to the biology of the study animals. Such hypotheses can stimulate a biologist to build mechanistic distributions, which *can* lead to new knowledge.

6.2 Introduction to Mechanistic Distributions

This chapter begins with an overview of the strengths of mechanistic distribution models and of how mechanistic distribution models work. This approach allows biologists to test their understandings of the ecologies, behaviors, and other aspects of the biologies of study animals. The chapter then uses examples from Moorcroft and colleagues (Moorcroft and Lewis 2006, Moorcroft et al. 1999, 2006) to introduce ways to use mechanistic models to build mechanistic distributions. Next come examples of testable, mechanistic distributions for meerkats and for long-tailed tits built to test hypotheses related to habitat specialization, conspecific avoidance, and group characteristics. The chapter then covers 2 modelling approaches that combine familiarity distributions with hypothesized attraction to resources to build quasi-mechanistic distributions. The chapter finishes with examples of mechanistic distributions for black bears used to test several hypotheses related to individual differences and related to the importance of food and to avoidance of neighbors, roads, and steep slopes.

Mechanistic distributions can be powerful tools for understanding animal behavior. Here is a quick overview of how the approach works. Imagine that I have been studying (hypothetical) green spotted cats ("green-spots") for many years. Because familiarity distributions for green-spots are stable over years, I hypothesize that when green-spots travel to the peripheries of their home ranges, they have a strong tendency to return toward the centers of their home ranges. Familiarity distributions for all green-spots have some patches with dense prey but my data provide little insight into whether total prey density or prey diversity is more important for understanding how green-spots use their environments.

To build mechanistic distributions, I start with the variables that are critical for my hypotheses, which are that 1) prey density drives how green-spots use their environments, 2) prey diversity drives how green-spots use their environments, or 3) both prey density and diversity drive how green-spots use their environments. Next, I develop models for those behavioral responses, which can then be applied on the landscape where each target animal lives, leading to a predicted distribution of movements across that landscape, which leads to the mechanistic distribution for that distribution of predicted movements. Each of these steps benefits from curiosity, imagination, and creativity.

Using maps of prey distributions, I build 3 mechanistic distributions: one that incorporates both a return-to-center model and a model for attraction to sites of different prey densities, another that incorporates the return-to-center model and a prey diversity model, and a third that incorporates the return-to-center model and both prey density and diversity models. I then rank how well the 3 mechanistic distributions describe my location data for green spotted cats. The highest ranked mechanistic distribution, whatever it turns out to be, provides strong insight into what characteristic of prey is most important to the green-spots. If the highest-ranking mechanistic distribution strongly resembles a familiarity distribution built from my location data, then that mechanistic distribution captures a very large proportion of the characteristics of the environment that are consequential for green-spots. If the highest-ranked mechanistic distribution does not resemble the familiarity distribution closely for each green-spot, then other characteristics of the environment are important to green-spots but have not been included in that mechanistic distribution, or I have not modelled the relationship between green-spots and prey completely correctly. Even in this latter situation, I have learned characteristics of prey that appear most important to green-spots and know that other aspects of green spotted cat biology affect their use of their environments more than does prey density or diversity.

Along this line, modelling resource selection, for example, and building mechanistic distributions has an advantage over using resource selection functions. The goodness of fit of the highest-ranked mechanistic distribution provides insight into whether characteristics of the environment not incorporated into the mechanistic distribution are consequential.

Once a mechanistic distribution has been tested and found to predict how animals use their environments, then variations on the underlying models can incorporate different levels of resource availability, allowing biologists to predict how animals will change use of their environments as resources change (such as food increase or decrease). These predictions have the potential to be quantitative, though they must be tested for their abilities to work across a range of values for a variable that is being adjusted. Information models, economic cost–benefit models, optimal foraging models, and game theoretic results can be incorporated into the rules for building and testing mechanistic distributions, thereby opening diverse and productive approaches to understanding home ranges. Known or hypothesized avoidance and attraction behaviors (such as avoidance of conspecifics, scent marking, and attraction to specific resources) can also be incorporated, as can loss of individuals or groups from a population (e.g. Moorcroft and Lewis 2006, Bateman et al. 2015).

The literature on mechanistic distributions disproportionately emphasizes territoriality (Potts and Lewis 2014 even state that territoriality is "ubiquitous"), but mechanistic models do not require territoriality. Indeed, the approach can be used to understand the biology of animals that have diverse social organizations. Some animals defend territories in some parts of the species' range but have broadly overlapping home ranges in other places (Powell et al. 1997, 2022). For these animals, the mechanistic approach has the potential to provide new insights into the conditions leading to variations of social organization.

Individual-based modelling is another strength of mechanistic distributions (Potts and Lewis 2014, Potts et al. 2014). Mechanistic distributions can be tested with location data for individual study animals, thereby detecting individual differences among animals in a population. Landscapes are heterogeneous and, therefore, individuals encounter different distributions of resources, different contacts with neighbors, and different dangers from predators. Individuals also have different personalities. All these potential differences may lead to the highest-ranked mechanistic distributions for different individuals to include different variables. Such results allow biologists to discern individual differences in resource requirements, foraging efficiencies, or predator detection. Biologists can use such information to understand better the relative importance of different resources for a population and to hypothesize reasons for individual differences. The better a biologist understands her study animals' cognitive maps, the more insights she has into potentially important variables to include in mechanistic distributions.

The rules used for animal interactions and movements in mechanistic distributions fall into 2 broad categories (Adams 2001) related to knowledge of the biology of the animals being modelled. First, if the biology being modelled is understood, then the biology should be incorporated into the rules, models, and mechanistic distributions. Second, if the biology is not well understood, alternative models can be built incorporating different potential distributions (rules) for model variables. As an example of the 1st category, energy expenditure increases linearly with distance travelled (Taylor et al. 1970, Schmidt-Nielsen 1971, Powell 1979b). In addition, area at a given distance from a home range center increases with the square of distance, increasing travel costs to travel to and from the center. A central place forager has increasingly more patches at increasing distances from her territory core, leading to travel costs that increase both linearly and with the square of the territory radius. Incorporating these 2 distributions into a testable, mechanistic distribution incorporates the known cost of animal movements. This type of mechanistic model is a "functional" model because it incorporates biological functions.

In contrast, as an example of the 2nd category, territory boundaries can be built from the condition that the importance of resource patches to a territory owner decreases with distance from the territory centroid. If that pattern of decrease is not well understood, it can be modelled using any of many distributions (linear, random normal, negative exponential, etc.). The distribution may be chosen for convenience of modelling as a decreasing pattern but may not be consistent with the as yet unknown biology of how animals visit resource patches at different distances. Building and ranking mechanistic distributions made using each of the different distributions provides insights into how the study animals actually use their home ranges.

Obviously, not all animals are central place foragers. Different animals tend to use the peripheries of their home ranges differently. Some make short, independent trips to peripheries, while others forage widely when at their peripheries (Broadbooks 1970, Hirsch et al. 2013). Choosing distributions and parameter values based on known biological and technological conditions, when possible, makes sense.

Mechanistic modellers have a history of using complex mathematics and statistics (from the perspective of many biologists) and mechanistic models have gained a reputation of requiring such math (Horne et al. 2008). The important characteristic of mechanistic models, however, is that biologically logical rules of animal behaviors are used to build testable mechanistic distributions. The models need not be mathematically complex if simple models capture the important biology. Using Ockham's Razor, the simplest models that capture the essence of the hypotheses being tested are the best models to use (Turchin 2023). What is critically important is that the models lead to testable mechanistic distributions that lead to new knowledge through use of the scientific method. That is the most powerful characteristic of mechanistic models and distributions. Biologists need not have extensive expertise with mathematics and statistics to build and to use mechanistic distributions productively.

Most mechanistic distributions are based on assumptions about memory: that animals remember locations of resources, encounters with aggressive neighbors, locations of scent marks from different neighbors, and so forth (e.g. Moorcroft and Lewis 2006, Van Moorter et al. 2009, 2016, Bateman et al. 2015, Merkle et al. 2014, Ellison et al. 2020, Collet et al. 2022). Direct tests have documented clearly that memory affects animal movements and establishment of home ranges (Ranc et al. 2020, 2021, 2022) and theoretical work has established the obvious that memory can be incorporated into mechanistic distributions (Potts and Lewis 2016). Nonetheless, although modellers have begun incorporating learning and memory into mechanistic distributions (e.g., Tan et al. 2002, Dalziel et al. 2008, Van Moorter et al. 2009, Smouse et al. 2010, Potts and Lewis 2016, Breed et al. 2017, Ranc et al. 2020 2021), the modelling has been inconsistent with the present understanding of animal memory and cognition (Eichenbaum 2017) and few have addressed explicitly how the value of information changes over time (Ranc et al. 2020, 2021).

The details of incorporating memory into mechanistic distributions can be awkward because memory is usually not the variable of interest. Memories do decay (Ebbinghaus 1885, Gleitman 1971, Mayes 1983), but the memories of mammals and birds last for years and decades (Garstang 2004, Tsalyuk et al. 2019, Dixon et al. 2022, Lewis et al. 2023), putting memory decay on a different timescale than the day-to-day interactions of animals. Consequently, the variables related to memory that are important for mechanistic models are what animals have learned about changes in behavior of neighbors, such as changes in aggression or rate of scent marking by a neighbor caused by decreased or increased importance of a resource patch along a territory boundary. As time passes, the quality of information about resources, neighbors, and predators changes and, therefore, the decay in information quality across time is consequential while the decay in memory is not. Other variables in mechanistic models also change with time. Bateman et al. (2015) incorporated into their mechanistic models a new meerkat group's changes in avoidance of its parent group as time passed. Mechanistic models and distributions for drifting home ranges or territories (Doncaster and Macdonald 1991) also require incorporating time.

6.3 Getting Started

Moorcroft and Lewis's (2006) book on *Mechanistic home range analysis* provides a good introduction to the process of developing testable, mechanistic distributions. Readers need not follow all the details of the mathematics presented to be able to understand the steps to follow.

Moorcroft and Lewis started with a discussion of random walks. An animal modelled to travel in a random walk moves at each step a distance drawn from a random distribution, with a change in direction also drawn from a random distribution. The result is a wandering path of connected line segments that tend to wander farther and farther from the starting point, though not always (Figure 6.1). If a large number of random walks starting at the same point are combined and the number of lines counted in each cell of a grid, the count builds a bell-shaped distribution centered on the animal's starting point.

Random walks are the ultimate null hypothesis for animal movement and are the basis for a large literature on animal movement (e.g. Bovet and Benhamou 1988, Blackwell 1997, Morales et al. 2004, Gautestad and Mysterud 2010, Breed et al. 2017, Michelot and Blackwell 2019, Michelot et al. 2019, Johnson et al. 2021). Acknowledging that no real animals actually walk randomly, Moorcroft and Lewis use modifications of random walks as ways to present hypotheses for behaviors or environmental conditions that affect animals' movements and animals' establishment of home ranges. The first hypothesis that Moorcroft and Lewis present is that an animal has an increasing bias to return toward her home range center the further she goes from that center: a localizing tendency (Figure 6.1; Dunn and Gipson 1977). Moorcroft and Lewis developed a mechanistic distribution incorporating such a bias; it was a radially symmetrical, bell-shaped distribution.

Chapter 5 showed that stable home ranges develop when animals return to resource patches at optimal times relative to information decay and resource renewal. Incorporating this result into a mechanistic distribution (rule category 1 above) is preferable to using the return bias used by Moorcroft and Lewis (rule category 2 above) but requires information on changes in resources over time and on rates of information decay, information that is not known by biologists for most ecological systems. The result is that the return bias is an acceptable fall-back model.

Moorcroft and Lewis (2006) tested this mechanistic distribution with data for red foxes in southern Minnesota, USA (Siniff and Jessen 1969) and with data for coyote packs on the Hanford National Environmental Research Park, Washington state, USA (Figure 6.2 A; Crabtree 1989, Moorcroft et al. 1999). Neither data for individual foxes nor for individual coyote packs matched well to

Figure 6.1 Three examples of random walks of 500 steps. All walks begin at the solid circle. The solid line ends in a triangle, the dotted in a diamond, the dot–dash in a square. The walk shown with dot–dash has a localizing bias to be more and more likely to turn toward the starting point the further the line deviates from the starting point.

the mechanistic distribution made from the center-return model other than centering the data to the center of the mechanistic distribution. This result is not surprising, since topography and distributions of resources, competitors and human activities (and much more) affect movements of animals and how they establish their home ranges. Nonetheless, the fox and coyote data did show the localizing pattern within other patterns.

In areas without wolves, coyotes live in packs that are family groups of parents and offspring from recent years. The coyote data used by Moorcroft and Lewis (2006, Moorcroft et al. 1999) were for 11 packs. The study site was in the rain shadow from the Cascade Mountains, had little topography, and was relatively homogeneously vegetated by a plant community of big sagebrush and blue bunch wheatgrass. Thus, the coyotes lived on a site with relatively little variation that might affect variation of familiarity distributions for packs. The coyote packs maintained territories that were roughly evenly spaced, suggesting similar access to food by all packs. Moorcroft et al. (1999), therefore, hypothesized that by adding an additional rule of mutual pack avoidance to the localizing rule would make a mechanistic distribution that would match pack locations better than had the central bias distribution alone. They modelled avoidance by incorporating scent marking behavior.

Upon encountering scent marks left by conspecifics, canids tend to leave their own scent marks and also to move toward the centers of their home ranges (Bowen and McTaggart-Cowan 1980, Macdonald 1983). Moorcroft et al. (1999) developed a mechanistic distribution incorporating these characteristics of scent mark avoidance for the Hanford coyote packs, and this mechanistic distribution fit the location data significantly better (using the Akaike Information Criterion) than had the mechanistic distribution with localizing movements only (Figure 6.2 B vs A). The model even allowed Moorcroft and colleagues to predict where the heaviest concentrations of scent marks would have been on Crabtree's study area.

The model for scent mark avoidance, however, did not match the locations of the coyote packs as

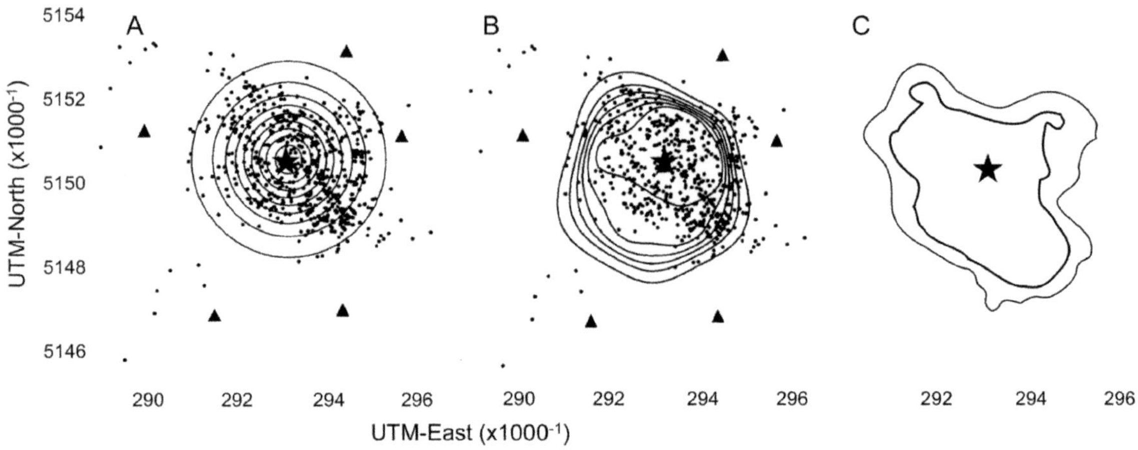

Figure 6.2 Distributions calculated for the Hopsage coyote pack studied by Crabtree (1989) on the Hanford National Environmental Park, Washington state, USA. A. Telemetry locations with the mechanistic distribution assuming random movements with a bias to move toward the telemetry centroid (star). Triangles show the centroids for surrounding packs. B. Telemetry locations with the mechanistic distribution assuming random movements with a bias to move toward the telemetry centroid and with increased bias toward the centroid after approaching scent marks of neighboring packs. This model described the data better than did the model without scent marking in panel A. C. Familiarity distribution calculated by Crabtree using the modified grid cell method of Rongstad and Tester (1969). (A and B adapted from Moorcroft et al. 1999; C adapted from Crabtree 1989.)

well as did grid cell familiarity distributions (Figure 6.2 C; Crabtree 1989, Rongstad and Tester 1969), which was to be anticipated. The familiarity distributions were, after all, built directly from the data for each pack while the mechanistic distributions were built to test a hypothesis. The mechanistic distributions with scent marking, by fitting the location data for coyotes better than did the localizing distribution alone, allowed Moorcroft et al. (1999) to accept (tentatively) that scent marking did, indeed, affect the spacing of the coyote's territories. The familiarity distributions could not provide that strong an insight into the coyotes' behavior. In addition, because the mechanistic distributions did not describe the location data as well as the familiarity distributions did, further mechanistic distributions could be built incorporating further models of the biology hypothesized to affect coyotes' movements. Tests of these mechanistic distributions would provide further knowledge of coyote biology.

Up to here, the mechanistic distributions developed by Moorcroft and colleagues (Moorcroft et al. 1999, 2006, Moorcroft and Lewis 2006) have highlighted no new information about coyotes. The exercises have, however, shown how to use mechanistic distributions to gain biological insight. Moorcroft and colleagues did gain new knowledge about coyotes by building mechanistic distributions to test hypotheses about the coyotes in the Lamar Valley of Yellowstone National Park, USA, before the reintroduction of wolves to the Park in 1995 (Phillips and Smith 1996). Compared to the Washington coyotes, the coyotes in Lamar Valley experienced considerable potential for increased energetic costs due to steep topography and they hunted prey whose distributions were affected distinctly by the patchiness of plant communities. Moorcroft and colleagues developed mechanistic models for coyote avoidance of steep slopes and for attraction to plant communities that supported high populations of small mammals, producing mechanistic distributions with different combinations of these ecological conditions: attraction to home range cores, avoidance of conspecifics, avoidance of steep topography, and attraction to habitats supporting small mammals.

The mechanistic distributions combining the center bias, conspecific avoidance, and prey availability matched the location data for the coyote packs better than did the mechanistic distributions with conspecific avoidance and topography. Both of the combined mechanistic distributions were superior to avoidance of conspecifics alone. These results suggest that prey availability is more important than the energetic costs of travelling up and down slopes. Unfortunately, the topography was not based on energetic or other actual costs of travelling across steep topography (rule category 2 above). Had the energetic cost of travel exceeded the cost incorporated into the mechanistic distribution, a revised mechanistic distribution for topography might have been found to be more important to the coyotes, as has been found to be the case for wolves and black bears (Sells et al. 2022b, this book Section 6.6).

Mechanistic distributions have an additional, extremely consequential characteristic. Models that

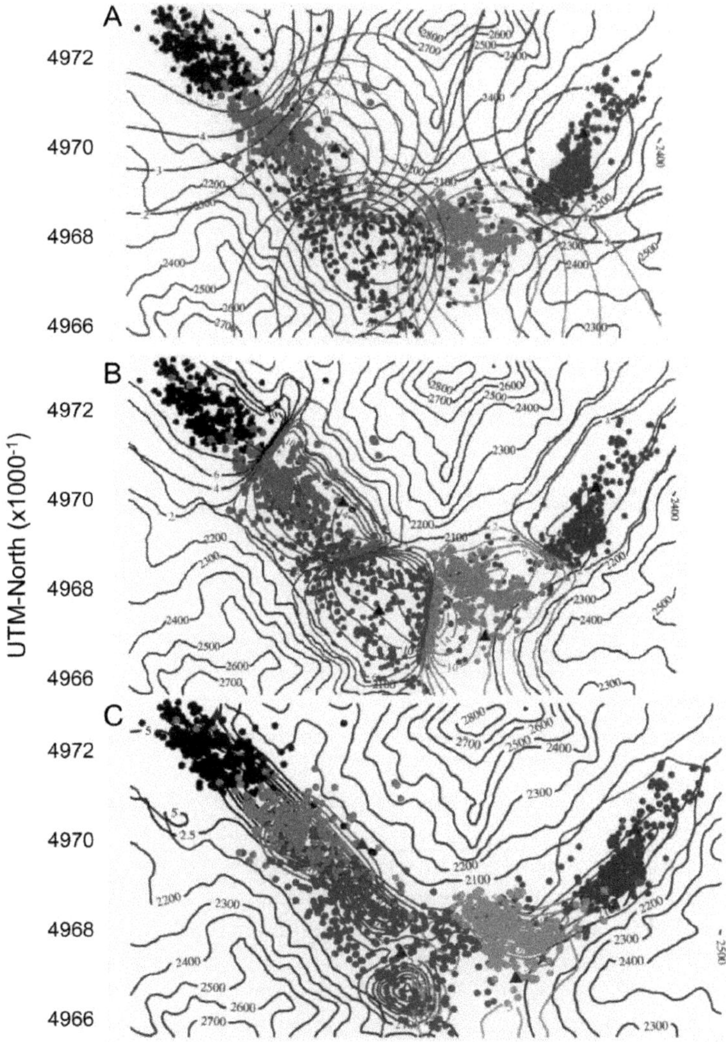

Figure 6.3 The testable, mechanistic distributions developed for 5 coyote packs in Yellowstone National Park before the wolf reintroduction of 1995. The coyote locations are identified to pack by shades of grey and the probability density functions by matching grey lines. The background lines represent topography; all 5 packs are centered in the Lamar Valley. A. Probability density function for the scent mark avoidance model. B. Probability density function for the model combining scent mark avoidance with avoidance of steep slopes. C. Probability density function for the model combining scent mark avoidance with attraction to prey habitat. This model described the data best. (Adapted from Moorcroft et al. 2006.)

produce mechanistic distributions can be modified to predict how animals will change their use of their home ranges when ecological conditions change. Moorcroft and colleagues predicted how coyote packs would change their territories as small mammal populations changed, and predicted how territories would change if one of the packs had a significant change in structure, such a losing one of the pack's breeders. They were able to test the latter prediction because one of the centrally located packs in Lamar Valley did, indeed, dissolve, allowing the neighboring packs to begin to use that pack's former territory. The set of mechanistic distributions incorporating conspecific avoidance and attraction to prey habitats predicted the changes in mechanistic distributions of the packs neighboring the dissolved pack (Figure 6.3).

6.4 Habitat Specialists, Conspecific Avoidance, and Group Characteristics

Mechanistic distributions have been used to clarify details of habitat preferences, group interactions, and the importance of group relatedness. Meerkats are highly social mongooses (Carnivora: Herpestidae) that live in territorial groups in sub-equatorial Africa, mostly on treeless plains. Whether meerkats' territories depended most heavily on overt defense or scent marks was not clear when Bateman et al. (2015) began research. Likewise, whether group size and habitat features affected territory maintenance were not clear. Bateman et al. developed a series of mechanistic distributions to test the importance of these behavioral and ecological effects.

Bateman et al. started with the null hypothesis of a biased random walk that drew members of each group back toward the center of the group's location data. They developed mechanistic distributions for all groups that also incorporated scent marking, scent mark abundance related to group size, and biased movement toward the territory center upon finding the scent marks of neighboring groups. Lacking data on prey availability and predation risk, Bateman et al. built mechanistic distributions incorporating avoidance of sand soil habitat (unprofitable), which correlated with river flats and extensive sour grass cover, and attraction to sand soil edge habitat (good habitat), which correlated with the interface between dune habitats and river flats. These mechanistic distributions incorporated slow movement in favorable habitat but quick movement through unprofitable foraging patches. Bateman et al. developed mechanistic distributions that incorporated direct, aggressive interactions between groups as an alternative to scent marking. And, finally, they did not always incorporate the centroid bias into their mechanistic distributions, allowing groups' territories to drift in response to avoiding other groups and in response to changing foraging habitat.

The mechanistic distributions that described best the stable use of their environment by meerkat groups (using Akaike and Bayesian Information Criteria) were the mechanistic distributions for scent marking without dependence on group size but with attraction to the sand soil edge habitat. Mechanistic distributions including scent marking described use of the environment consistently better than did mechanistic distributions incorporating direct, aggressive interactions, though the very best of the several mechanistic distributions including aggressive interactions produced results similar to the best scent marking mechanistic distribution. Using this approach of mechanistic distributions, Bateman et al. learned that both scent marking and habitat selection were consequential contributors to how meerkats used their environment and that use of good habitat was associated with slow movements while use of unprofitable habitat was associated with rapid movements. Using resource selection functions would not have yielded as much information about meerkat use of habitats or yielded the additional information about meerkat biology.

Bateman et al. documented a new group budding from an established group. The mechanistic distribution that best described the use of the environment by the new group included territory drift and attraction to sand soil edge habitats. The new group moved its territory from starting within the territory of its parent group to having minimum overlap. A post hoc mechanistic distribution incorporating increasing avoidance of the parent group through time described best the territory drift of the new group. All groups fluctuated in size over time, assumed to be in response to changing prey availability, while territory sizes remained reasonably stable over time. The patchiness of the habitat may have prevented meerkat groups from shrinking their territories when prey availability increased. Alternately, meerkat groups may in all years maintain territories of a size required during food shortages.

Bateman et al. did not know exactly how well their partial differential equation models mimicked the actual behavior of the meerkats (rule category 2 above). Mechanistic distributions more anchored on biology could well have provided even more information than did their research, which was, nonetheless, highly productive.

Research on long-tailed tits provides another example of using mechanistic distributions to clarify habitat preferences, group interactions, and the importance of group relatedness. Long-tailed tits are territorial passerine birds of northern Eurasia that are year-round residents in England, where they are woodland specialists. Ellison et al. (2020) studied long-tailed tits outside of Sheffield, where the tits lived in flocks with non-overlapping territories during the non-breeding season. The flocks appeared to avoid other flocks passively, without known displays or calls (Ellison et al. 2020). Ellison et al. (2020) hypothesized that long-tailed tits avoided places where they saw other flocks foraging, demonstrating that territoriality does not require calls, scent marking, aggressive defense, or other advertising of a territory, and does not require a central den or nest site, two major characteristics of the mechanistic distributions for territories presented by, and known widely from, the work of Moorcroft and Lewis (2006).

Before testing their hypothesis on territory maintenance, Ellison et al. needed to establish in what way the long-tailed tits were attracted to woodlands, as that could affect how flocks distributed themselves within woodlands. Therefore, they built mechanistic distributions to test whether tit flocks were drawn to woodland edges vs woodland interiors and whether tit flocks were drawn from areas close to the woodlands vs from across the whole landscape. In addition, they built a null mechanistic distribution that lacked attraction to woodlands. Once they had evidence for the best of the 5 woodland attraction mechanistic distributions (drawn to woodland interiors from the whole landscape), they added to that mechanistic distribution the avoidance of conspecific flocks. Lastly, Ellison et al. modified the avoidance to include whether another flock was larger or smaller than a target flock and whether another flock contained tits related to members of the target flock.

The research team live-trapped and banded most members of the study population. The team outfitted some members of each flock with global positioning system (GPS) transmitters programmed to locate themselves every minute. Flocks were located using recognizable calls and then by identifying banded individuals. Global positioning system locations began when a flock was identified and the flock was followed until it was lost, yielding an average duration of about 20 minutes per observation. Mechanistic distributions were ranked using Bayesian information criteria.

Adding avoidance of other flocks built a mechanistic distribution with stronger predictability (Figure 6.4). And stronger yet were mechanistic distributions incorporating flock size and relatedness. Small flocks avoided larger flocks while large flocks were less affected by small flocks, and these effects diminished as flock relatedness increased. Thus, Ellison et al. clarified mechanisms of habitat preference and mechanisms of group interactions and, in addition, provided evidence that territories can be maintained passively by avoidance of places where competitors are known to forage.

6.5 Other Landscapes

Shepard et al. (2013) suggested the concept of building energy landscapes, where the value of any place is the energy required to traverse that place. Shepard et al. listed a series of drivers that can affect cost

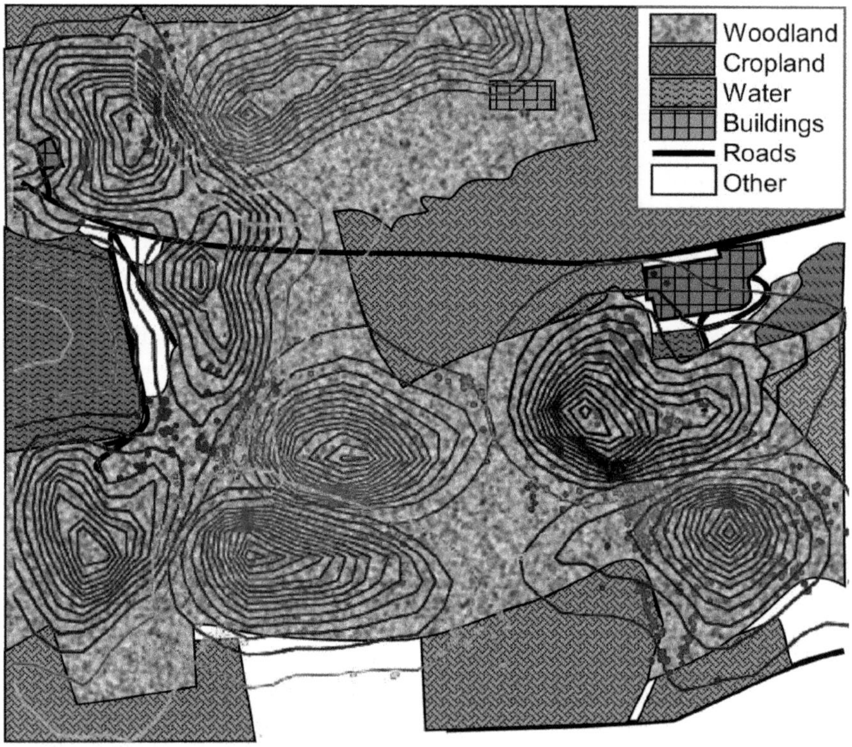

Figure 6.4 The testable, mechanistic distributions developed for 6 long-tailed tit winter flocks. These distributions included avoidance of other flocks and preference for interiors of woodland habitats, and described the location data better than other mechanistic distributions. The tit locations are identified to flocks by shades of grey and the distributions by matching grey lines. The flock located the farthest to the west (left) had a distribution that extended nearly across the whole side of the figure from north to south (top to bottom). The background patterns show habitats. (Adapted from Ellison 2019.)

of transport, including ground and aerial vegetation, ground compaction (or lack thereof), weather conditions, and tides and other water currents. They built such cost-of-transport landscapes for a black vulture and for soaring birds in general, incorporating wind speed and direction and how ground topography and vegetation affect wind. For birds, a mechanistic cost of transport distribution would be important to test for many species. For mammals, a cost-of-transport mechanistic distribution could incorporate density of vegetation and physical complications for travel, such as coarse woody debris on the ground, rocky terrain, and soft soil.

Brown et al. (1999) highlighted the "ecology of fear" (also e.g. Charnov et al. 1976, Ylönen 1989, Jędrzejewski et al. 1993, Ylönen 1994), and Laundré et al. (2001) introduced the concept of a "Landscape of Fear" for prey animals. Since then, Joel Brown and his colleagues developed the experimental use of "giving up density" (GUD) of food as a way to measure a Landscape of Fear for diverse animals (Kotler et al. 1991, Brown et al. 1992, Brown 1999, Olsson et al. 2001, Brown and Kotler 2004, van der Merwe and Brown 2008, Abu Baker and Brown 2014, Emerson and Brown 2015, and others). A Landscape of Fear, or perhaps its inverse, a Landscape of Safety that would highlight attractive patches of a landscape, could be used to build a mechanistic distribution that could be exceedingly important for prey animals, especially those with diverse predators. Such a mechanistic distribution would allow biologists to include fear or safety in a ranking of environmental variables important to individuals. Building separate mechanistic fear or safety distributions for several individuals would provide an understanding of how fear or safety differs among individuals and how their environments differ correspondingly.

6.6 Combining Mechanistic and Familiarity Distributions

Given that time spent in an area may not be the best currency for building familiarity distributions, combining environmental and behavioral characteristics of animals with familiarity distributions has the potential to provide insight into what resources are of great or small importance to an animal, and how that importance changes with time and with the animal's condition or status. For example, a currency built from limiting nutrients for a moose can produce a mechanistic distribution that provides insight into the importances of different resource patches. Likewise, using fear as a currency (Laundré et al. 2001, van der Merwe and Brown 2008) produces a mechanistic distribution for chipmunks that incorporates more information about ground squirrels' cognitive maps than does using locations alone. Matthiopoulos (2003b) and Horne et al. (2008) have presented approaches to combining mechanistic and familiarity distributions.

6.6.1 "Model-Supervised" Smoothing

Matthiopoulos (2003b) incorporated resources, behavior, environmental constraints, and other components of cognitive maps with familiarity distributions, making modified mechanistic distributions. For each cell of a grid covering a landscape, his "model-supervised" smoothing combines the value of an animal's familiarity distribution for that cell with the value for that cell of a "synthetic model" incorporating important co-variates, each weighted by its importance. Matthiopoulos specified using a kernel estimator to build the familiarity distribution but other estimators work. He also used extensive modelling of potential movement paths to explore habitat preference and resource accessibility. These models did not, however, incorporate changes of information over time.

The model-supervised value of the modified mechanistic distribution at site x, $s(x)$, is

$$s(x) = wu_x + (1 - w)v_x,$$

and

$$u_x = f(w_{1...i,x}u_{1...i,x}),$$

where w is the weighting parameter for the synthetic model, u_x is the value at site x for the synthetic model, v_x is the value of the familiarity distribution at site x, $w_{1...i,x}$ are the weighting factors for resources 1 through i available at site x, and $u_{1...i,x}$ are the values for resources 1 through i available at site x. u_x is given as a function of resources i available at site x since resources can be complementary, supplementary, facilitative, and so forth. Values of u_x, v_x, and $s(x)$ must each sum to 1 across all x to make a probability distribution.

Matthiopoulos used maximum likelihood cross validation to calculate w in simulations, though other approaches could be used, such as using important biological information, using objectives of a given research project, or using technological limitations of data. By incorporating information about consequential resources and behaviors into simulations, the final modified mechanistic distribution matched the "true" utilization distribution better than did the familiarity distribution built from a kernel density estimator using only location data (Matthiopoulos 2003b). Even rough models of consequential resources and behaviors can improve estimates of use of an environment. The weighting parameters, w and w_i, can change with season and with resource harvest by a target animal and with competitors. A model can incorporate a target animal's "state," such as its nutritional state or its understanding of location-dependent dangers of predation. Mathiopoulos tested his model using data for a British population of grey seals.

If the co-variates can be modelled from the perspective of how an animal perceives his environment in maintaining his cognitive map, then the modified mechanistic distributions have the potential to provide insights into why the animal moves through his home range as he does. In addition, u_x can even be tested as a mechanistic distribution on its own, without combining it with the familiarity distribution. Biologists can use this approach to develop several biologically pertinent mechanistic distributions, using different combinations of co-variates and different values for weighting parameters, and using Akaike's Information Criteria for small sample sizes (AICC), or other statistics, to rank how well each model describes the data. If numbers of locations in an area indexes importance, then ranking of models provides insight into the relative importance of the modelled resources.

6.6.2 Synoptic Model

The "synoptic model" of Horne et al. (2008) builds mechanistic distributions but calls them by another name. Horne and colleagues started with a null mechanistic distribution, $f_0(x)$, for use of an environment that incorporates attraction to an animal's center of activity. They allowed the null mechanistic distribution to incorporate a bivariate normal distribution, or an exponential power distribution (Horne and Garton 2006a), or a kernel of higher order than 2 (equation 11.2), which represents more uniform use of an environment than do the other 2 options. The synoptic model then adds distributions of resources (considered broadly to include environmental constraints, behaviors such as fear, and so forth), weighted by their importance, as co-variates to the null mechanistic distribution with center bias. The value of the synoptic distribution at site x, $s(x)$, is

$$s(x) = \frac{f_0(x)\prod_{i=1}^{k}(1+\beta_i H_i(x))}{\int_x \left[f_0(x)\prod_{i=1}^{k}(1+\beta_i H_i(x))\right]},$$

where $H_i(x)$ is the measure at site x for resource i of k resources and β_i is an estimated selection parameter controlling the effect of resource i on $s(x)$; that is, β_I converts $H_i(x)$ (which could be resource density, nutritional value of a plant, escape cover, etc.) to its importance to a target animal. The denominator for the synoptic model integrates over space to be a constant that ensures that $s(x)$ integrates to 1, making it a probability distribution. Horne et al. also presented a time-dependent synoptic model, $s(x, t)$, noting (the obvious) that the models of co-variates could change with time.

Horne et al. assume that the importances of resources $(\beta_I \cdot H_i(x))$ are multiplicative, which is often not true (Ries et al. 2004). Horne et al. recommended the general approach used for research using mechanistic distributions: develop several biologically pertinent mechanistic distributions and then rank them by how well each describes location data for the animals studied.

The "synoptic model" of Horne et al. could be modified to use the familiarity distribution for a given animal instead of the null model of a center bias for movements, $f_0(x)$. If using the familiarity distribution fits the data better than the null mechanistic distribution, then $s(x)$ does not include all of the consequential resources or models them poorly (β_I or $H_i(x)$ or both are faulty).

One strength of both the "synoptic model" of Horne et al. and Matthiopoulos's (2003b) "supervised smoothing" is that alternate modified mechanistic distributions can be built by varying values for the weighting factors (β_i, w_i), allowing testing for relative importances of different resources.

6.7 Mechanistic Distributions for Black Bears

Female black bears are good subjects for testing behavioral hypotheses using mechanistic distributions because: 1) resident female bears have well-defined familiarity distributions that have consistent characteristics by age, sex, and breeding classes within a region (Powell et al. 1997); 2) bears live in heterogeneous, patchy habitats and move among patches containing varying food resources on a daily and seasonal basis while producing annual familiarity distributions that are generally stable from year to year (Powell et al. 1997); 3) female black bears appear to structure their home ranges based on the productivity of food resources, especially plant foods and ants, but also based on avoidance of humans (Amstrup and Beecham 1976, Alt et al. 1977, Young and Ruff 1982, Powell 1987, Noel 1993, Powell et al. 1997, Mitchell et al. 2002); 4) the food resources that might structure a bear's home range are fixed in space and fixed seasonally in time; and 5) energetic demands (e.g. lactation; Mauritzen et al. 2001, Moen and Boomer 2005) and the ability to sequester resources (e.g. because of social dominance based on age; Powell et al. 1997) vary among bears and should influence size, shape, and resource content of familiarity distributions. These ecological characteristics are not unique to black bears but represent the conditions under which many animals establish and maintain home ranges (Powell 2000). Familiarity distributions of male black bears, in contrast, do not vary with food supply (Powell et al. 1997, Mitchell et al. 2002) and, therefore, I use data from females only.

I directed a research team that studied black bears in the Blue Ridge Mountains of western North Carolina, USA for 22 years, 1981–2002. During 3 of those years, we collected several hundred very high frequency telemetry (VHF) locations on each of several female bears with overlapping or adjacent home ranges. These data could be used to test mechanistic distributions developed for the bears. The literature and my team's past analyses suggested building distributions that included up to 6 rules of behavior: 1) bias to return toward the centroid of the telemetry locations; 2) avoidance of neighboring females; 3) attraction to close food patches; 4) attraction to rich food patches; 5) avoidance of steep slopes; and 6) avoidance of roads. Building mechanistic distributions for these individual female bears can provide insights into how cognitive maps differ among individual bears.

From previous analyses, I knew that these female bears maintained stable and overlapping familiarity distributions and, therefore, that they had cognitive maps with massive amounts of local information on the environment and on other bears. Because their movements were not confined within territory boundaries (Powell et al. 1997), I hypothesized that their movements would have an inherent bias to move toward the centroid of their activities after having moved to distant resource patches (Hypothesis 1 and Rule 1 for building a mechanistic distribution). In addition, previous analyses (Powell et al. 1997) suggested that some individual pairs of females avoided each other even though the overall social pattern for these bears was a lack of territoriality. From this information, I knew that the bears' cognitive maps kept track of the movements of other bears. Consequently, I hypothesized that mechanistic distributions could document individual variation and would identify females that avoided other females (Rule 2).

Black bears are predominantly vegetarian (Noyce and Garshelis 1997, Powell et al. 1997, Rettler et al. 2021), so these females' cognitive maps would have information on the richness of food patches and renewal of food patches after harvesting. I hypothesized that the movements of these females would be more attracted to food patches as richness increased (Rule 3) and that the attraction would increase as a bear moved closer to a rich food patch (Rule 4).

Past analyses showed that ridge tops correlated with boundaries of familiarity distributions for the bears and that some bears avoided steep slopes (Powell and Mitchell 1998; Chapter 1). Clearly, the bears' cognitive maps included details of the topography of their home ranges. In contrast, Zimmerman's (1992) model of habitat quality for black bears in the Blue Ridge Mountains included attraction to steep slopes as a component for escape cover from humans. Tests of the habitat model, however, showed that the model predicted use of the study area by black bears better when the escape cover component was removed (Zimmerman 1992,

Mitchell et al. 2002). This result also suggested that the bears' cognitive maps included details of the topography of their home ranges. I hypothesized that the bears avoided steep slopes and built mechanistic distributions including avoidance of steep slopes (Rule 5).

Analyses by Reynolds-Hogland and Mitchell (2007b) showed that the bears avoided roads, which were therefore included in their cognitive maps. Zimmerman's habitat model also included avoidance of roads in its escape cover component, which did not contribute positively to the overall model. I hypothesized that bears avoided roads (Rule 6). As with steep slopes, I foresaw that by building mechanistic distributions including attraction vs avoidance of roads by individual bears could clarify the contradiction.

Most of the research I have cited in this chapter used partial differential equations to build distribution kernels that were then used to build mechanistic distributions. I built distribution kernels directly from independent data collected in the field on food patches and from independent data on topography and roads. I modelled attraction to the centroid of the location data distribution by multiplying a bivariate normal distribution by a K_4 kernel distribution (see equation 10.2, using exponent 4 instead of 2) and then transformed the result so that its volume equalled 1.

For avoidance of neighboring bears, I set the value at 1 for places beyond the overlap of movements for each other individual bear, then decreased the value toward 0 inside that overlap. I modelled bears' avoidance of roads to increase as distance to roads decreased and attraction to food to increase as distance to food patches decreased. I modelled avoidance of slopes to increase with steepness. I tested the abilities of the mechanistic distributions to describe the locations of each bear, starting with the complete model with 6 variables (the 6 hypothesized rules). I used Type II Sum of Squares to remove variables that did not contribute to the significance of a mechanistic distribution's ability to describe a bear's locations. Ultimately, I obtained mechanistic distributions that incorporated only the variables that contributed to significance for each bear (Figure 6.5).

The centroid bias and avoidance of roads contributed to the final mechanistic distributions for all 15 bear-years, slope avoidance contributed to 13, distance to food patches to 8, food quality to 7, food combined to 11, and avoidance of neighbors to 4 (Table 6.1). These results highlight individual differences among bears that were not available from past analyses. For example, clearly, avoidance of neighbors was not widespread but was, nonetheless, significant for a few pairs of bears. Past analyses of neighbor avoidance included apparent mutual avoidance of female bears 61 and 87 but those analyses could not be tested for significance. Of the 4 final distributions that did include neighbor avoidance, 2 were for bear 87 in the 2 years with data for her, and a third was for bear 61 in one of the 2 years when she and bear 87 were both tracked. Both of these females were well into adulthood (8 to 11 years old) and had overlapping familiarity distributions in areas with good food patches, where each could consider herself to be dominant.

The rules included in each final mechanistic distribution appear to highlight consequential characteristics of the environments within each bear's home range. Lactating females were represented in 6 bear-years. Five of these females responded to food quality or food distance and the only one who did not was lactating in a year of abundant hard mast in the autumn. This result suggests the commonsense hypothesis that lactating female black bears are more sensitive to food availability than are non-lactating and juvenile females.

In the year with the poorest hard mast crop, distance to food patches was important for all 3 females followed in that year and food quality

Table 6.1 The 6 models incorporated into testable distributions for black bears in the Southern Appalachian Mountains of North Carolina, USA, and the numbers of final distributions that incorporated each of the 6 variables.

Rules related to Behavior	**Number of bears including model**
Bias to move toward center	15
Avoidance of neighbor	4
Distance to food	8
Quality of food	7
Combined food	11
Avoidance of steep slopes	13
Avoidance of roads	15

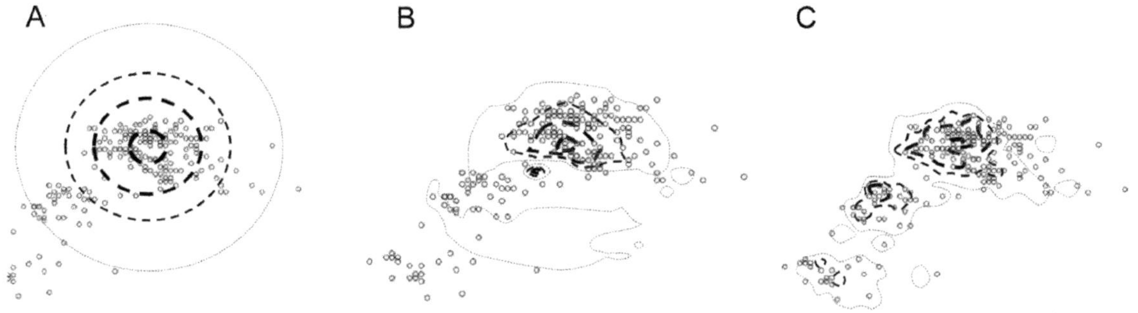

Figure 6.5 Telemetry locations for female black bear 61 in the Blue Ridge Mountains of North Carolina, USA, in 1985, plotted with 2 testable, mechanistic distributions and a familiarity distribution calculated using a kernel estimator. A. Mechanistic distribution for a random walk starting at the bear 61's centroid (study area cell with the most telemetry locations) combined with a bias to return toward the centroid. B. Mechanistic distribution including a bias to return to the centroid plus avoidance of neighbors, attraction to patches of high food quality, attraction to patches close to food, avoidance of steep slopes, and avoidance of roads. This testable familiarity distribution described the data better than did the all other mechanistic distributions built using other combinations of the behavior rules. C. Kernel generated familiarity distribution with h = 250 m and the k_2 kernel. The contour lines represent the 5%, 35%, 65%, and 95% familiarity distributions.

was important for 1. These results, though limiting, suggest that in different years with different abundances of food the bears' cognitive maps valued food resources differently and the bears used that information when planning movements.

The universal avoidance of roads and the near universal avoidance of steep slopes suggest that Zimmerman's model of habitat quality should be tested with the escape cover variable included but modified to retain distance from roads but not steep slopes. The intermediate importance for quality of and distance to good food patches suggests that the bears in the study population were not food limited, consistent with lack of territoriality and consistent with mortality rates (predation by humans), not reproduction, limiting the population (Powell et al. 1996, Brongo et al. 2005).

After doing the above modelling, I combined, using Matthiopoulos's (2003b) "supervised" approach, the final mechanistic distribution for each bear and year with the matching familiarity distribution, built using a fixed kernel estimator with k_2 and h = 250 m. In all cases, combining the mechanistic and familiarity distributions, weighted equally, improved the fit to the distribution of location data compared to that for the final mechanistic distribution alone (ΔAIC>2). This result is not surprising because the familiarity distribution was built from location data alone. But, importantly, for all but 1 bear in 1 year, the combined mechanistic distribution fit the data as well as or better than did the familiarity distribution alone, adding strong support to the final choices of the mechanistic distributions chosen for each bear.

6.8 Cautions

As is the case for all research and management, a method is only as good as the biological knowledge used to choose and to apply the method. For example, brown capuchins and ring-tailed coatis have very similar diets of fruits and invertebrates (Hirsch et al. 2013). Members of the 2 species, however, have very different foraging behaviors: capuchins travelling in the canopy in straight lines from fruit tree to fruit tree, harvesting fruits in the trees and eating arboreal insects opportunistically; coatis travelling in tortuous routes among fruit trees, foraging for fruits on the ground, visiting some trees multiple times in a day, and seeking sites where they can forage on soil invertebrates. To build mechanistic distributions for the 2 species based only on diets could yield spurious results when testing hypotheses related to partitioning of resources.

Bateman et al. (2015) noted that all of their meerkat groups fluctuated in size from year to year but territory sizes were relatively stable. If meerkat groups maintain territories of a size required during years of food shortage, they use memory not just for resource locations but also for past resource availability. The hypothesis to test is that meerkat groups maintain territory size needed in bad years. Ranc et al. (2021) showed that spatial memory (resource locations) decays more slowly than does what they called "attribute memory" (resource abundance), which is really information about resources (Chapter 5). Animals may, however, associate some aspects of resource abundance, such as minima, with resource locations and, if they do, they could maintain minimum sizes for territories based on spatial memory. If they do not remember minimum abundances of foods, then the hypothesis to test is that patch distributions and sizes are more consequential for establishing territory sizes than patch richness in the year of territory establishment. These 2 hypotheses for territory size can lead to different mechanistic distributions and, therefore, can be tested. And, if mechanistic distributions are built for each meerkat group, testing can show whether different groups use a different rule. These hypotheses should be tested with other animals.

CHAPTER 7

Economic and Cost–Benefit Models and Home Ranges

Just about all models of animal behavior explore how benefits from a behavior are related to costs. Information theoretic models compare how benefits and costs change over time leading to optimal times to forage in different food patches. Game theoretic models compare pay-offs to constraints and find the greatest benefit. Thus, many modelling approaches incorporate economic analyses, and usually cost–benefit analyses.

This chapter deals only with models that use cost–benefit analyses directly to provide insight and to stimulate hypotheses related to behaviors and home ranges. The chapter deals mostly with territories: when to maintain a territory vs when to accept home range overlap. The first example examines why North American red squirrels maintain individual territories instead of maintaining territories shared by mated pairs. Then the chapter presents cost–benefit criteria for when a territory will be profitable and when home range overlap is more profitable. The chapter ends with some modifications of a model for when to maintain a territory and applies the modified model to black bears, comparing 2 populations that contrast with respect to territoriality vs overlap of familiarity distributions.

Animals are territorial only when they have a limiting resource, that is, a critical resource that is in short supply, and when maintaining a territory is economical (Chapter 2; Brown 1969). By decades ago, this relationship between resources

Home Ranges of Animals. Roger A. Powell, Oxford University Press. © Roger A. Powell (2025). DOI: 10.1093/9780191949142.003.0007

and territories had been studied extensively and established, both theoretically and empirically, primarily using economic analyses of fitness trade-offs between benefits gained from resources and costs of defending them (Stenger 1958, Brown 1969, Gill and Wolff 1975, Carpenter and McMillen 1976, Kodric-Brown and Brown 1978, Ebersole 1980, Hixon 1980, Schoener 1981, 1983, Powers and McKee 1994). Members of some species maintain territories within some populations but not others, or during some seasons but not others (Chapter 2; some mammals, Grant et al. 1992, Powell et al. 1997, 2022; many birds).

7.1 Individual Territories of Red Squirrels

North American red squirrels are predominantly mammals of northern conifer and mixed conifer–hardwood forests (Young 1999). They defend individual territories using diverse vocalizations and chasing, and each territory has a central cache or caches of conifer cones, dried mushrooms, and other items that are food for winter (Smith 1968, 1981, Webber et al. 2023, and much literature between and since). The squirrels' cognitive maps know in detail the trees in their territories, the values of the cones of different species and individual trees, what to expect from each annual cone crop, and the diverse routes a squirrel can take to reach cones on the trees. Red squirrels harvest cones and other foods and transport them to their caches. Thus, each storage trip involves a trip from the cache to the food item or items and then a return trip to the cache. Smith (1968) built a simple model comparing the benefits and costs of individual territories vs territories shared by a mated pair of squirrels. Mated pairs of red squirrels do not share territories, though mated pairs of many birds and mammals do. Smith hypothesized that, because mated pairs could share territories, individual territories must have advantages and the first place to look would be at simple cost–benefit analyses.

If one assumes that each individual territory is circular with radius r and with its cache at its center (Figure 7.1), then caching each item involves a trip of distance $2R$ from the cache to the food item and back, where R is the distance from the cache to the food item ($R \leq r$; Smith 1968). For simplicity, assume that the resident squirrel fetches food items close to the cache first, working its way out through concentric rings of radius R and some narrow width ΔR. And assume that the squirrel does not have to climb trees or do anything else other than to travel on the ground to all the food items. The number of trips at each radius will be proportional to the circumference of a circle of radius R, or $2\pi R$. For radius R, the total distance travelled will be $2R \cdot 2\pi R \cdot \Delta R$. Compressing the concentric rings leads to integrating $4\pi R^2$:

$$\int_0^r 4\pi R^2 dR = 4/3 \cdot \pi \cdot r^3$$

which is the total distance travelled by a resident squirrel in an individual territory to fetch all food items to its cache. If the circle of radius r and area πr^2 supplies food for 1 squirrel, then a circle with twice that area is needed to provide for 2 squirrels sharing a territory: area $2\pi r^2$ and radius $\sqrt{2} \cdot r$ Total distance travelled by each squirrel is

$$1/2 \cdot 4/3 \cdot \pi \cdot (\sqrt{2} \cdot r)^3 = \sqrt{2} \cdot 4/3 \cdot \pi \cdot r^3,$$

which is $\sqrt{2} \approx 1.414$ times the total distance travelled by a squirrel maintaining an individual territory. This result is logical because the extra area to be foraged by 2 squirrels all lies outside the area that would be foraged by 1 squirrel (Figure 7.1). The longer total distances travelled for 2 squirrels maintaining a shared territory would also lead to greater time spent foraging and greater energy expenditure.

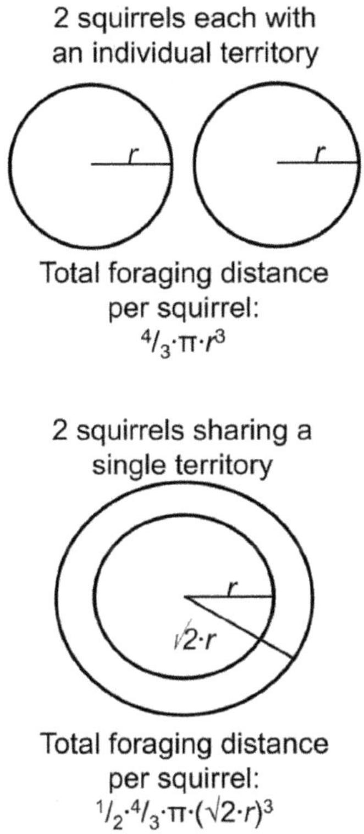

Figure 7.1 The relative distances travelled to store winter food under idealized conditions for 2 potential social organizations for North American red squirrels. Red squirrels maintain individual territories only, consistent with the shorter distances travelled and consequent smaller total foraging times and energy expenditure.

Smith's simple model is not realistic in detail. Red squirrel territories are seldom truly circular, squirrels do climb trees to cut cones loose and to retrieve mushrooms left to dry on tree limbs, no 2 territories have exactly the same density of foods, and so forth. Nonetheless, both squirrels of a pair would have those same extra sources of energy expenditure. Smith's model captures the essence of the question and stands as a strong biological null hypothesis for territory ownership. Biologists must understand how the benefits of a shared territory compensate for the extra costs.

This result begs the question, then, why any territorial animal would share a single, larger territory with a mate or other conspecifics. The answer is, obviously, that other benefits more than offset the added foraging costs. Most songbirds and many mammals need both parents to contribute to raising offspring. Cooperative breeders, like red-cockaded woodpeckers, beavers, and woodland voles get help from offspring of previous reproductive cycles to raise young, to maintain dams and lodges, to build food caches, and to maintain tunnel systems. Even if the juveniles do not help, the parents benefit from tolerating them, which helps them to reach reproductive age and to find places to reproduce. Either way, the benefits of cooperation exceed the costs for these species.

7.2 Resource Productivity, Territory Benefits, and Territory Costs

Carpenter and MacMillen (1976) showed theoretically that an animal should be territorial only when productivity of its food (or another limiting resource) is between certain limits. When productivity is low, the costs of maintaining a territory are not returned through exclusive access. When productivity is high, requirements can be met without exclusive access. Carpenter and MacMillen's model incorporates clearly the limiting conditions for territoriality to exist, as opposed to simply assuming that territoriality is economical, which has been the usual case going back decades (Ebersole 1980, Kodric-Brown and Brown 1978). Understanding limiting conditions provides insight regarding how animals use information from their environments to make decisions about how to use resources within their environments. Those decisions, therefore, depend on the animals' cognitive maps, which house the critical spatial knowledge needed for foraging and for otherwise making use of limiting resources.

Carpenter and MacMillen argued that an animal should defend a territory only if the animal can not meet its basic, energetic cost of living (E) with the share of the limiting resource it can obtain when it is not territorial (a fraction a of productivity P). Thus, an individual should be territorial only when

$$aP < E, \quad \text{or} \quad P < E/a.$$

If P is above this limit ($P > E/a$), then an animal should not maintain a territory but should share resources. In addition, an individual should defend a territory only when the basic cost of living (E) plus the cost of being territorial (T) is less than the expected yield to a non-territorial individual (aP) plus the extra yield gained from being territorial (an additional fraction b of productivity P), which can be written:

$$E + T < aP + bP, \qquad \text{Eq. 7.1}$$

or

$$\frac{(E+T)}{(a+b)} < P.$$

If P is less than $(E + T)/(a + b)$, then the animal can not maintain a territory or home range and must travel elsewhere to establish a home range. Thus, an animal should be territorial only when:

$$\frac{(E+T)}{(a+b)} < \frac{P}{a}. \qquad \text{Eq. 7.2}$$

Equation 7.1 states that territorial behavior can be exhibited only when

$$\frac{(E+T)}{(a+b)} < \frac{E}{a}, \qquad \text{Eq. 7.3}$$

which is equivalent to

$$T < \frac{Eb}{a}. \qquad \text{Eq. 7.4}$$

Equation 7.2 sets the conditions for territorial behavior. Members of species with large species ranges can live under different conditions in different parts of their ranges, leading to differences in territorial behavior by populations in different parts the species range. My research team showed that, for 2 black bear populations (Section 7.7), the female bears in a population living where food productivity was low maintained territories, while the female bears living where food was far more productive did not maintain territories and tolerated extensive home range overlap (Powell et al. 1997, 2022). Rogers (1977, 1987) documented nomadic behavior by black bears in northern Minnesota in autumn when berry crops failed, indicating that berry production had fallen below the lower limit permitting territorial behavior (left side of equation 7.2). Hummingbirds vary territorial behavior on a daily basis as productivity of nectar in flowers moves across the upper limit for territoriality (Hixon 1980). Hummingbirds with the level of resource productivity appropriate for maintaining territories may experience enough territory trespass that T becomes so great as to violate equation 7.1, leading to territory shrinkage or territory abandonment to find a territory elsewhere with lower T (Hixon 1980, Schoener 1981, Schoener and Schoener 1982).

An assumption for equations 7.1, 7.2, and 7.3 is that the limiting home range productivity for a non-territorial individual is the same as that for a territorial individual (Carpenter and MacMillen 1976). This assumption is unrealistic, but it can be relaxed with little effect on the general conclusions. (What follows has been adapted from a model I presented earlier; Powell et al. 1997.)

Consider an animal that lives on a landscape with required resources in patches separated by non-habitat, which may be areas used only for travel between suitable resource patches. For simplicity, let all habitat patches be the same size and have the same resource productivity. Let P be the productivity of a patch and let a territory have N patches. Equation 7.2 now becomes

$$\frac{(E+T)}{(a+b)} < PN < \frac{E}{a}. \qquad \text{Eq. 7.5}$$

What happens if the productivity of the patches increases: berries have a bumper crop, or the vole population explodes in numbers, or early successional trees sprout densely after a forest fire producing rich browse? Let P increase by an increment of p such that

$$(1+p)PN > E/a. \qquad \text{Eq. 7.6}$$

This condition is consistent with ceasing to maintain the territory and allowing home range overlap. But what if an animal that ceases to maintain a territory then must maintain a non-territorial home range that is larger than the territory it used to maintain,

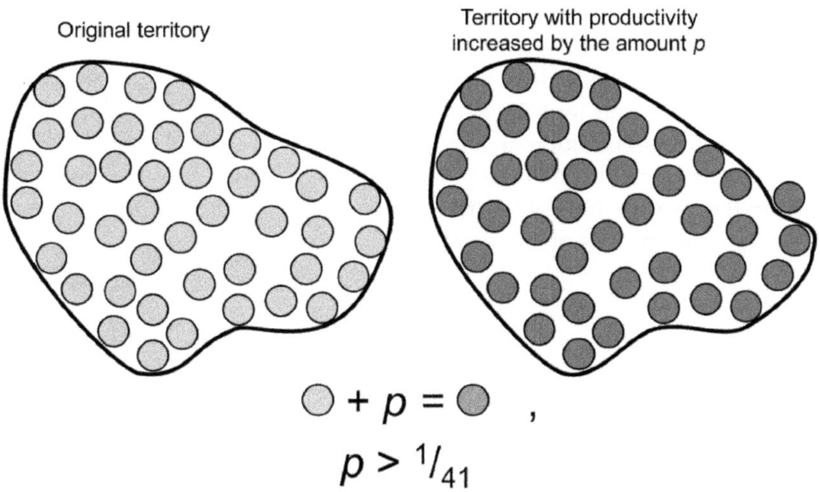

Figure 7.2 When productivity of the limiting resource increases for an animal maintaining a territory having many small habitat patches (in this example, 42), the animal should shrink his territory instead of maintaining a larger home range that overlaps with home ranges of conspecifics. Here, productivity of the limiting resource must increase at least by the amount $p = 1/41$ of the total resources in the original territory.

with larger travel costs and lower resource availability because of competition (only aPN available now instead of $(a+b)PN$; Grant et al. 1992)?

Let N_T be the number of patches in a territory and N_H be the number in a non-territorial home range; $N_H > N_T$. Assume that travel costs and territory maintenance costs are both proportional to some extent to home range or territory area (which equals the number of patches). We can now break the cost of living (E) into basal metabolic (B) and movement (M) costs. Equation 7.1 now becomes

$$\frac{(B + (N_T)\,M + (N_T)\,T)}{(a+b)} < PN_I < \frac{(B + (N_H)\,M)}{a}. \quad \text{Eq. 7.7}$$

The condition in equation 7.5 for abandoning territoriality now becomes

$$(1+p)PN_T > \left(B + N_H M\right)/a. \quad \text{Eq. 7.8}$$

The potential exists that

$$(1+p)P\,(N_T - 1) < \left(B + N_H M\right)/a. \quad \text{Eq. 7.9}$$

The only difference between equations 7.8 and 7.9 is a decrease of 1 resource patch in the animal's territorial home range. If equation 7.9 holds, then the animal should continue to maintain a territory but a smaller territory than she maintained before the increase in productivity. For this to occur, the following must also hold:

$$(1+p) > N_T/(N_T - 1), \quad \text{Eq. 7.10}$$

and

$$(1+p)\,P(N_T - 1)\,(a+b) > aPN_H. \quad \text{Eq. 7.11}$$

If N_T is large and each habitat patch is small, then $N_T \approx (N_T - 1)$ and $N_T \approx N_{\underline{H}}$. If we substitute these approximations into equations 7.8 and 7.9 we get

$$1+p > 1 \quad \text{and} \quad (1+p)\,(a+b) > a,$$

both of which are always true. Thus, when productivity of the limiting resource increases for animals maintaining territories with many small habitat patches, the animals should shrink their territories instead of maintaining larger home ranges that overlap with home ranges of conspecifics (Figure 7.2). This is not a novel insight.

But, what happens if a territory consists of few, large patches? Then $N_T > (N_T - 1)$ and $N_T > N_{\underline{H}}$, the inequalities in equations 7.10 and 7.11 may not be true, and the animal should cease maintaining its territory. Losing one large patch from the territory would make the territory uneconomical. Alternately, the animal could maintain a territory reduced in size (lacking the one large patch) and continue to forage in the large patch that was formerly within its territory but that is now shared with conspecifics. For this latter situation, let us return to equations 7.5 and 7.6 and let N be small (and P appropriately large). The condition can exist such that an increase in productivity could be proportionately less than the decrease in total territory productivity stemming from the loss of one patch. In this situation

$$(1+p)PN > E/a,$$
$$(1+p)P(N-1) < E/a, \quad \text{and}$$
$$\frac{(E+T)}{(a+b)} > (1+p)P(N-1), \qquad \text{Eq. 7.12}$$

which means that eliminating a patch from an animal's territory would make the territory unsustainable, while retaining that patch leads to a territory much larger than it needs to be. As an example, let $p = 0.2$ (a 20% increase in productivity), $P = 10$, and $N = 4$. Here,

$$PN = 40,$$
$$(1+p)\,PN = 48, \quad \text{and}$$
$$(1+p)\,P\,(N-1) = 36.$$

If $40 < E/a < 48$ and $36 < (E+T)/(a+b) < 40$ (both of which are true), then the animal is stuck with looking for a new territory, maintaining a territory larger than needed, or using a patch that she does not defend but is willing to share. These latter conditions constitute partial territoriality. Some coyote packs (Person and Hirth 1991) and eastern chipmunks (Getty 1981) generally adopt this pattern of social organization.

If, however, patches vary in size and productivity, then a territory owner might be able to trade a big patch for a small patch and continue to maintain the slightly changed territory. If the large patch in question contains an essential resource, then the animal must continue to include it in its territory (Figure 7.3).

Other ecological conditions exist that may cause animals to cease maintaining territories. If an increase in productivity is local, not landscape-wide, the increase could lead to increased trespassing by conspecific neighbors, increasing the costs of territory maintenance tremendously and,

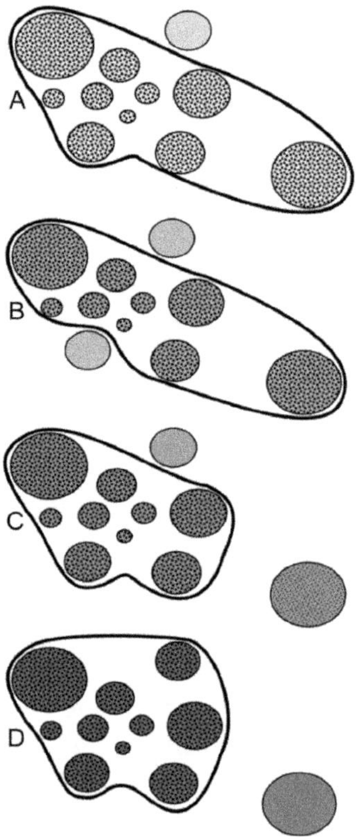

Figure 7.3 Territories are shown within black outlines. Circles are patches of good habitat. Productivity of habitat patches increases with darker grey color. Patches used exclusively by the territory owner have heavy internal patterns, those shared have medium patterns, and those not used at all have no internal patterns. A. Original territory. B. Productivity has increased to the point that the territory owner can abandon a small habitat patch (as in Figure 7.2). If travel cost is high to use the large habitat patch to the far right and separated from the other patches, the territory owner may be forced to use the strategy shown in panel C. C. Productivity has increased to the point that the territory owner has stopped maintaining a large patch within her territory but still forages in that patch. D. Productivity has increased enough that the territory owner can cease to forage within the large patch formerly within her territory as long as she can absorb a small patch to make up the difference.

potentially, leading animals with augmented productivity to abandon territoriality. This is exactly what happened when Carpenter and MacMillen (1976) supplemented with nectar the flowers within some but not all territories of their Hawai'ian honeycreepers.

The ecological and social conditions modelled in this section provide fodder for building mechanistic distributions. The productivity of a resource patch in an animal's cognitive map incorporates resources lost to competitors, which delay resource renewal, and incorporates information loss since a target animal last visited a resource patch. Incorporating these sorts of information into a cost–benefit model can be done by incorporating equations for model components presented in Chapter 5 or building successive models for different times. Results of tests, in turn, provide further insight in how animals use their cognitive maps when making foraging decisions. The conditions modeled are nothing that animals do not handle with their cognitive maps daily, and understanding that animals do deal with these conditions provides new hypotheses to test.

7.3 An Example of Black Bears

Black bear project colleagues John Zimmerman and Erran Seaman and I collected data from our black bear project in the Blue Ridge Mountains of North Carolina, USA (Powell and Seaman 1990, Powell et al. 1997), and from bear biologists in Minnesota's southern boreal forests (Rogers 1977, 1987, Arimond 1979, Noyce and Garshelis 1997, Rettler et al. 2021; Dave Garshelis, Minnesota Department of Natural Resources, personal communication) to test the model of Carpenter and MacMillan (1976) for when animals should maintain territories. We used data from adult female black bears only. (What follows is abridged from Powell et al. 1997.)

7.3.1 Area of Familiarity Distributions

Blue Ridge: Mean area for familiarity distributions of adult female black bears was 16.9 ± 11.7 km^2 (Powell et al. 1997). We also estimated that the approximate mean minimum familiarity distribution size that an adult female black bear maintained when food productivity was high was 9 km^2. Familiarity distributions overlapped broadly, each female bear's familiarity distribution overlapping with 2, 3, or more familiarity distributions of other female bears.

Minnesota: The familiarity distributions of adult female bears studied by Rogers averaged 34.7 ± 12.0 km^2 and those studied by Garshelis averaged 33.3 ± 26.6 km^2. These areas were almost exactly twice that for the adult female bears in the Blue Ridge and, therefore, we estimated the minimum familiarity distribution size for northern Minnesota bears to be twice that for the Blue Ridge bears, or 18 km^2. Familiarity distributions for the bears did not overlap much at all.

7.3.2 Food Production

Blue Ridge: Bear corn (a saprophyte on oak roots) dominated as the early summer food, diverse berries dominated from mid-summer through early autumn, and diverse hard mast dominated across the rest of autumn. We estimated food production to average 1.7×10^5 metabolizable kcal/km^2/day during summer and 6.3×10^4 metabolizable kcal/km^2/day during autumn.

Minnesota: Berries dominated as summer foods in Minnesota and autumn foods were berries and hard mast, which included nuts on shrubs, such as beaked hazel, yielding 2.5×10^2 metabolizable kcal/km^2/day in summer and 1.7×10^3 metabolizable kcal/km^2/day in autumn.

7.3.3 *a* and *b* Blue Ridge and Minnesota

Given that some individual food plants are found in places bears can not reach, that other individual food plants actually produce little food for bears, and that many bear foods are also eaten by birds, other mammals, and insects, estimating the amount of food production actually available to bears is complicated. We assumed that 10% of the food produced was available to bears, that on the average the equivalent of 4 bears used every resource patch in both study areas (1 adult male bear, 1–3 adult female bears, some with young and some without), and that a female bear, through territoriality, can reduce the bear use to the equivalent of 1–2 other bears. These assumptions lead to $a = 0.025$ and $a + b = 0.075$.

7.3.4 Energy Expenditure

An animal loses oxygen as water and as CO_2 but loses hydrogen only as water. Water turnover rate can be estimated by injecting the bears with doubly labelled water (water labelled with heavy hydrogen, deuterium, and heavy oxygen, $D_2{}^{18}O$)

and measuring deuterium loss, allowing CO_2 to be calculated as that oxygen not lost as water. Daily energy expenditure can be estimated from CO_2 production using a reasonable respiratory quotient (Nagy 1980, 1987, Speakman and Racey 1988, Speakman 1993, Speakman et al. 1993).

We obtained 3 estimates of daily energy expenditure for 2 adult female black bears in the Blue Ridge Mountains in 1990. Bear 178 had 3 cubs at heel and we estimated her energy expenditure during a week in early June and again during a week in August. Bear 214 was a young adult female who was in estrus for her first time and we estimated her energy expenditure during a week at the end of July. Rates of energy expenditure for these 2 female bears ranged from 61.8 to 79.3 kcal/hr (1483–1903 kcal/day) but we also needed to know the rates at which these bears expended energy for basic body maintenance (basal metabolic rate, BMR), that is, the energy that the bears burned while in a resting, thermoneutral, non-reproductive, post-absorptive (i.e. not digesting food) state (Robbins 1983). Kleiber (1975) estimated empirically for mammals that:

$$\text{BMR} = 3W^{0.75}\ \text{kcal/hour}, \quad \text{or}$$

$$\text{BMR} = 72W^{0.75}\text{kcal/day},$$

where W is a mammal's weight.

Using these equations, the estimated mean BMR for the 2 female bears was 52.5 kcal/hr, or 1260 kcal/day. Consequently, their measured rates of energy expenditure were approximately 20–60% above basal. Female 178, who had cubs at heel, expended 61.8 kcal/hr in early June and increased her rate of energy expenditure to 79.3 kcal/hr by August. During that same period, her mean travel rate increased from an average of 0.26 ± 0.23 km/hr ($N = 23$) to 0.29 ± 0.41 km/hr ($N = 16$); not a statistically significant increase in travel, but it does parallel the increase in her daily energy expenditure. Female 214 traveled 0.56 ± 0.51 km/hr in August. We estimated travel rate from the straight-line distance between consecutive location estimates recorded 2 hours apart, so these estimates are biased low.

We have no direct estimates for energy requirements of adult female bears in Minnesota. Bears in northern North America average heavier than bears in the Blue Ridge and the estimated BMR of a 75 kg (appropriate for Minnesota) female black bear is 78 kcal/hr or 1900 kcal/day. This estimate is 1.5 times that for a 45 kg female bear in the Blue Ridge. We therefore estimated energy expenditure for Minnesota bears as 1.5 times that for bears in the Blue Ridge.

Bears' actual food requirements exceed their daily energy expenditures, however, because they must gain weight for denning. To predict territorial vs non-territorial behavior, we used 1500 kcal/day as the daily energy requirement for an adult female Blue Ridge bear during summer and 1900 kcal/day for early fall.

7.3.5 Completed Model

For Blue Ridge bears, $P > E/a$ both in summer, $1.1 \times 10^6 > 3.0 \times 10^5$, and in autumn, $5.8 \times 10^8 > 3.8 \times 10^5$. Thus, Carpenter's and MacMillan's model predicts that Blue Ridge black bears should not maintain territories, and they do not.

For Minnesota bears, $P = 4.5 \times 10^3$ in summer and 3.1×10^4 in autumn. $E/a = 3 \times 10^5$ in summer and 3.8×10^5 in autumn. The left side of equation 7.2 can not be calculated because the cost of maintaining a territory is unknown. Using the equivalent of equation 7.7 from Powell et al. (1997), female bears in Minnesota should be territorial if $T < 1088$ kcal/day, which is about 55% of a Minnesota female's other energy requirements, or about 35% of total daily requirements ($E + T$). Actual estimates of costs of territory maintenance for diverse animals, including North American red squirrels, spotted hyaenas, red foxes, harriers in California, USA, lesser short-tailed bats in New Zealand, chimpanzees, and even *Anolis* lizards (Smith 1968, Henschel and Skinner 1991, White and Harris 1994, Amsler 2010, Ord 2021, Temeles 2022) are considerably less than this 35% upper limit. Consequently, the completed model predicts that black bears in Minnesota should maintain territories, and they do.

7.3.6 Cognitive Maps and Home Range Alternatives

The cognitive map of a territorial animal will differ from that of a conspecific who lives where the productivity of the limiting resource is higher and who tolerates home range overlap. The cognitive map of the territorial animal keeps track of foraging patch renewal and does so with information loss caused by competitors of other species but not conspecifics. Competitors of other species may affect food resources in ways that may not affect the territorial animal as much as would resource removal and information loss caused by conspecifics. The cognitive map of the territorial animal keeps track of when and where she has advertised her territory to conspecifics and keeps track of who and where her neighbors are. In contrast, the cognitive map of the non-territorial animal tracks food resource renewal but deals with information loss about resources caused both by interspecific and conspecific competitors. Thus, her cognitive map keeps track of the movements of conspecifics and their potential effects on the resources that she also tracks. Her cognitive map need not track territorial displays but does track information that communicates when conspecifics are in the vicinity, just as their cognitive maps track her. This information may stimulate conspecifics (or her) to go elsewhere to forage, and her cognitive map tracks this potential effect on information loss. Thus, the cognitive maps of both territorial and non-territorial individuals track information about conspecifics, which affects their decisions regarding where to forage next, but their cognitive maps plan travels differently due to the different effects of conspecific neighbors.

CHAPTER 8

Foraging Optimally for a Home Range

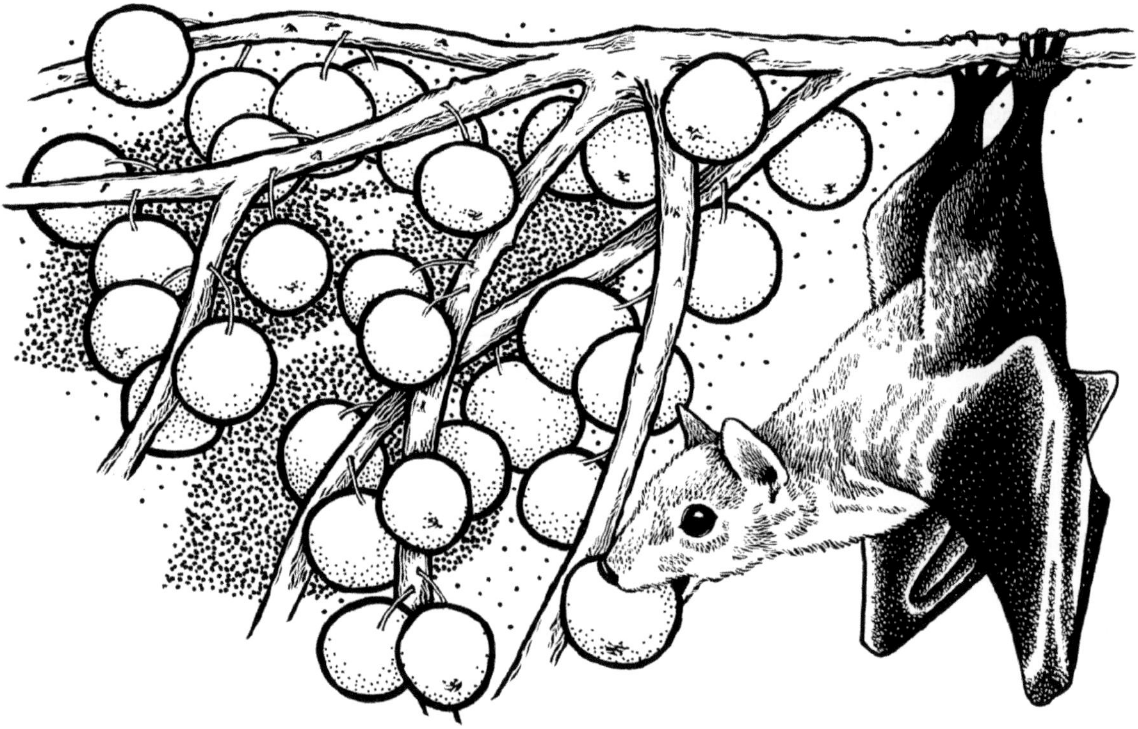

My black bear research crew was spread throughout an exceptionally rich site for high bush blueberries, diligently picking every ripe berry in the site to be weighed but not eating a single one. With my plastic ice cream bucket hooked to my belt, I would pull down a berry branch, hold it with one hand so that a berry bunch was over my bucket, then strip the berries into the bucket with my other hand. I got to thinking about how bears must pick berries. A bear must reach to enclose a bunch of berries in her mouth, close her lips but not her teeth, then pull her head back from the branch to let the berries come off in her mouth. That seemed to me a much more efficient way to pick berries than what I was doing. I walked out of the 10 m radius of the sampling site, took a blueberry branch that was at head height and directed the bunch into my mouth. It was a big bunch dense with berries but I fit it into my mouth. Then I closed my lips but not my teeth and pulled the branch. The berries stayed in my mouth but the leaves and green berries stayed on the branch, leaving only ripe berries in my mouth.

"Wow!" I thought. "*That* is cool!" I took another berry bunch and filled my mouth again. Then another bunch. The time I took to grab a new branch and to get it to my mouth was almost, but not quite,

Home Ranges of Animals. Roger A. Powell, Oxford University Press. © Roger A. Powell (2025). DOI: 10.1093/9780191949142.003.0008

the time I needed to chew and swallow the berries in my mouth. I was on a roll. Within 10 minutes, I was so stuffed with berries that I had to sit down and bend my waist to give my stomach some room to expand.

Thinking more about bears and berry patches: a bear must want to incorporate berry patches like this patch into her home range. So, how might a bear select an optimal home range that concentrates the best food patches? This was not a question about how a bear might best forage within a food patch, or travel between patches, but about how a bear might choose, or forage for, the best set of patches to make an optimal home range. Could that be done?

Optimal foraging models were first developed in the 1960s to guide behavioral studies (Emlen 1966, MacArthur and Pianka 1966). The concept was to develop good null hypotheses for how animals behave. If animals have evolved to behave most efficiently, thereby allowing the most energy to be put into reproduction and survival, then optimality models should be able to predict such optimal behavior. But, if important aspects of an animal's biology are not known, then optimality models will not include that missing information and will make wrong predictions. Wrong predictions lead to new research to fill the gaps in knowledge.

Most of the early optimal foraging models related to diet breadth, foraging strategies, and use of patchy environments (e.g. Emlen 1966, MacArthur and Pianka 1966, Charnov 1976a, b). The importance of food as a limiting resource has been cited in home range studies going back more than three quarters of a century (Hayne 1950, Brown 1969, Carpenter and MacMillen 1976, Lindzey and Meslow 1977, Dill 1978, Harestad and Bunnell 1979, Hixon 1980, Lindstedt et al. 1986, Litvaitis et al. 1986, Jones 1990, Holzman et al. 1992, Joshi et al. 1995, Duncan et al. 2015), particularly for females (Young and Ruff 1982, Ims 1987b, Powell et al. 1997). Most research into how food and other resources affect home ranges relate to home range size and to understanding when animals will maintain territories vs when they maintain home ranges that overlap with those of other conspecifics. Little was known about how animals might choose their home ranges with respect to potentially limiting resources, though Fretwell and Lucas (1970) and Fretwell (1972) presented a null, theoretical prediction that animals would spread themselves across a landscape so that all have equal access to required resources. The common assumption was, and still is, that an animal's life requisites are satisfied by the resources available in the environment delineated by an observed familiarity distribution. Modelling and research were needed on optimal foraging approaches for establishing home ranges (Mitchell 1997, Mitchell and Powell 2004).

This chapter begins with an overview of optimal foraging models applied to animals searching for home ranges, explores linear programming approaches, and then explores resource-maximizing and area-minimizing models. Resource-maximizing and area-minimizing models are then applied to black bears living in the Blue Ridge Mountains of North Carolina, USA. The area-minimizing models are then expanded and applied to the wolf populations in Montana, USA. The chapter ends with an overview of applications of optimal foraging models applied to choosing home ranges.

8.1 Optimal Foraging for a Home Range

All home ranges, not just territories, should be structured primarily by the fitness-driven need for efficient accumulation of resources required for survival and reproduction. A home range should be a function, among other things, of the availability and distribution of potentially limiting resources, with availability discounted by the costs of resource acquisition.

The guiding hypothesis is that an animal optimizes necessary resource accrual per unit area of her home range through the selection of resource-bearing patches to include in her home range (Mitchell 1997, Mitchell and Powell 2004, Powell and Mitchell 2012), analogous to optimal foraging for food items in a diet (Charnov 1976a, Stephens and Krebs 1986, Krebs and Kacelnik 1991). The approach focuses on the spatial distributions and on the importance of resources that structure an animal's movements (that is, the animal's cognitive map; Peters 1978, Spencer 1992, 2012, Powell 2012a). In this way, this approach is consistent with animals using cognitive maps to guide resource acquisition and differs at its core from the common view of a home range being the sum of an animal's movements (Worton 1987, Loehle 1990, White and Garrott 1990, Gautestad and Mysterud 1993, 1995, Bascompte and Vilà 1997).

Modelling optimal foraging for patches to be included in a home range provides insight into how resource depletion and depression by competitors affect resource availability and, therefore, how an animal's cognitive map handles perceived resource quality. Resource depression can be modelled as the combination of resource renewal with information loss following foraging (Chapter 5). Although the approach does not model animal movements directly, it assumes that an animal has developed a cognitive map incorporating local resource patches, along with the myriad relationships among patches (location, topography, complementary and supplementary qualities, state of renewal, and potential travel routes with their predation risks and their metabolic costs), and then chooses resource patches to include in her realized home range.

Optimal foraging models for choosing home ranges emulate the selection of patches over time by dispersing or immigrating animals but do not depict the animals' day-to-day movements. In this way, the models explain development of a cognitive map, but not the map's use in day-to-day foraging decisions.

8.2 Linear Programming Models

During the mid-20th century, livestock producers used linear programming to optimize cattle diets for growth or milk production at low cost. Westoby (1974) appears to have been the first biologist to use linear programming in optimal foraging models. Belovsky highlighted the approach with research on diets of moose, beavers, and other mammals (Belovsky 1978, 1984a, b). Most linear programming models in biology have dealt with diet selection, though publications about diets of wild animals have dwindled since research on optimal foraging went out of fashion in the late 20th century. Most recent publications using linear programming have related to diets of humans, especially to improve diets across the world and to develop strategies for dealing with climate change (e.g. Colombo et al. 2023, Donkor et al. 2023). Recent publications on wild animals are widely spaced in time (e.g. Amano et al. 2004, Barcia et al. 2007, Pretorius et al. 2012, Oudman et al. 2015).

The linear programming approach is easy to adapt to choosing resource patches to include in a home range as opposed to foods to include in a diet. The approach is logical and straightforward. To use linear programming to model optimal foraging, the target animals must have a natural history that lends itself 1) to modelling a goal of maximizing or minimizing the values of limiting variables, and 2) to modelling constraints that affect being able to reach that goal. Constraints may require extensive modelling. Modelling of food requirements may involve study of activity patterns, foraging returns, resource distributions, and digestion. Constraints can also be sizes of vacant patches of suitable habitat or physical structures such as sites for reproductive dens or complex structures on the forest floor that facilitate foraging.

Here is a simple outline of the approach. Imagine an animal dispersing across a landscape that has resource patches of 2 types, P_1 and P_2, with similar distributions of sizes and total area. If the animal can meet its carbohydrate requirements by foraging in an area of 75 units of P_1 patches only, or in 25 units of P_2 patches only (Figure 8.1A), then (with P_1 and P_2 representing the combined areas of all P_1 and all P_2 patches)

$$P_1/75 + P_2/25 \geqslant 1$$

is required for the animal to meet its carbohydrate requirements. If the animal can meet its protein requirements by foraging in 25 units of P_1 patches

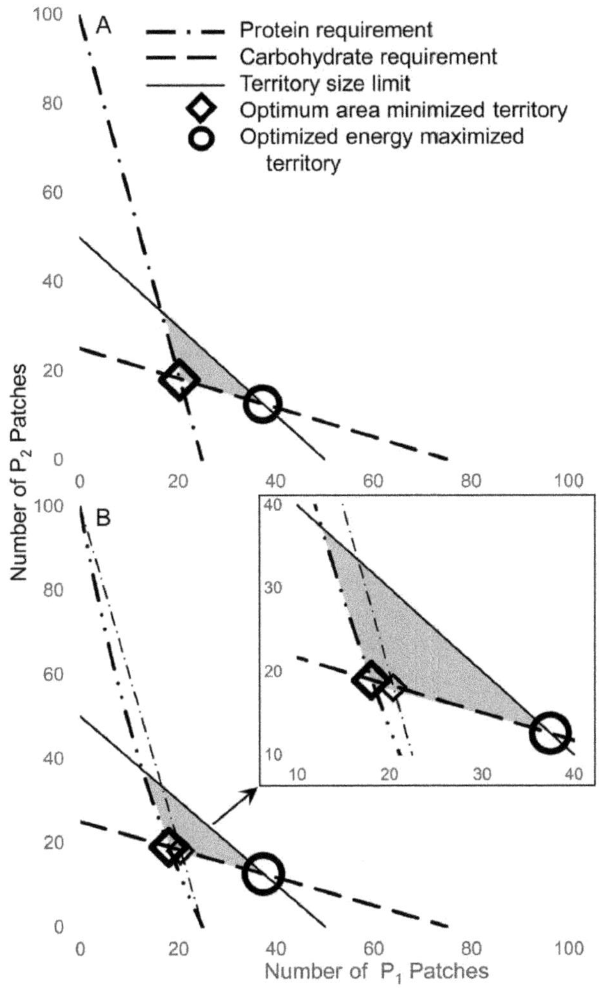

Figure 8.1 Using linear programming to hypothesize optimal home range sizes on a landscape with patches of 2 resource types, P_1 and P_2. The constraints for home range size are that P_1 and P_2 have different concentrations of carbohydrates and proteins. A. To meet his carbohydrate requirement, an animal must choose for his home range a combination of P_1 and P_2 patches that falls above and to the right of the line for carbohydrate requirements. To meet his protein requirement, the animal must choose for his home range a combination of P_1 and P_2 patches that fall above and to the right of the line for protein requirements. To be able to maintain his territory in the face of trespass by conspecifics after resources, the combination of patches must fall below the line for the territory size limit. The diamond shows the patch combination with the smallest territory size satisfying all constraints. The circle shows the patch combination with the highest available energy. B. If P_1 and P_2 patches have different and somewhat complementary proteins, then the protein constraint will bow somewhat to the left (bold dot–dot–dash line), changing the area-minimizing patch combination (now the large diamond) but not the energy-maximizing combination.

only, or in 100 units of P_2 patches only, then

$$P_1/25 + P_2/100 \geqslant 1.$$

The intersection of these equations is at $P_1 = 20.45$ and $P_2 = 18.18$, shown by the diamond in Figure 8.1A.

Thus, the area of patches of each type available must be in the area above and to the right of both the carbohydrate requirement line and the protein requirement line in Figure 8.1A. If the goal of our dispersing animal is have the smallest territory possible (an area-minimizer; Mitchell and Powell 2004), then he should establish a territory with area having 20 to 21 P_1 patches and 18 to 19 P_2 patches.

A territory may have a maximum size that can be maintained without excessive trespassing by conspecifics (Figure 8.1A), putting a constraint on territory size. If our animal's goal is to maximize the energy available in his new territory (a resource-maximizer) while meeting both carbohydrate and protein requirements, then he will maximize the energy available within a territory equal to or smaller than the maximum territory size. Because carbohydrates and proteins can be metabolized to gain much the same amount of energy, P_1 patches are slightly more energy concentrated. Consequently, the circle in the lower right of the triangle in Figure 8.1A made by the 3 constraints (proteins, carbohydrates, territory size) shows the characteristics of the optimum energy-maximizing territory. If resident territory owners have left no spaces between territories that are as large as the territory size limit, our disperser might be forced to establish a territory smaller than the maximum size but that still meets carbohydrate and protein requirements.

The concept of linear programming does not require that all relationships be linear. If the proteins available in P_1 and P_2 patches differ and complement each other, combinations of the 2 patch types have the potential to meet our animal's protein requirements in combinations of patches that fall below the protein requirement line, as shown in Figure 8.1B. In this example, the optimal area-minimizing territory shifts to slightly less area of P_1 patches and slightly more area of P_2 patches. The optimal resource-maximizing territory is unchanged.

Multidimensional models are also possible: a model in 3 or more dimensions, for example, with proteins, carbohydrates, and fats.

8.3 Resource-Maximizing and Area-Minimizing Models

Optimal diet models presented the 2 options for an animal: maximizing his energy acquisition or minimizing the time needed to meet his energy requirements (Charnov 1976a). These 2 options can be converted to optimal patch choices for a home range as maximizing resources available within a home range vs minimizing home range area required to have sufficient resources to meet requirements. Modelling these 2 alternatives provides clear, alternative scenarios for characteristics of home ranges and their distributions on a landscape.

Choosing a patch is not always a simple function of the resources in the patch. Social interactions (reproduction, territoriality, dominance) and depletion of resources by other individuals in a population (and individuals of other species) can depress the perceived and real values of patches, thereby affecting patch selection by an animal establishing a home range and, consequently, affecting how animals distribute themselves on the landscape. Consequently, the value to an animal of resources in a patch depends on: 1) the inherent quantity and quality of the patch's resources; 2) the number of other animals using that patch and who they are; 3) the extent to which those animals depress or deplete the value of resources; and 4) the average cost of travelling to the patch (Charnov et al. 1976, Ylönen 1989, 1994, Jędrzejewska and Jędrzejewski 1990, Ylönen and Ronkainen 1994, Mitchell and Powell 2004).

A real dispersing animal must choose where on its landscape to establish a home range, which he does by choosing patches to incorporate into his home range. The animal presumably values a patch based on how the patch's net benefit (benefit discounted by cost) contributes to his fitness. Once a home range is established, the animal will spend a disproportionate amount of time in his home range core (an area used disproportionately often) because, if for no other reasons, travelling to resources leads the animal to move through the core more than through peripheries. Therefore, the cost of travel discounts the value of patches, so one can model the realized value of patch n, V_n', to be the value of its resources, V_n, discounted in some way by distance, D, from the center of the animal's home range. The discounting function could quantify the energetic cost of travel (Taylor et al. 1970, Schmidt-Nielsen 1971), but could also include the rate of resource renewal after a patch is visited, the loss of information about a distant patch stemming from its being visited less often than patches central to the home range, and so forth (Chapter 5; Vander Wall et al.

2006, Spencer 2012, Cao et al. 2018, Dimitri and Longland 2022).

Mitchell (1997, Mitchell and Powell 2004) modelled 2 alternative, individual-based approaches for choosing patches for a home range: 1) maximizing the realized resource value within a model home range; and 2) accumulating resources sufficient to satisfy a pre-set minimum resource threshold, representing resource requirements. If an animal chooses every patch he enters on a random walk, he will, on average, accrue resources for his home range at a rate proportional to the mean value of resources per patch on the landscape (*Random selection* line in Figure 8.2A). In contrast, a selective animal (Charnov 1976a, Krebs and Kacelnik 1991, Stephens and Krebs 1986) chooses patches in decreasing order of patch value V'_n for his home range and, as he does so, the accumulated value of selected patches

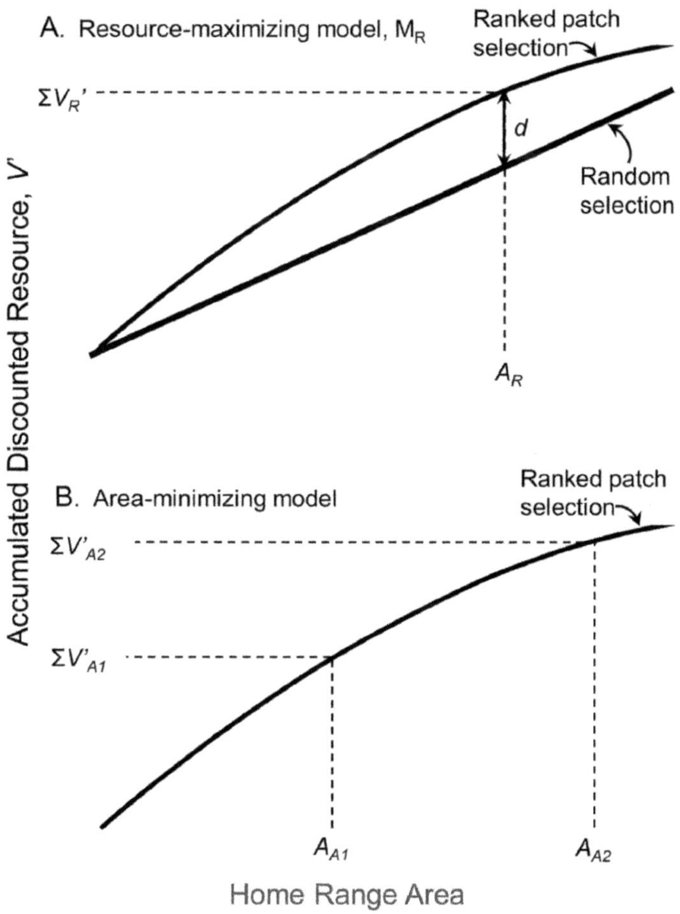

Figure 8.2 Graphical models for constructing optimal home ranges based on selecting patches containing high quality resources. In both the resource-maximizing (A) and the area-minimizing (B) models, an animal selects patches in order of their resource value discounted for travel costs required to reach the patches (V'). A. Under the resource-maximizing model, M_R, an animal stops selecting patches when the difference between random and selective choice of patches, *d*, is maximized, representing the maximal balance between costs and benefits of patch ownership that can be obtained from the landscape. The resource-maximizing home range defined by resource accumulation of EV'_R area A_R. B. Under the area-minimizing model, M_A, an animal stops selecting patches after reaching the minimum resource threshold necessary for survival. Two possible area-minimizing home ranges are defined by resource thresholds EV'_{A1} and EV'_{A2}, yielding home ranges of sizes A_{A1} and A_{A2}. Note that area-minimizing home ranges can range from smaller to larger than the optimal resource-maximizing home range. (After Mitchell and Powell 2004.)

initially increases in manner exceeding that of random selection but in a decreasing manner until the only patches left have values below average (Figure 8.2A). This approach maximizes the difference (*d* in Figure 8.2A) between selective and random choice and is equivalent to Charnov's (1976a) energy-maximizing optimal diet. This home range choice model is a Resource-Maximizing model, M_R. Mitchell (1997) showed that multiple potential functions for discounting patch value, V_n, by distance, *D*, resulted in similar patterns of patch choices and, therefore, he chose, for simplicity, to discount V_n by dividing by *D*, which decreases V_n' at a decelerating rate, with *D* in units of cell size.

Alternatively, animals might choose patches for their home ranges so as to accumulate only the minimum amount of resources required to survive and to reproduce. When a selective animal (again choosing patches in decreasing value of V_n') has chosen enough patches to meet his resource requirements, he has chosen his Area-Minimizing home range, M_A (Figure 8.2B), the equivalent of Charnov's (1976a) time-minimizing optimal diet.

Over time, resource values on a real landscape change. Prey populations change from season to season and year to year; winter weather and late frosts affect vegetative foods. Changing resource values lead to contrasting predictions for resource-maximizing and area-minimizing strategies. When resource productivity decreases, for example the resource curves in Figure 8.2 are lowered as shown in Figure 8.3. Resource-maximizing home ranges maximize the difference between the random and selective accumulation curves, leading to no change in the optimal home range area, A_R, because both curves drop by the same proportion. Nonetheless, an animal with a resource-maximizing home range accumulates fewer total resources $V = \Sigma V_n'$ when resource productivity decreases (Figure 8.3A). For an animal with an area-minimizing home range, in contrast, accumulated resources dictate how big a home range must be. Consequently, accumulated resources, $V' = \sum V_n'$, have the same threshold but, with low resource productivity, an area-minimizer needs a larger home range (Figure 8.3B).

On a real landscape, animals within a population have been resident for different periods of time. Animals with overlapping home ranges deplete and depress resources for other residents (hereafter, for convenience, simply "depress"; Charnov et al. 1976). If competition increases, the curves for random choice and selective choice (Figure 8.2) become shallower but not necessarily proportionately so, because resource depression is patch-specific.

Mitchell added resource depression to models M_R and M_A, making models M_{RD} and M_{AD}. A long-term model resident, assumed to be dominant, perceives no or little resource depression but depresses resources in all patches of her home range for all other model animals. A more recent colonizer perceives resource depression by longer-term residents but also depresses patches for the most recent colonizers. The most recent colonizers perceive some cells to have been depressed by several longer-term residents.

Mitchell simulated 100 home ranges for model animals on each of 5 landscape types that had distributions of resource patches ranging from evenly distributed to highly clumped. All landscapes had the same mean resource value per patch (meanV_n) and standard deviations as similar as possible. Individual patches ranged in value from 0 to 1. For models without resource depression, a model colonizer initially chose a random location and then chose patches sequentially from highest to lower V_n' until reaching her resource-maximizing or area-minimizing threshold. From the chosen patches, the patch with the highest resource value (V_n) was chosen as the home range center and the home range was rebuilt. Mitchell set area-minimizing thresholds at 25%, 33%, and 50% of the mean for the resource-maximizing home ranges.

For models with resource depression, a colonizer explored the model landscape using a moving window of mean home range size to locate the neighborhood with maximum total resource value and then chose patches. Mitchell explored 3 levels of resource depression, all of which exhibited the same patterns and, therefore, he used just the medium level of depression, 0.15. All patches within each new home range depressed V_n for all subsequently simulated home ranges. Patch values could not go below 0 and no patch could be included within >4 home ranges.

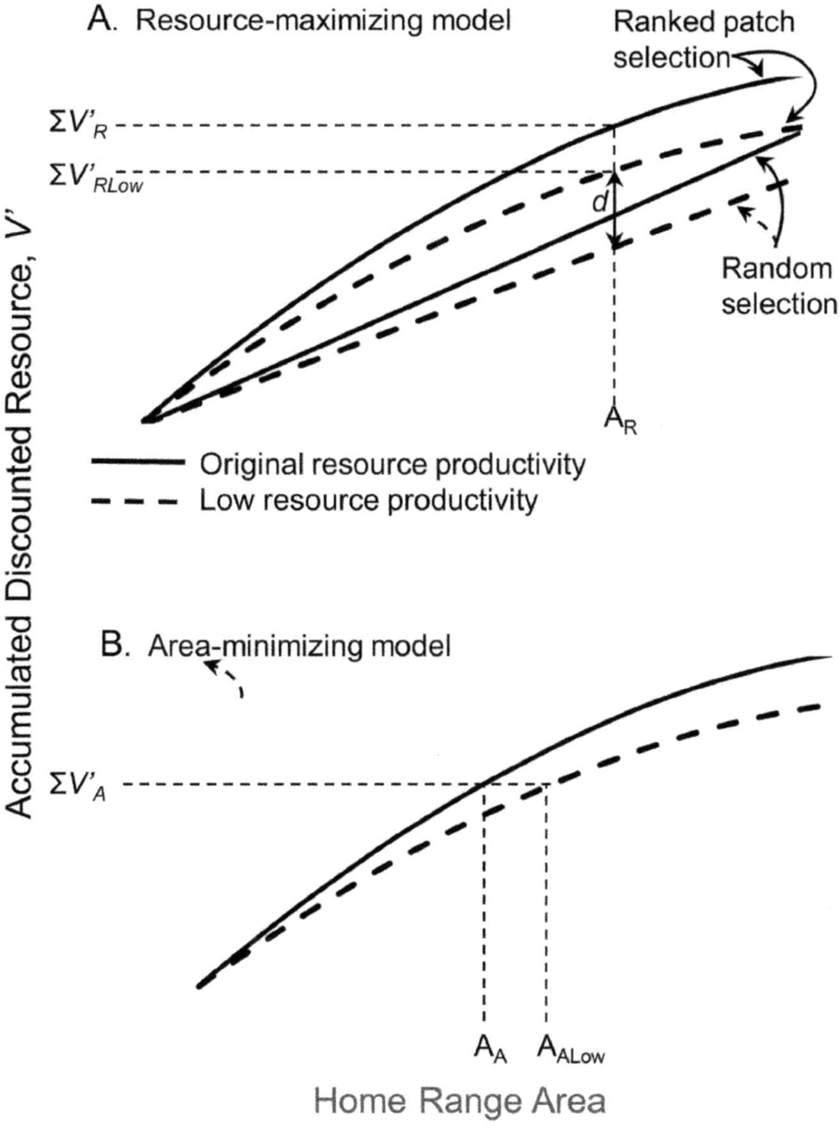

Figure 8.3 How decreased resource availability affects the selection of patches for home ranges under two models of optimal patch selection, as in Figure 8.2 except that the dashed lines show the condition of proportionally decreased resource availability. A. Because resource-maximizing home ranges maximize the difference between selective and random patch choice, the point at which d is maximized does not change with proportional changes of resource availability. Home range area, A_R, does not change but accumulated resources decrease from EV'_R to EV'_{Rlow}. B. For area-minimizing home ranges, accumulated resources, EV'_A, do not change because this is the animal's threshold no matter what the level of resource availability; but home range area increases from A_A to A_{ALow} because more area is needed to reach the resource threshold. (After Mitchell and Powell 2004.)

Mitchell characterized home ranges by number of patches, overlap, mean V_n per patch, total $V(V_n$ summed over all selected cells), and fragmentation (ranging from 0 to 1).

8.3.1 Results of Resource-Maximizing and Area-Minimizing Models

The resource thresholds for area-minimizing models (M_A) affected the number of patches selected

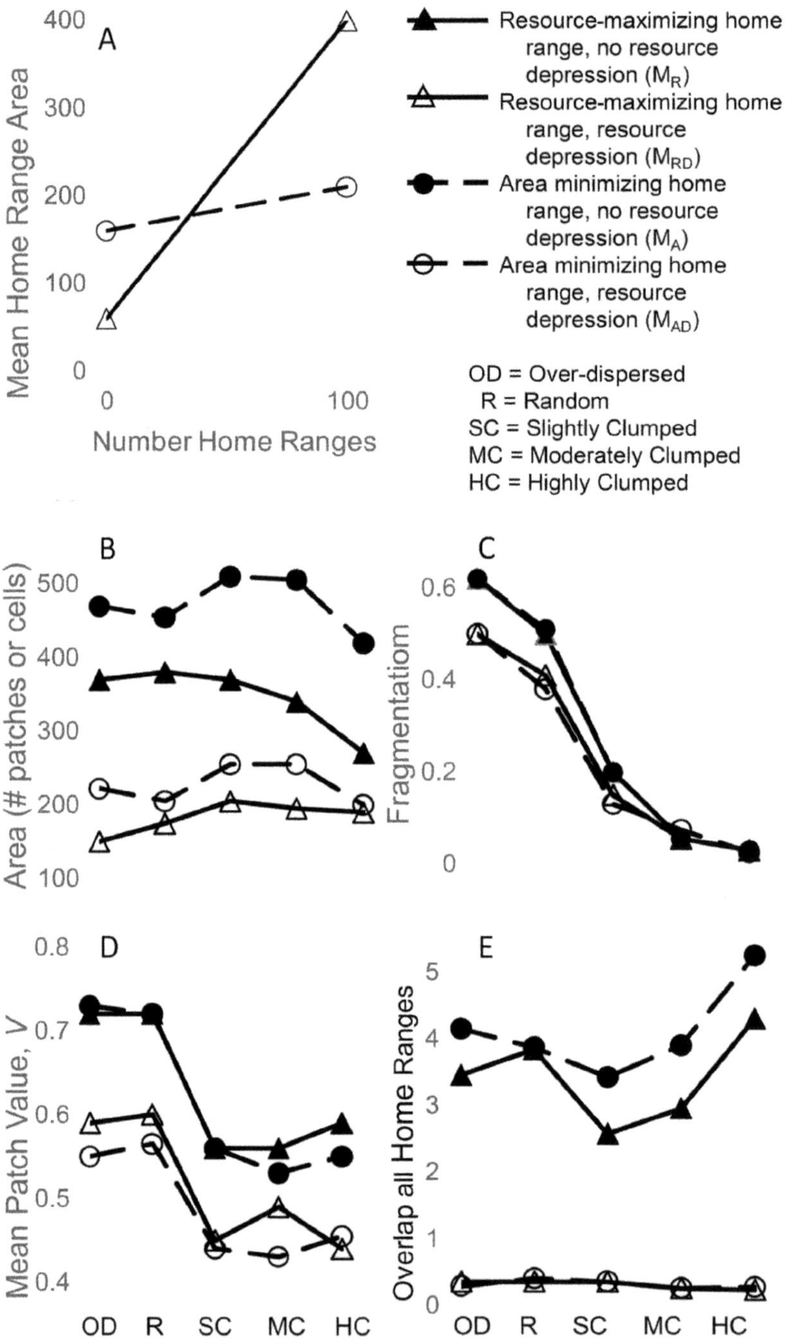

Figure 8.4 Characteristics of 100 simulated home ranges on each of 5 simulated landscapes, created using patch selection designed to maximize resource density, Model M_R, or to minimize home range area, Model M_A, either with or without resource depression, Models M_{RD} and M_{AD}. Resources on the simulated landscapes varied from being over-dispersed (OD) to highly clumped (HC). A. Mean area of home ranges (numbers of patches selected, all patches of equal size) as number of home ranges increased on the landscape. B. Mean area of home ranges on different landscapes. C. Mean fragmentation of selected patches. D. Mean quality of patches (mean *V*). E. Mean overlap among all home ranges. For models with resource depression overlap is always near 0. (After Mitchell and Powell 2004.)

(home range area) for model home ranges on all resource distributions across the landscapes (evenly distributed to highly clumped). Mean home range areas increased as successive home ranges were added using models with resource depression (M_{RD}, M_{AD}; Figure 8.4A). Areas of home ranges incorporating resource depression were larger than those without resource depression on all 5 landscapes (Figure 8.4B). For models including resource depression, averaged across landscapes, home range area increased for both resource-maximizing and area-minimizing models as more and more animals were added to the landscapes; the increase was far steeper for area-minimizing home ranges. An increase in area for resource-maximizing home ranges is not predicted for a landscape that decreases uniformly in quality but resource depression does not decrease quality uniformly, causing a very gradual increase in area for resource-maximizing home ranges with increasing numbers of home ranges on a landscape.

Fragmentation of home ranges decreased with increasing resource clumping but was similar for resource-maximizing and area-minimizing home ranges with and without resource depression (Figure 8.4B). On landscapes with evenly distributed resources, all good patches were separated from other good patches by patches of low value. Landscape patterns had greater effects on fragmentation than did numbers of home ranges.

Mean V_n was lower for home ranges on clumped landscapes and lower for home ranges that incorporated resource depression (Figure 8.4C). Clumped landscapes had high-quality patches close together, reducing travel costs, which caused less discounting of resource values. Consequently, model animals could choose close patches with modest V_n that had higher V_n' than did distant patches with higher V_n but lower V_n' due to travel costs. Likewise, resource depression decreased realized values for patches (V_n').

For home ranges not incorporating resource depression, overlap increased as landscapes became more clumped (Figure 8.4D). On clumped landscapes, patches with abundant resources were close together, leading to overlapping home ranges. Home ranges experiencing resource depression, however, had very little overlap. Resource depression caused by dominant, long-term residents led newer, subordinate residents to avoid patches already colonized.

The models define the optimal choice of resource patches for inclusion in a home range but such a home range may differ from a familiarity distribution based on movement data, particularly where the models choose disjunct patches to be in a home range. The models exclude from a home range those areas without resources through which or over which (birds, bats, insects) a target animal might be forced to travel to reach its selected patches (Mitchell and Powell 2008). These home ranges will not match familiarity distributions built from animal locations. If, however, animals must choose travel routes to minimize predation, for example, then the routes are a resource and will be included in animals' home ranges, and should be included in familiarity distributions.

8.4 Resource-Maximizing and Area-Minimizing Models Applied to Black Bears

Mitchell's models can be used to build mechanistic distributions. Consequently, Mitchell (1997, Mitchell and Powell 2007) compared mechanistic distributions to familiarity distributions for female black bears on our study area in the Blue Ridge Mountains, North Carolina, USA. In Chapter 6, section 6.6, I summarized why female black bears are good subjects for testing general hypotheses, which include hypotheses related to optimal foraging for a home range.

Mitchell built familiarity distributions for 104 adult female black bears followed between 1981 and 2001 using a fixed kernel estimator with kernel K_2 and h = 250 m, and used the 95% familiarity distributions for his comparisons. He used those familiarity distributions to test against his mechanistic distributions that he built based on the distribution and characteristics of resource patches in the study area. He tested the following hypotheses.

1) Adult female black bears choose optimal home ranges, as approximated by their familiarity distributions tested against mechanistic distributions built from optimal foraging models.

2) Resource depression by competing female bears affects home range choice.
3) Familiarity distributions are efficient with respect to the spatial distribution of limiting resources; mean resource content of bears' familiarity distributions exceeds the average availability of resources in the study area. Mitchell defined "efficiency" as how much larger the resource content, total$V$, of the familiarity distribution is than would be realized from random choice (i.e., the magnitude of d in Figure 8.2A).

To understand the importance of energy requirements and social dominance on patch selection, Mitchell built the mechanistic distributions using his models with and without resource depression for each familiarity distribution of an adult female bear. To build the mechanistic distributions, he quantified resource quality (V) using the life requisite variable for food from Zimmerman's model of habitat quality for black bears (Chapter 6; Zimmerman 1992, Mitchell et al. 2002). He started building each mechanistic distribution for each real bear at the location with the highest intensity of use. He compared similarity of patch choices by real bears to those built into mechanistic distributions using each model, thereby exploring how interactions among optimal foraging strategy, resource threshold, and resource depression explained patch choice.

To build mechanistic distributions for patch selection models M_A, M_{AD}, and M_{RD}, values must be included for two biological parameters: 1) a resource threshold (the minimum resource requirement for models M_A and M_{AD}); and 2) the extent to which a bear depresses resources within her home range (how many bears can share a given patch for M_{RD} and M_{AD}). No data existed to estimate these parameters for black bears, so Mitchell built mechanistic distributions using a realistic range of levels for the resource threshold and for resource depression in models M_A, M_{AD}, and M_{RD}. For resource thresholds, he divided the distribution of summed V for the familiarity distributions of female bears into 6 equal quantiles representing 6 hypothetical resource thresholds.

Overlap of familiarity distributions was extensive in the study area (Powell et al. 1997), suggesting that levels of resource depression among the female bears were relatively low compared to resource levels. Therefore, Mitchell set 0.20 as the semi-arbitrary, maximum amount that a simulated bear would depress resources in patches that are within her familiarity distribution. To test for levels of resource depression < 0.20, he included 3 lower levels in simulations. Six levels of resource thresholds and 4 of resource depression generated 24 mechanistic distributions for each familiarity distribution using modelM_{AD}; M_A generated 6 mechanistic distributions, M_{RD} 4, and M_R 1.

The mean quality (mean V_n) for real bears' 95% familiarity distributions, 0.60($CI_{95\%}$ = 0.59 – 0.61), was significantly greater than the mean quality for the study area, 0.54($CI_{95\%}$ = 0.53 – 0.55), consistent with the bears selecting patches to include in their home ranges based on patch quality and maintaining quality of patches in their cognitive maps. The familiarity distribution for each real bear was approximated reasonably well by at least one mechanistic distribution and the mean similarity value with the best mechanistic distribution compared to the familiarity distribution was 0.75 ± 0.12 (Figure 8.5), reinforcing a conclusion of bears selecting patches to include in their home ranges based on patch quality. In ΔAIC_C rankings of the mechanistic distributions' abilities to describe the real bears' familiarity distributions, M_A ranked highest (model weight = 0.25, likelihood = 1.00) and M_R ranked second (model weight = 0.24, likelihood = 0.96; Table 8.1). Distributions incorporating resource depression (models M_{RD}, M_{AD}) had lower similarity to the familiarity distribution, only the mechanistic distribution with minimal resource depletion having $\Delta AIC_C \leqslant 4.0$ (ΔAIC_C = 3.89, model weight = 0.04, likelihood = 0.18). In contrast, 6 area-minimizing mechanistic distributions without resource depletion had $\Delta AIC_C \leqslant 4.0$ (combined model weight = 0.73, likelihoods ranging from 1.00 to 0.14). Mechanistic distributions built using models M_A with the 3 highest resource thresholds and no resource depression ranked highest. Only 2 mechanistic distributions built using models M_{AD} had $\Delta AIC_C \leqslant 4.0$ and those models had the lowest level of resource depression. The results for resource depression are consistent with the tests of mechanistic models for these female bears presented in Chapter 6 and

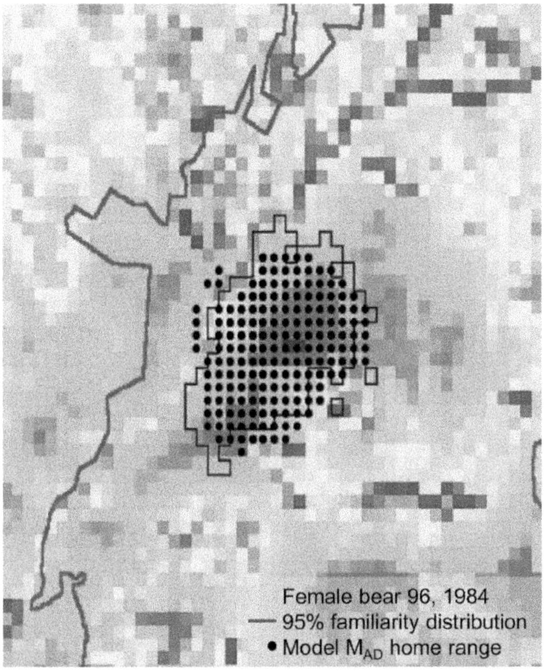

Figure 8.5 The optimal mechanistic distribution (dots) built using resource-maximizing model M_R and the 95% familiarity distribution (outline) for female bear 96 in 1984, Blue Ridge Mountains, North Carolina, USA. The optimal mechanistic distribution was generated based on the underlying distribution of resources depicted by the Food component of a habitat suitability index for black bears. Dark background represents low food value; light background represents high food value. The familiarity distribution was built with a kernel estimator using kernel k_2 and h = 250 m. (After Mitchell and Powell 2004.)

with the results of other research showing that the population was regulated by mortality (Powell et al. 1996, Brongo et al. 2005). Although the bears maintained information about other bears in their cognitive maps, that information appears not to be central in their minds when they plan future movements. This result is also consistent with the results presented in Chapter 6.

Importantly, mean mechanistic resource-maximizing distributions (M_R) consistently and significantly predicted higher resource quality (total V_n) and were consistently larger (numbers of patches selected) than the familiarity distributions built for real bears.

Female black bears in our study area in the Blue Ridge Mountains appeared to maintain area-minimizing home ranges with low levels of resource depression. Resource depression sets a maximum to the number of home ranges that a landscape can support and suggested that Mitchell's models could be used to estimate carrying capacity, K, for the population (Mitchell and Powell 2012). To that effect, Mitchell added mechanistic area-minimizing distributions sequentially onto the study area. Mechanistic distribution areas increased as the simulated population grew, ultimately reaching the point where no more mechanistic distributions could be added (Figure 8.6). The estimated carrying capacity was 52 female bears, which he and I found to be a credible number (Mitchell and Powell 2012). Areas of the familiarity distributions for the bears averaged in the mid-teens (km^2). Our best estimate of the numbers of female bears living in the study area each year, using capture–recapture, minimum number of alive, and "ghost" home ranges (places that appeared to have a resident female that we could not catch) was in the low 20s, consistent with population size estimated in Figure 8.6.

Table 8.1 AIC results of similarity between home ranges simulated using different optimality models for patch selection and distributions of adult female black bears in the Pisgah Bear Sanctuary, North Carolina, during 1981–2001.

Model	Resource threshold	Resource depression	AIC	ΔAIC	Model weight, w_t	Model likelihood
$extM_{extA}$	5	0	1098.095	0.000	0.25	1.00
$extM_{extR}$	—	0	1098.174	0.078	0.25	0.96
$extM_{extA}$	4	0	1098.193	0.097	0.24	0.95
$extM_{extA}$	6	0	1100.470	2.374	0.08	0.31
$extM_{extAD}$	4	0.05	1100.584	2.488	0.07	0.29
$extM_{extAD}$	3	0.05	1101.427	3.376	0.05	0.19
$extM_{extRD}$	—	0.05	1101.544	3.448	0.04	0.18
$extM_{extA}$	3	0	1101.987	3.891	0.04	0.14

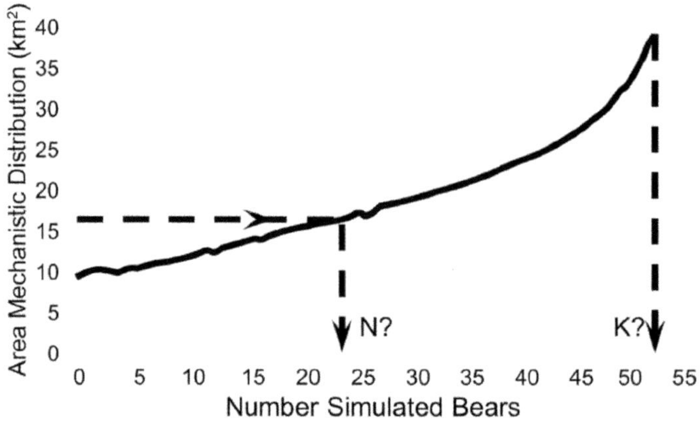

Figure 8.6 Change in area of simulated, area-minimizing home ranges for female black bears in the Blue Ridge Mountains, North Carolina, USA, as the population increases. Simulations were of sequentially established optimal home ranges with moderate resource thresholds and low resource depression, and based on the food component of a habitat suitability index for bears. Areas of familiarity distributions for real bears averaged in the mid-teens over the years of the project, estimating population size to be in the low 20s (N), which is consistent with trapping data for the population. After the number of simulated, area-minimizing home ranges reached 52, no new home ranges could be added, making 52 the estimated carrying capacity (*K*) for the study area.

8.5 Area-Minimizing Home Ranges Remodelled

Sells (Sells 2019, Sells and Mitchell 2020) built on Mitchell's models of foraging optimally for a home range. Her model of patch choice can be used to produce mechanistic distributions for a wide range of animals. Specifically, once applied to real animals, Sells's model predicted wolf territory sizes accurately and predicted accurately how wolves' territories would decrease in size as prey density increased and as competitor numbers and sizes of neighboring packs increased. In addition, her model predicted accurately how much territory size would be inversely related to density of roads and human residences (Sells et al. 2021, 2022a).

To evaluate her model, Sells simulated optimal, area-minimizing home ranges on simulated landscapes that ranged from having the limited resource (modelled to be food) distributed evenly to clumped. (Although Sells consistently used only the term "territory," some of her models generated home ranges with such extensive overlap that exclusive use was limited to only some 50% of the home ranges. Consequently, I use the term "home range.") She simulated variation in food production from low to high and she simulated predator distributions (predators on target animals) ranging from even to clumped with predator densities from low to high.

For Sells's models, the discounted (Sells used the term "net") value V_n' of patch n is

$$V_n' = B_n - C_\Sigma - T_\Sigma - P_\Sigma,$$

where Bn is the benefit (food) to be gained by the simulated animal from patch n, C_Σ is the decrease in benefit caused by competition from conspecifics in patch n, T_Σ is the cost of travel from the home range center to patch n, and P_Σ is the local predation pressure in patch n. Sells noted that other limiting resources, such as burrows or den sites, could be the driver for home range or territory establishment.

Sells set

$$C_\Sigma = \sum_{i=1}^{n} \frac{C_{cost} \cdot \sum C_i}{\sum (C_i + C_A)},$$

where C_{cost} is the relative cost of competition compared to benefits, C_i is the competitive ability of conspecific i who has already incorporated patch n into her home range, and C_A is the competitive ability of the model animal who is establishing or adjusting her home range.

Sells set

$$T_\Sigma = D \cdot T_{cost},$$

where D is the number of patches between the simulated home range center and patch n and T_{cost} is the relative cost of travel compared to benefits.

And Sells set

$$P_\Sigma = \Sigma P_i \cdot P_{cost},$$

where P_i is the predation risk in patch i and along the route from the home range center to patch n, and P_{cost} is the cost of predation risk compared to benefits.

Each model run began with a simulated animal choosing a neighborhood of patches with high benefit and low predation risk, as in Mitchell's model. The animal then added new patches to her home range in the order of decreasing $V_n{}'$ until the home range reached the resource requirements, V'_{total}, set for the model run. As other simulated animals were added, they chose neighborhoods based on $V_n{}'$, with C_Σ adjusted by competition from conspecifics that have already chosen patches within the neighborhoods. In this way, Sells's model mimics Mitchell's resource depression. After all animals had established home ranges, each animal adjusted her home range if subsequent animals had chosen patches she had chosen before them. Sells also explored the effects of variable predator densities (low to high) on the number of patches selected by model animals and the carrying capacity of model animals (= number of home ranges at the end of each run).

Sells's results were consistent with those of Mitchell (Mitchell 1997, Mitchell and Powell 2004) and extended beyond Mitchell's results by adding predation risk and modelling competition differently (Figures 8.7, 8.8). Numbers of patches selected, and numbers of patches selected plus patches used for travel, decreased with increasing food clumping and food abundance, and with decreasing food requirements (Figure 8.7). Home range overlap increased with increasingly clumped resources, especially at low population densities with low food requirements. Increasing predator densities (number of predators on a landscape) decreased carrying capacity slightly for simulated animals but the effect was much smaller than the effects of food requirements of simulated animals and of food abundances (Figure 8.8). Similarly, predator distributions (from even to highly clumped) and predator densities (numbers of selected patches that have resident predators) had only modest effects on numbers of patches selected by target animals. These were results consistent with a long history of modelling and empirical data (e.g. Carpenter and MacMillen 1976, Harestad and Bunnell 1979, Hixon 1980, Moorcroft and Lewis 2006, Rich et al. 2012, Lesmerises et al. 2012, Ausband et al. 2017).

The result is that Sells developed a mechanistic model of home ranges with realistic outputs that could be used to build mechanistic distributions and tested against data. Though the model's equations for travel, competition, and predation produce realistic behavior summaries, the equations do not mimic actual biological characteristics of real animals. The same was true for Mitchell's models. Sensitivity or elasticity analyses would have strengthened the conclusions of both Mitchell and Sells.

8.5.1 Area-Minimizing Models Applied to Wolves

Sells and colleagues have used Sells's models (Sells 2019, Sells and Mitchell 2020) to explain characteristics of territories (Figure 8.9), and the distribution and population size for wild wolves

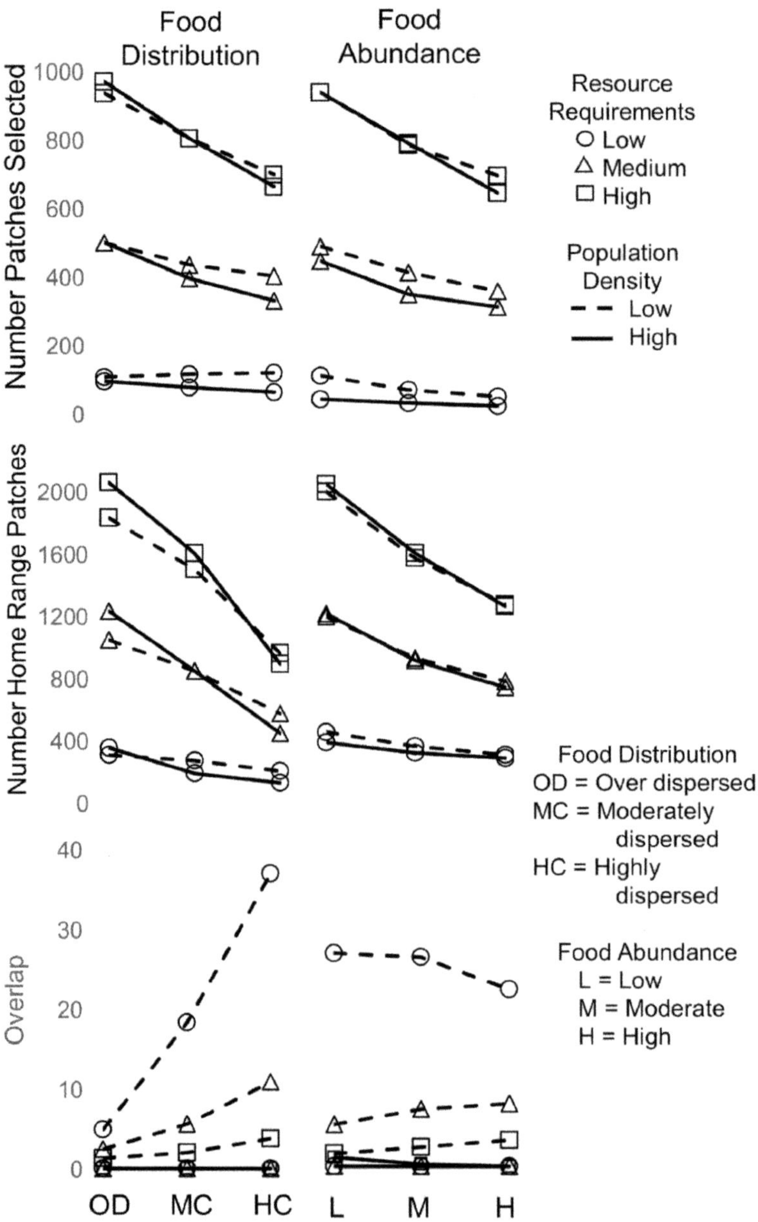

Figure 8.7 Mean characteristics of home ranges of simulated animals as food distribution and abundance varied over model runs. Number Patches Selected was the mean number of patches selected by simulated animals based on their discounted value, $V_n{}'$. Number Home Range Patches was the number of patches selected plus patches in travel corridors to selected patches. Overlap was the percentage of home range patches shared with home ranges of other simulated animals. (After Sells and Mitchell 2020.)

in western Montana, USA, studied in 2008–2009 and 2014–2019 (Sells et al. 2021, 2022a, b). Specifically, mechanistic distributions generated by her model predicted accurately the characteristics of the familiarity distributions for wolf territories in western Montana, including 2nd and 3rd order use of landscapes (Johnson 1980); the territory sizes and the way that territories would

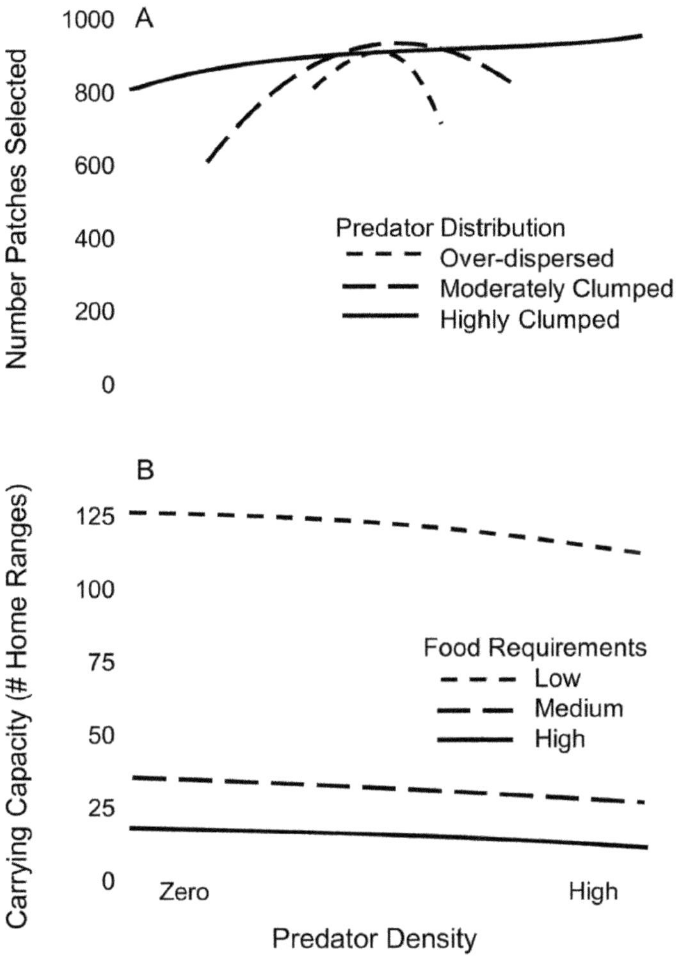

Figure 8.8 Effects of predator density on (A) mean number of patches selected for a simulated home range and (B) mean carrying capacity (number of home ranges on the landscape) for simulated animals on a landscape with moderate food clumping. Patterns for over-dispersed and highly clumped landscapes showed similar patterns. (After Sells and Mitchell 2020.)

respond to variation in prey populations, terrain, and dangers of human hunters; the distribution of wolf territories (Figure 8.10); and the total wolf population size (Sells et al. 2021, 2022a, b).

In addition, by combining the mechanistic distributions with occupancy modelling and an empirical model of pack size, Sells et al. (2022b) showed that their estimates of wolf abundances were more accurate than previous, field estimates while also requiring less monitoring effort. First, based on environmental variables and 14 years of wolf data, Sells and colleagues used occupancy modelling to estimate the wolf distribution in Montana. They used Sells's mechanistic distributions, including prey densities (benefit), terrain ruggedness (travel cost), and human density (predation on wolves) to predict areas of familiarity distributions for wolf packs. Together, these mechanistic distributions predicted the number of packs and showed that pack sizes correlated positively with local densities of packs and negatively with terrain ruggedness, local wolf mortalities, and the intensity of the management of the wolf harvest. Wolves' cognitive maps appear to maintain information about locations and activities of other packs, about local terrain, and about

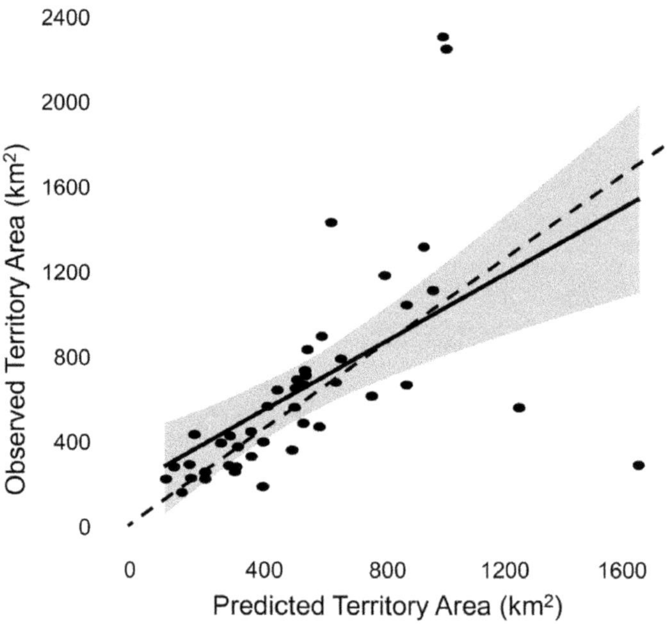

Figure 8.9 Territory sizes for wolf packs in western Montana, USA, predicted by Sells's home range models and observed for GPS-collared wolf packs (n = 9, 2008–2009; n = 26, 2014–2019). Observed territory areas correlated significantly with predicted territory areas. Solid line shows regression of observed on predicted. Dotted line is at 45°, perfect predictions. Shaded area shows the 95% confidence interval. (After Sells 2019.)

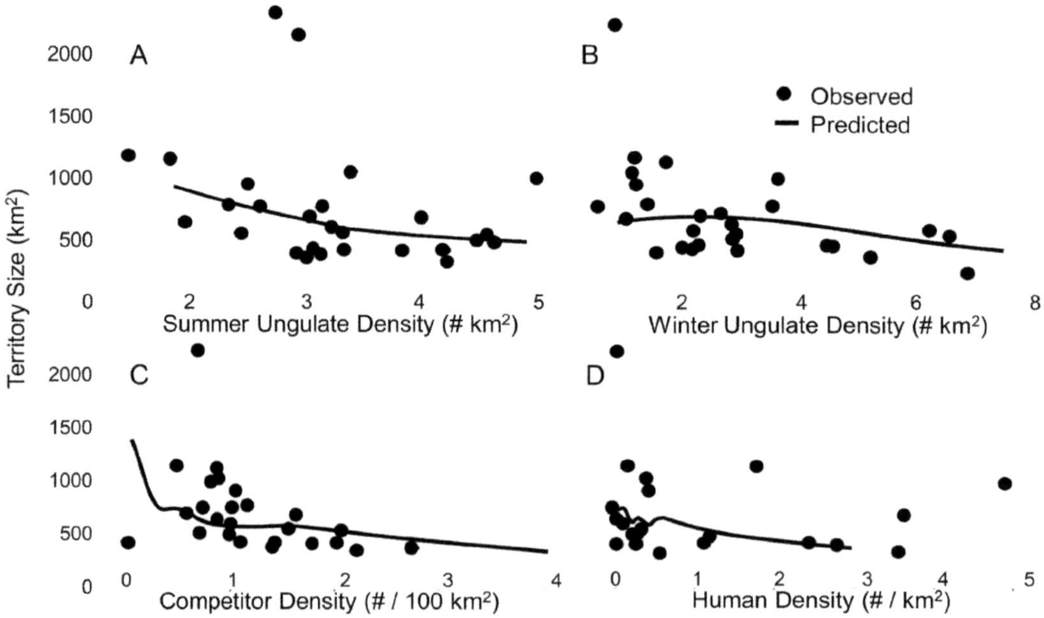

Figure 8.10 For 26 wolf packs in western Montana, USA, in 2014–2019, observed territory areas and smoothed, predicted territory areas taken from mechanistic distributions plotted against (A, B) seasonal densities of ungulates, (C) competitors, and (D) human densities (3 outlier observations had densities of 26, 30, and 72). (Adapted from Sells et al. 2022a.)

when and where they can anticipate danger from humans.

The applications of Mitchell's and Sells's models to real populations (Mitchell 1997, Mitchell and Powell 2007, Sells 2019, Sells and Mitchell 2020, Sells et al. 2021, 2022a, b) suggest strongly that Mitchell's optimal foraging approach to home range selection and Mitchell's and Sells's models capture major components of home range behaviors. The models highlight the importance of incorporating critical biological variables into models and of modelling those variables in ways that mimic, as much as possible, true biological functions, or at least the outcomes of those functions. The application of the models demonstrates that territoriality, and home ranges in general, are emergent behaviors that connect how animals use landscapes given the distributions of resources and constraints on those landscapes. The benefits (food resources) and the costs of travel, competition, and predation provide Sells's and Mitchell's models with considerable flexibility to include diverse environmental and social conditions. Sells and her colleagues demonstrated that predation on wolves by humans could be indexed using road densities, human densities, and direct human-caused mortality. With appropriate data, competition with pumas or grizzlies could be incorporated in the competition variable.

Of course, resources are not limited to food. For other cooperatively breeding animals, such as red-cockaded woodpeckers and woodland voles, the limiting resources that stimulate territorial behavior are not food but sleeping cavities for red-cockaded woodpeckers and tunnel systems for woodland voles (Walters et al. 1988, 1992, Powell and Fried 1992). For these situations, benefits could be modelled to include the costs of cavity construction or tunnel construction that are avoided by maintaining territories.

8.6 Overview of Optimal Patch Choice and Home Ranges

Mitchell's (1997, Mitchell and Powell 2004, Mitchell and Powell 2012) were the first mechanistic home range models based on optimal choice of resources distributed on a landscape. Mitchell's (1997) and Sells's (2019) models differ significantly from other mechanistic home range models in not being based on animal movements (e.g. models of Holgate 1971, Gautestad and Mysterud 1993, Moorcroft and Lewis 2006, Börger et al. 2008, Van Moorter et al. 2009, Wakefield et al. 2013, Riotte-Lambert et al. 2015, Beardsell et al. 2023, Rohwäder and Jeltsch 2022, and Xu et al. 2022 are all based on movements). Most mechanistic models build mechanistic distributions that are static (that is, do not change with time; a model of Bateman et al. 2015 (see Chapter 6) is an outlier) even though the models are based on movements. Many mechanistic distributions have been based on correlated random walks structured according to animal locations and biased toward home range centers (Chapter 6). Thus, such models are based on a null model for movements.

The models of Mitchell and Sells, however, and the mechanistic distributions built using them, are not based on movements but on optimal selection of resource-bearing patches, presumably the foundation (i.e. the cognitive map) underlying the movements of animals. In this way, Mitchell's and Sells's models provide different insights in animal's cognitive maps by emphasizing establishment, and updating of resources and other characteristics, of home ranges (building cognitive maps) rather than the use of cognitive maps to plan foraging bouts and routes. Mitchell's and Sells's models emphasize that daily movements are emergent from home ranges.

Further, Mitchell's and Sells's models are general and can be used to build mechanistic distributions for any taxonomic group for which the spatial distributions of resources or habitat quality can be quantified, and their models are particularly well suited for application to species that have independently tested models of habitat quality. Mitchell's and Sells's models are not unique on this latter point, however (Van Moorter et al. 2009). Mitchell's and Sells's models are mechanistic and founded on ecological theory (Charnov 1976a, Stephens and Krebs 1986, Krebs and Kacelnik 1991). Their models allow the development of biologically credible hypotheses that can be tested empirically, leading to new insights into how animals' view their home ranges, their cognitive maps. And: the models work.

Application of Mitchell's and Sells's models to real animals showed that, for black bears and especially for wolves, distributions of food were consequential for choice of patches to include in home ranges and, therefore, distributions of food must have importance in black bears' and wolves' cognitive maps. Even though food may not be limiting for some black bears that Mitchell studied in the Blue Ridge Mountains (Chapter 6), food, nonetheless, was central enough for organizing home ranges in the bears' cognitive maps. Using topography to discount travel cost for wolves showed that topography had a critical place in their cognitive maps, as did including indices of humans such as roads and density of residences. The mechanistic models applied to black bears in Chapter 6 showed that topography was also important for most bears, suggesting that incorporating topography into travel costs for those bears might have provided more insights to the importance of topography in bears' cognitive maps. Including roads and human densities to index predation on wolves proved to be important, showing that landscape characteristics related to humans must be incorporated into wolves' cognitive maps. Hunting and poaching mortality appeared to be the controlling factor for the bear population that Mitchell studied (Powell et al. 1996, Brongo et al. 2005). Given that the distribution of roads contributed to all bears' choices of home ranges (Chapter 6), including roads and trails as a predation component in Mitchell's optimal foraging model might also have improved it.

Nonetheless, Mitchell's and Sells's models have been cited but not adopted generally to gain insight into the biologies of other species. Seidel and Boyce (2016), for example, did test and show that elk based their selection of patches on patch quality. They did not address, however, whether elk developed resource-maximizing (resources exceeding requirements) vs area-minimizing home ranges or whether elk home ranges maximized quality of familiarity distributions.

Landscape structure has the predominant influence on both resource-maximizing and area-minimizing model home ranges. If this result is true of most real animals' home ranges, then animals on a given landscape and practicing either strategy (resource-maximizing, area-minimizing, or perhaps any other) will have familiarity distributions that are similar in many ways, unless individuals have very small or large resource requirements relative to the productivity of the landscape. Nonetheless, the 2 strategies, resource-maximizing and area-minimizing, differ in some predictions that can be tested for populations of real animals. First, a uniform decrease (or increase) in resource quality across a landscape will not affect size for resource-maximizing familiarity distributions greatly but will lead to larger (or smaller) sizes for area-minimizing familiarity distributions (Figure 8.3). A large literature illustrates this result, leading to the hypothesis that most animals maintain area-minimizing home ranges.

Second, for animals that maintain area-minimizing home ranges, total resources contained within familiarity distributions should be similar for all animals within a class within a population (class based on sex, reproductive status, age, etc.), while this should not be the case for resource-maximizing home ranges. Third, when the resource threshold for area-minimizing home ranges matches the optimal home range size for resource-maximizing ones the two types of home ranges will be much alike until resource abundances change. In contrast, however, when the area-minimizing threshold does not line up with the optimal size for resource-maximizing, then areas and total resources will not match and may be very different (Figure 8.2).

Finally, because the resource-maximizing model incorporates no biological parameters (other than seeking to maximize resources without a biological reason to do so), predicting only the optimal use of a resource distribution, it represents a null model for evaluating hypothesized spatial behaviors that do incorporate biological parameters. The relationships between animals and their limiting resources change over time, due to seasonal changes in resource requirements (such as reproduction), due to seasonal and annual variations in the availability of resources, due to population fluctuations, and due to deaths of high-ranking individuals within the population. Indeed, an animal might maximize resources during part of a year but meet a minimum resource threshold at other times. If the resource-maximizing strategy yields more resources than the area-minimizing strategy during the reproductive season, resource-maximizing might be easier to maintain than planning ahead for increasing resource requirements as young grow and have increasing nutritional and energetic requirements. If the biology is understood, building a home range model that combines the two strategies should be straightforward.

Real animals do compete with other animals for resources and, therefore, mechanistic distributions incorporating competition with conspecifics and also with members of other species will often match best the familiarity distributions of animals and provide insights into the animals' cognitive maps. Likewise, many real animals cooperate with group members to obtain resources and, therefore, mechanistic distributions incorporating cooperation will often match best the familiarity distributions of animals and provide insights into the animals' cognitive maps. Mitchell's and Sells's results suggest that, at low population densities, home ranges of animals that consume or protect resources will be similar to those of animals that do not, but be more evenly dispersed on the landscape. For high population densities, the proximity and overlap of home ranges will depend on the distributions and qualities of resources and depend on the number of animals living in the area. An area-minimizing strategy including resource depression could result in home ranges distributed on a landscape roughly according to an ideal free distribution (Fretwell and Lucas 1970, Fretwell 1972).

CHAPTER 9

Game Theory and Home Ranges

Of a winter evening, I sat reading in the living room of the Kawishiwi Lodge, a historic log cabin in the woods made of huge white pine logs by the Civilian Conservation Corps back during the Depression. During those winter days, I was live-trapping fishers in northern Minnesota, USA, and shared the lodge with other graduate students and with forest service researchers. Across the room, I saw a white short-tailed weasel stick his head out from under an overstuffed chair. Quickly, he snuck back under the chair, then snooped around the walls of the room. I walked into the kitchen and cut a knuckle-sized chunk of raw venison from a haunch that we had in the refrigerator.

Back in the living room, I sat on the floor holding the venison at the end of my extended arm. Before too long, Mr Short-tail ventured out from under his chair, one jerky step at a time, his tail fluffed, till he was close enough to touch the venison. He did not grab the venison and pull, as I had anticipated. Instead, he hit my index finger with a good punch from his muzzle. Surprised, I dropped the venison and he grabbed it, zipping back under his chair, trilling.

I felt a bit foolish, so I fetched another piece of venison. Sure enough, Mr Short-tail came back, less cautious this time. Again, he approached the venison and muzzle-butted my index finger. Again I dropped the venison and he grabbed it, running back under his chair, trilling.

Now, I really felt foolish. With a third piece of venison, I sat on the floor again, holding out my hand with the venison. I relaxed. I waited for the fine fellow who had twice proven himself more intelligent than I. This time when he gave my finger the muzzle punch, I held tight. He tried again. Again I held tight. Then he did grab the venison and tug, but I still held on. Finally, in the middle of the living room, he settled and ate the venison from my fingers. When he had eaten so much of the venison that

Home Ranges of Animals. Roger A. Powell, Oxford University Press. © Roger A. Powell (2025). DOI: 10.1093/9780191949142.003.0009

I could no longer hold it firmly, I let him have it. He became a regular visitor in the living room during evenings.

Mr Short-tail and I played a game. We contested over a piece of venison. Our game strategies can be modelled using Game Theory.

9.1 Game Theory Background

For understanding contests and cooperation between animals, game theory is one of the most productive approaches. Game theory has been used to generate hypotheses related to finding mates, to obtaining food, to understanding the selfish herd, and to maintaining territories. It provides methods for building models of behaviors that researchers can use to develop hypotheses related to the evolution of those behaviors. Hypotheses developed from game theoretic analyses can be incorporated into mechanistic distributions and tested in the field, increasing the power of game theoretic analyses.

This chapter provides an introduction to game theory with simple yet, I believe, insightful examples. After an introduction to the jargon used in game theoretic models, the chapter introduces the simple model that is used almost universally to introduce game theory. This model is then made more realistic and modified to apply to territory holders and their neighbors. Then the chapter explores some models for wolf pack behaviors, including the conditions that could lead to invasion of the territory of a neighboring pack. It examines the conditions that lead to home range sharing and to applications of the War of Attrition game to territory maintenance. Finally, the chapter looks into why animals might adopt landmarks as territory boundaries.

Game theory is an active area of research with a huge literature providing large numbers of hypotheses in need of field testing (for example: Mesterton-Gibbons and Sherratt 2016, Varga et al. 2020, Noonburg and Anderson 2021, Saraiva Menezes and Oliveira-Santos 2021, Asgharnia et al. 2022). The literature, however, is dominated by sophisticated modelling that has generated hypotheses that are unavailable to many field biologists who do not read journals that emphasize modelling. Game theoretic modelling, however, does not require complicated mathematics, as the models used as examples in this chapter show. Biologists of all flavors should consider developing simple game theoretic models when their research is appropriate, thereby building hypotheses for testing. Hypotheses related to how animals use their environments and how information affects animals' cognitive maps can be incorporated into mechanistic distributions. The opportunities for learning new biology are vast.

Game theory was first formalized by von Neuman and Morgenstern (1953) to understand human economic behavior. The early biological applications were used to model animals playing games vs the environment to minimize the probability of extinction (e.g. Lewontin 1961), and more recently to understand the ecological consequences of climate change (Sutherland 2006). Maynard Smith (1982) provided a solid, introductory overview of game theory for biologists, which Hinsch and Komdeur (2017) have supplemented with a review of game theory and territoriality. McNamara and Leimar (2020) have more recently provided an overview of sophisticated uses, showing the breadth of applications of game theory to biology.

Because game theory explores interactions among animals, and even among plants, game theory modelling has largely ignored uncontested home ranges, except for some work on sharing resources that can be applied to sharing home ranges. Hinsch and Komdeur (2017) proposed that game theoretic models fall into 3 categories related to territories, without noting that other situations exist: 1) A territory invader (usually an animal lacking a territory) strives to oust a territory owner and take over the territory for herself. 2) An invader (usually a neighbor with her own territory) strives to move the boundaries between territories. And 3) an invader strives to pilfer resources without otherwise changing territories or ownership. Although Hinsch and Komdeur (2017) note correctly that these 3 situations are, indeed, different, they do not show clearly that territory owners (real animals, not modelled animals) respond differently to the 3 situations. Nor do they document that these 3 most-modelled situations are the 3 situations most experienced by real animals. Smith (1968; Chapter 7), for example, showed that, for North American red squirrels, visits to the peripheries of their territories are expensive moves, devaluing such moves. Consequently, red squirrels may not contest small changes in boundaries much. This chapter shows that red squirrels recognize calls of known neighbors and respond little to them because they are familiar neighbors and are not threatening.

For game theoretic models of home range and territorial behaviors, the outcome of a contest is understood to have been incorporated into the contestants' cognitive maps, though not stated in those terms. The contestants' cognitive maps have information on locations and values of uncontested and potentially contested resources and how those values change with information loss. Thus, game theoretic approaches are consistent with our best understanding of animals' home ranges.

Before game theory became a common approach to animal behavior, Fretwell and Lucas (1970) introduced the idea of an "ideal free distribution" of home ranges, which uses an application of game theory. Fretwell and Lucas hypothesized that if every animal has perfect information about habitat quality on a landscape and is able to move to the area where its fitness is maximized, the "ideal free distribution" results with each individual having equal fitness, though not necessarily equally sized home ranges nor fitness ≥ 1. If animals maintain territories, local densities may be limited because some animals can not settle in their preferred locations, resulting in the "ideal dominance distribution." In this case, fitness should be ≥ 1. Myriad variants of Fretwell's and Lucas's approach have been developed, incorporating animals of different sizes, with different competitive abilities, different abilities to maintain territories, different abilities to use resources, and living on landscapes with different distributions of resources and floaters who do not maintain home ranges (e.g. Sutherland 1996, Milinski and Parker 1991).

Game theory works largely with phenotypes, not genotypes, for which fitness depends on the frequencies of behaviors within a population. If a phenotype is at least in part heritable, behaviors (including flexibility of behaviors) can be selected without having to know the genetic basis for such selection (as Darwin showed). Most behaviors are probably polygenic with the genetics unknown, also making modelling phenotypes easier than modelling genotypes.

Most game theoretic models deal with the behavior of 2 interacting individuals within the context of their population. A common approach is to assume that the individuals' behaviors affect their individual, or sometimes inclusive, fitnesses. If all or almost all members of a population have a particular behavior related to a situation (the "resident behavior" or "resident strategy"), such as defending a territory, game theoretic modellers ask whether a new member of the population (through immigration or birth) that uses a different behavior in that situation can introduce the new behavior to the population such that the new behavior becomes the predominant behavior ("invasive behavior" or "invasive strategy"). Such a behavior is stated to have "invasive fitness." Seldom, however, do game theoretic models deal directly with fitness, but, instead, work with proxies for fitness (McNamara and Leimar 2020, Alif et al. 2022), such as energy or ownership of a resource required for fitness (a reproductive nest cavity, for instance). In many models, competitors seek to gain the value of victory, "V," which is a proxy for fitness; V is the value for the contested resource that the competitors have in their cognitive maps.

Behaviors modelled using game theory can be considered broadly, from fighting over possession of a territory, to choosing mates, and even to "decisions" not made consciously, such as allocating sex to offspring and allocating resources toward growth vs reproduction. "Strategies" are the rules that specify behaviors of model animals and behaviors can be state dependent: the owner of a territory behaves in ways that contrast with the behavior of an individual who does not have a territory. A model animal may seek to maintain a territory in one environment while a model animal in a different environment may have nothing to gain from maintaining a territory (Chapters 7, 8). Thus, a strategy can be a "pure strategy" (always do x) or a mixed strategy (do x in situation A and y in situation B). A strategy that is better than or equal to all other strategies is at a "Nash equilibrium," after John Nash, who developed and proved the concept (Nash 1950). A strategy that is better than all other strategies and that can not be displaced is called an "evolutionarily stable strategy," or an ESS.

Suppose an animal adopts (that is, has evolved genetically or culturally the phenotype that causes that animal to adopt) strategy I in a given situation and meets an opponent who adopts strategy J in that situation. Imagine 2 dispersing beavers of the same sex arriving at a promising, uninhabited pond on the same night. Or imagine 2 male red-winged blackbirds, 1 breeding for the first time who has established a breeding territory and the other a returning adult who owned that very territory the previous year; each believes that he has legitimate ownership of the territory. The payoff (the fitness proxy) to the animal who adopts strategy I and whose opponent adopts J is written as $E_J(I)$.

Alternately, if I is the resident strategy for a stable population (this is the strategy that all or almost all individuals in the population adopt), then the number of offspring that one such individual is expected to contribute in its lifetime to the population at equilibrium can be written $\lambda(I, I) = 1$. $\lambda(I, I)$ must equal 1 because the population is assumed to be at equilibrium. The fitness of an immigrant individual who adopts the invasive strategy J in this population is $\lambda(J, I)$.

For strategy I to be an evolutionarily stable strategy, it must meet 2 criteria:

$$1: \lambda(J, I) \leq \lambda(I, I) \text{ or}$$
$$E_I(J) < E_I(I),$$

which is the condition for a Nash equilibrium; and

$$2: \text{if } \lambda(J, I) = \lambda(I, I) \text{ then}$$
$$\lambda(J, J) < \lambda(I, J), \text{ or}$$
$$\text{if } E_I(J) = E_I(I) \text{ then}$$
$$E_J(J) < E_J(I).$$

Note that $\lambda(J, I)$and $E_I(J)$provide equivalent information except that the former emphasizes fitness while the other emphasizes another, more immediate, payoff.

These criteria state that for I to be an evolutionarily stable strategy: 1) An animal that adopts strategy J when surrounded by animals that have adopted resident strategy I obtains lower or equal payoffs (in fitness or proxies) than do all those animals adopting strategy I who meet others adopting strategy I; and 2) If the animal adopting strategy J gets equal payoffs to those adopting I, then those adopting I do better when surrounded by animals adopting J than do those adopting J surrounded by colleagues also adopting J.

This chapter provides several examples of building game theoretic models to develop hypotheses related to home ranges and territories that can be tested in the field. These can be a springboard for biologists to use game theoretic models to guide their research.

9.2 The Boxer–Dancer Games

The standard game used to introduce game theory has habitually been called the Hawk–Dove game, wherein the Hawk strategy is "always fight" when confronting a conspecific over a resource that can not be shared, while the Dove strategy is "always display." For many reasons, the Hawk and Dove names for strategies are inappropriate. Hawks almost never fight conspecifics, having better ways to resolve conflicts, while doves are not always peaceful. In addition, the names that humans apply to situations affect their unconscious understanding of both the situations and the names, potentially leading non-biologists to think that hawks might always fight and that doves never have aggressive conflicts.

I prefer the name Boxer–Dancer game. Boxers do only fight when confronted with an opponent in the ring. And dancers do not fight, not even when enacting a fight: Think the Montagues and Capulets in Act I, Scene I of Prokofiev's Romeo and Juliet. In addition, dancers can cooperate and, thereby, share the payoff: Think Fred Astaire and Ginger Rogers.

The payoff matrix for the Boxer–Dancer game is in Table 9.1, where the entries are the payoffs to the player adopting the strategies in the left-hand column for meeting an opponent from columns 2 or 3.

Here, V is the value of victory or, more accurately, the value of the resource that is the source of the conflict. C_W is the cost of being wounded (or otherwise beaten) when a Boxer meets a Boxer. We assume that when 2 Boxers meet, each has a 50:50

Table 9.1 Payoff matrix for the Boxer–Dancer game. Boxers always fight in a conflict over a contested resource and Dancers always display. V is the value of the resource, C_W is the cost of being wounded, and C_D is the cost of dancing (displaying). For 2 Boxers meeting, each is assumed to win half the time. For 2 Dancers meeting, each is assumed to win half the time or they share V each time. The Boxer strategy is an evolutionarily stable strategy.

	Boxer	Dancer
Boxer	$1/2\,(V - C_W)$	V
Dancer	0	$1/2V - C_D$

chance of winning, each gets on average ½V, and if being beaten, each pays, on average, −½C_W. When a Boxer meets a Dancer, the Boxer always gets the resource. When a Dancer meets a Boxer, the Dancer always loses. When 2 Dancers meet, the value of the resource is discounted by the cost of dancing (displaying), C_D, which is, logically, much smaller than C_W. Each Dancer may get the resource half the time or, alternately, they share; either way, each gets ½V, though each pays the cost of Dancing.

If $V > C_W$, being a Boxer is a pure strategy (a strategy without alternatives) and being a Boxer is an evolutionarily stable strategy, because

$$E_B(D) < E_B(B);$$
$$0 < 1/2\,(V - C_W).$$

If, however $C_W > V$, then no pure strategy is an evolutionarily stable strategy. A population of Dancers can be invaded by a Boxer, who does well as long as the majority of his opponents are Dancers. But, when Boxers become the majority of the population, Dancers who meet Dancers have the highest fitness. Given that being wounded can hinder an animal from obtaining resources for a long period after a confrontation $C_W > V$ is not unrealistic.

Now, we must all admit that the Boxer–Dancer game is hard to align with most realistic behaviors of animals, such as contesting over territory ownership. Suppose, though, that Boxers can become disabled for significant periods of time, that disability occurs with probability q, that disabled Boxers try to Dance, while Dancers have learned to Box (at least a little) with probability p ($p + q = 1$), then one can solve for probabilities p and q that lead to an evolutionarily stable strategy for these mixed strategies. This situation is not far-fetched, actually. Animals do get injured. A stoat wounded by attacking a rabbit must avoid confrontations with other stoats and may not be able to catch more rabbits until its wounds have healed (Allen 1938). Bighorn sheep occasionally break horns when fighting, immediately affecting dominance status and fitness.

A variation on the Boxer–Dancer game is the Boxer–Dancer–Retaliator game, in which Retaliators always dance first but will box if attacked. This also resembles reality. Few, if any, animals fight immediately when meeting a competitor, even in the presence of a contested resource. Animals want not to be wounded and, therefore, they test each other, often with some complex displays. If an animal is attacked, he will defend himself, at least a bit, even if he runs ultimately. The classic table of payoffs for the Boxer–Dancer–Retaliator game (Table 9.2 A) is complicated by the payoffs to Dancers and Retaliators being the same when they meet. The consequence is that Dancers can invade a population of Retaliators because Retaliators receive the same payoffs as Dancers. But, because Dancers and Retaliators do not have the same payoffs when confronting Boxers, Boxers can invade a population that has a mix of Dancers and Retaliators. One can add a small, realistic change to the payoff matrix

Table 9.2 A. The payoff matrix for the Boxer–Dancer–Retaliator game. Boxers and Dancers are as in Table 9.1. Retaliators always start by displaying but fight if attacked. V, C_W, and C_D are as in Table 9.1. This game has no evolutionarily stable strategy.

	Boxer	Dancer	Retaliator
Boxer	$1/2\,(V - C_W)$	V	$1/2\,(V - C_W)$
Dancer	0	$1/2V - C_D$	$1/2V - C_D$
Retaliator	$1/2\,(V - C_W)$	$1/2V - C_D$	$1/2V - C_D$

Table 9.2 B. As in Table 9.2 A except that "A" represents a minor advantage for Retaliators over Dancers. Retaliator is an evolutionarily stable strategy.

Boxer	Dancer	Retaliator
$1/2\,(V - C_W)$	V	$1/2\,(V - C_W)$
0	$1/2V - C_D$	$1/2V - C_D - A$
$1/2\,(V - C_W)$	$1/2V - C_D + A$	$1/2V - C_D$

by assuming that Retaliators, given that they are willing to fight if attacked, have a slight advantage over Dancers because they are willing to shove a little bit when dancing (Table 9.2 B; "A" in the payoff matrix represents the small advantage of Retaliators). This small change makes Retaliating an evolutionarily stable strategy.

The games introduced so far have been symmetrical in that all individuals have the same options available to them and receive the same payoffs for choosing each option. Most confrontations of animals over resources have some asymmetries. One contestant may be significantly larger, the contestants may be of different sexes with different responsibilities for offspring, one contestant may own a territory while the other not, or the contestants may seek different rewards. The latter was the case in my game with the short-tailed weasel.

9.3 The Short-tailed Weasel Game

Table 9.3 shows the payoff matrix for my face-off with Mr Short-tailed weasel in the log lodge. This game is not symmetrical in that the payoffs to Mr Short-tail and to me were not the same and neither of us could adopt any of the other's strategies.

If I dropped the piece of venison when Mr Short-tail punched my fingers with his muzzle, Mr Short-tail gained the value V of the food and I gained some pleasure P for having interacted with him and for having given him some food. If I did not drop the venison and Mr Short-tail stayed and ate, he

Table 9.3 Payoff matrix for my interaction with Mr Short-tailed weasel. Mr Short-tail's behavior options are given in the far left column and my options across the top. Mr Short-tail's payoffs are given before the "/" and my payoffs after the "/." "Punch" is an evolutionarily stable strategy for Mr Short-tail and "Hold" is an evolutionarily stable strategy for me.

	Roger Drop	Roger Hold
Punch	V/P	$V - D_1/P^*$
Retreat		$0/P$
Bite	$V - D_2/-B$	

gained V but also paid a cost, D_1, for exposing himself to potential danger, while I gained P^*, a greater pleasure because I got to watch Mr Short-tail eat the venison. If I held but Mr Short-tail retreated without eating, he gained nothing while I did gain the pleasure of interacting with him. If Mr Short-tail had bitten me, I would have dropped the venison. Then Mr Short-tail would have obtained the venison but would have been subject to a danger, D_2, from being frightened by my jumping when bitten. The payoff to me was highest for Mr Short-tail's "Punch" option and my "Hold." Had Mr Short-tail bitten me to get me to drop the venison, my option of "Hold" was not possible. Without repeats of the game, Mr Short-tail's option "Punch" yielded his highest payoff if I "Dropped" but my highest payoff came if I "Held." Either way, Mr Short-tail's best option was always "Punch" as long as $D_1 < V$. Given that Mr Short-tail was familiar with the log building and, therefore, with living around people, and given that he had interacted with me with only friendly responses, he would have known that D_1 was small.

The game, however, did allow repeats if Mr Short-tail did not "Bite" and if I did not "Hold." Thus, with repeats, my choosing "Drop" yields a better total payoff for both Mr Short-tail and for me. The maximum payoff comes, however, on the repeat that satiates Mr Short-tail. If I "Hold" on that repeat, my total payoff is maximized but Mr Short-tail's is not. My total will be $P^* - P$ more and Mr Short-tail's $-D_1$ less than the maximum. Again, no matter what I choose, Mr Short-tail's best option was still "Punch." I could not, however, have been able to identify the satiating repeat without Mr Short-tail giving me some (unintended) signal. Not giving such a signal was his best strategy.

Now, let us explore an asymmetrical game with a territory owner and a potential territory thief.

9.4 Territory Defense and Maintenance

9.4.1 Territory Owners and Thieves

Assume that this game has our old options of Boxing and Dancing morphing into an Owner, who can fight or display when her territory is invaded by a conspecific, and a Thief, who wants to steal the territory and who can also fight or display (Table 9.4). This situation differs from Boxer–Dancer games in that an entire territory is the source of conflict, whereas individual resource patches could be the source of conflict for Boxer–Dancer games (Hinsch et al. 2013). Boxing (fighting) from the start is appropriate in this situation because the Thief is immediately a threat to the Owner. In addition, the value of the territory is greater for the Owner ($V_O > V_T$) because the owner's cognitive map includes information on where resources are and what their values and availabilities are. In addition, the cost of being wounded is smaller for the Owner because she knows how to take advantage of her territory to avoid being wounded in a conflict and, if wounded, she knows how to find easy food while she heals (Crofoot et al. 2008). Finally, being without a territory may have led the Thief to be in poorer physical condition than the Owner.

As long as $V_O > V_T$, Owner–Boxer is an evolutionarily stable strategy. If, however, the value of the territory is actually greater for the Thief, then Thief–Boxer strategy becomes an evolutionarily stable strategy if the cost of being wounded is not too great.

A Retaliator can be added to this game with the same results as for the Boxer–Dancer–Retaliator game (Table 9.2 B). In this new game, the Retaliator can represent a Neighbor who is known to the Owner and who has an understanding with the Owner that neither is a threat to the other under every-day conditions. If a Neighbor trespasses, the Owner will remind the Neighbor but restrain aggression because the Neighbor is not a Thief who would steal her territory. Innumerable examples exist of territory Owners recognizing the advertisements of their Neighbors. Neighbors' identities and behaviors are included in an Owner's cognitive map. For both, Owner–Neighbor is an evolutionarily stable strategy.

Schelling (1960) noted that territorial neighbors face a "coordination problem." Each territory Owner might like to have a somewhat larger territory but his Neighbor is unwilling to yield such desired land. In the end, and regardless of where the boundary actually lies, the outcome may be worse (outright fighting and the potential for being wounded) if no agreement can be made or if reaching an agreement requires a costly process. Such

Table 9.4 An Owner owns a territory and a Thief wishes to steal, or otherwise obtain, the territory. Under each Thief column, the payoff to the Owner is on the left and the payoff to the Thief is on the right. Variables are as in the Boxer–Dancer game except that subscripts "O" and "T" refer to Owner and Thief. Owner and Thief payoffs are proportional to the relative values of V_O and V_T. Owner–Boxer is an evolutionarily stable strategy if $V_O > V_T$ but Thief–Boxer is an evolutionarily stable strategy if $V_T > V_O$.

	Thief–Boxer		Thief–Dancer	
Owner–Boxer	$\frac{V_O(V_O-C_{OW})}{(V_O+V_T)}$	$\frac{V_T(V_T-C_{TW})}{(V_O+V_T)}$	V_O	0
Owner–Dancer	0	V_T	$\frac{V_O(V_O-C_{OD})}{(V_O+V_T)}$	$\frac{V_T(V_T-C_{TD})}{(V_O+V_T)}$

aspirations to expand a territory are not universal, as shown by examples in Chapters 4–8. Nevertheless, Owners and Neighbors usually adopt the solution that is mutually obvious: leave the boundary where it is (Schelling 1960).

Some game theoretic approaches to wolf territoriality follow. Whether these examples apply to other animals that maintain territories, especially other cooperative breeders such as beavers, woodland voles, and red-cockaded woodpeckers, needs to be tested.

9.4.2 Wolf Packs and Neighbors

Wolves live in packs that are extended family groups, usually 2 parents and their offspring from the previous 2–3 years (Mech and Boitani 2003; Chapter 2). They maintain territories, advertising ownership using, at least, urine marks (which are also visual marks when snow is on the ground), defecation, foot scrapes on the ground (which are both visual and scent marks), and howls (Peters and Mech 1975, Harrington and Asa 2003). Territories contain adequate food and other resources that packs need, all of which are part of the wolves' cognitive maps. Wolf behavior conforms basically to the Owner–Neighbor strategy. When a wolf pack travels near a territory boundary, pack members scent mark. When a pack finds a scent mark from an adjacent pack, frequency of scent marking increases during subsequent travel, causing pack boundaries to have a higher density of scent marks than found elsewhere in either territory. A pack leader sometimes trespasses a short distance into the neighboring pack's territory to scent mark, perhaps trying to obtain a somewhat larger territory. A leader of the neighboring pack reciprocates. Thus, territory boundaries are nudged back and forth but generally remain where they are. The boundaries and past scent marking are also incorporated into the wolves' cognitive maps, affecting foraging moves.

Now consider a wolf pack (or a cooperative group of another species) that maintains a territory in an area that experiences a drastic decrease in the prey population such that the pack can no longer support itself with the prey in its territory. The value of its territory is now some portion 'b' ($0 \leq b < 1$) of its previous value, V. The pack can continue to maintain its present territory, leading to mortality or dispersal of pack members until the pack size drops to a size that the territory can support. Or, it can abandon the Owner–Neighbor strategy and take an Invader strategy by invading a neighboring pack's territory and fighting the resident wolves, trying to add part or all of that neighboring territory to its own, inadequate territory and maintaining pack size without mortality from starvation (if it wins). Although rare, such invasions do occur among wolf packs (Mech 1977a, Peterson 1977, Meier et al. 1995, Cassidy et al. 2015, Vucetich 2021), and also among coyote packs (Gese 2001), packs of free-ranging dogs (Boitani et al. 1995), capuchin monkey groups (Crofoot et al. 2008, Crofoot and Gilby 2012), white-browed sparrow-weaver groups (York et al. 2019), and groups of many other species.

In this potential situation, 2 packs interacting under the rules of the Owner–Neighbor agreement retain the full value of their territories, which is bV for the potentially invading pack (Table 9.5). When a pack invades the territory of a pack that uses the Owner–Neighbor strategy, assume that each pack wins half the time. Consequently, the payoff for the Invading pack if it wins is

$$bV + \frac{1}{2}(1 - b)V - C_W :$$

bV because the Invaders retain their depleted territory whether they win extra land from the invaded pack or do not;

Table 9.5 Payoffs shown for the strategies in the left-hand column for a potential Invader pack whose territory has decreased significantly in quality to a level of bV, where V is the value that its territory has had in the past and $0 < b < 1$ is the proportion of the old level and is not enough to support the pack. If the potential Invader pack maintains using its Owner–Neighbor strategy, it does not have adequate food. Invading a neighboring pack's territory that has value V is an evolutionarily stable strategy if b and C_W are both small enough. If the Invader pack learns information "a" about the neighboring territory and pack that helps it to win in a contest, the higher the value of "a," the larger the range of values of variables that lead to Invading being an evolutionarily stable strategy.

	Owner–Neighbor
Owner–Neighbor	bV
Invader	$bV + 1/2(1 - b)V - C_W$
Augmented Invader	$bV + 1/2(1 + a)(1 - b)V - C_W$

½(1 − *b*) *V* is one half times the value of the extra space they need;

and C_W is the cost of being wounded, which the Invading pack pays whether it wins or loses.

To learn what information needs to be in the Invaders' cognitive maps to calculate whether the Invasion strategy (*I*) is an evolutionarily stable strategy as opposed to the Invaders maintaining the Owner–Neighbor strategy (O), we must learn whether $E_0(I) > E_0(O)$:

$$bV + \frac{1}{2}(1 - b)V - C_W > bV, \qquad \text{Eq 9.1}$$

which can be rewritten as

$$V(1 - b) > 2C_w, \qquad \text{Eq 9.2}$$

which means that the smaller the cost of having pack members wounded, Cw, relative to the value of a good territory, *V*, and the more depauperate the Invaders' territory (the smaller '*b*' is), the more likely that the Invader strategy is an evolutionarily stable strategy and wolf packs with depauperate territories should try to expand their territories to obtain more resources (Table 9.5, Figure 9.1A). In this scenario, Cw is likely to be large, seldom making Invasion an evolutionarily stable strategy.

Most often, animals know their territories far better than do neighbors. Consequently, Owners are more likely to win in a confrontation because they know how to take advantage of the local environment (e.g. Crofoot et al. 2008). If this is the case regarding a wolf invasion, Owner packs will win more often than 50% of the time and the "2" in equation 9.2 will be a larger number, making the Invader strategy less likely to be an evolutionarily stable strategy. Thus, *b* must be very small for the Invader strategy to be worth while, especially given that Cw is likely large.

Now, consider the possibility that a couple members of a potential Invader pack trespass into the territory of a neighboring Owner pack and learn important new information about that Owner pack, such as pack size, pack age distribution, and health of pack members, and learn a bit about the lay of the land. From this information, the potential Invader pack can judge better its chances of winning a fight with the Owner pack. We can modify our assumption that each pack wins half the time by inserting a variable "*a*" ($0 \leq a \leq 1$) that increases the probability

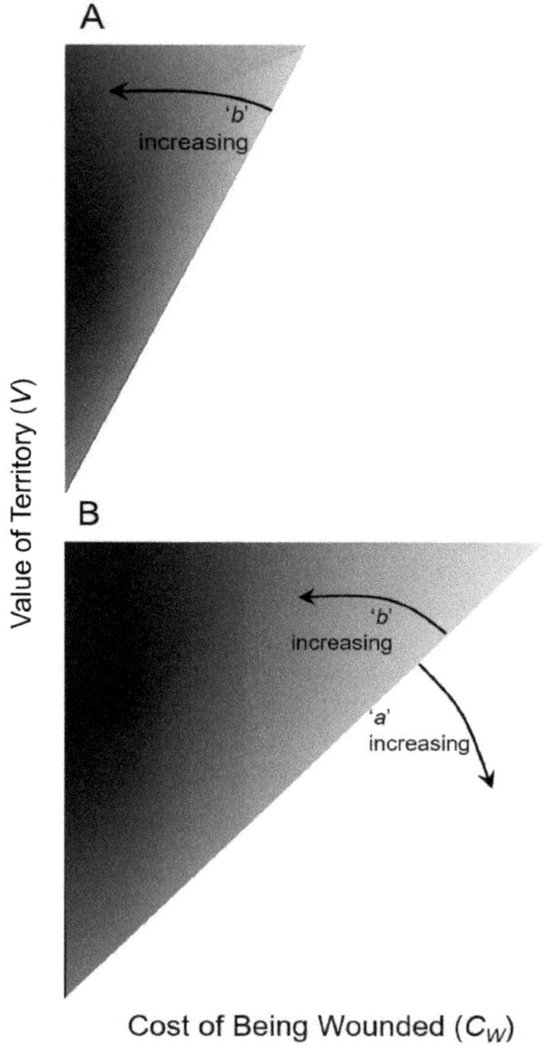

Figure 9.1 A. The Invader strategy is an evolutionary stable strategy in the grey area. As "b" increases, the range of values for key variables that lead to the Invader strategy being an evolutionary stable strategy become more constrained. B. Same as 9.1A except that the Invader pack has learned an amount "*a*" about the pack whose territory it will invade, such as pack size and pack health, that will help it win a contest with the invaded pack. This information increases the range of values for key variables that lead to the Invader strategy being an evolutionary stable strategy. Dark shading shows decreasing chances of the Invader strategy being an evolutionary stable strategy.

that the Invader pack will win. Now, equation 9.1 becomes

$$bV + \frac{1}{2}(1 + a)(1 - b)V - C_W > bV \qquad \text{Eq 9.3}$$

which can be rewritten as

$$V(1 - b) > 2C_W/(1 + a).$$

Under these conditions, the more that the potential Invader pack can learn suggesting that it can win, the more likely that the Invader strategy will be an evolutionarily stable strategy (Figure 9.1B). Nonetheless, unless a is very large, Invasion will still seldom be an evolutionarily stable strategy.

Finally, consider a truly dire situation for the Invader pack. Pack members judge that their territory is so poor that many of them will die if they do not obtain more resources and judge that if many pack members do starve or disperse then the pack will likely disintegrate. Under these conditions, members of the potential Invader pack may choose to invade the Owner's territory without consideration of wounding: they might as well chance dying trying to make conditions better as opposed to dying later from starvation. If potential wounding does not affect behavior of invaders, then $C_W = 0$ and the Owner–Neighbor strategy can never be an evolutionarily stable strategy for the Invaders.

Information included in all these simple models (the value of resources in a territory, the costs of wounding, locations of known neighbors) is available to animals through their cognitive maps including their knowledge and physiological responses to their nutritional and physical status. The models lead to the following hypotheses regarding neighbors invading territories. 1) The lower the value of a potential invader's territory, the more likely an invasion. 2) The more information known about neighboring packs by a potentially Invasive pack and the more that information implies weakness of the Owner pack, then the more likely an invasion. 3) A potential invader in dire straits is most likely to invade. These are not novel insights but they show how building simple models can lead to hypotheses that can be tested in the field, perhaps not with wolves but very possibly with North American red squirrels or Hawai'ian honeycreepers (Chapter 7), whose ecologies can be manipulated. For individuals having territories, the strategy of dispersing to establish a new territory should be added to the payoff matrix. Hypothesis 3 above is consistent with behavioral data on other animals in dire straits that choose to gamble on finding an unlikely jackpot rather than choose to stick with a guaranteed food source that is inadequate (Real 1981a, b, Harder and Real 1987, Dukas and Real 1993).

9.5 Home Range Sharing

In Chapter 7 the question of partial territoriality arose. The situation was of a territory owner whose supply of the limiting resource has grown such that the owner could shrink her territory, except that the only resource patches that could be abandoned are so large that to abandon one would leave the territory without enough resources. Whether the territory owner continues to maintain a patch in her territory that is not completely needed depends on how neighbors react, which is a situation for game theory.

Suppose 3 strategies exist for the territory owner regarding patches only partially needed: Defend, Share, Abandon. Assume that the territory owner and her neighbors use the Owner–Neighbor strategy. And assume that the total value of the patch in question is V but that no animal in this game needs more than ½V of the resources in the patch (Table 9.6). An Owner who Defends always prevents a Neighbor from using the patch in question, no matter what the Neighbor's strategy. A Defender will maintain ownership of a complete patch that is not completely needed and will acknowledge the Owner–Neighbor agreement, even though she will

Table 9.6 A territory has increased in value such that the owner no longer needs all resource patches in the territory. The only patches that it can eliminate from its territory, however, have values so large that eliminating one decreases the value of the entire territory below the level to meet requirements. The territory owner can continue to Defend all patches, Share one patch with neighbors, or Abandon one patch and suffer not meeting its requirements. Neighbors can adopt Invade, Share, and Ignore strategies. Payoffs shown are for the owner's strategies in the left column.

	Invade	Share	Ignore
Defend	$1/2V - C_W$	$1/2V - C_W$	$1/2V - C_A$
Share	$1/2V$	$1/2V$	$1/2V$
Abandon	0	0	0

be defending unneeded resources in the resource patch and even though her payoff is only ½V. Given that the Owner–Neighbor agreement leads "defense" to be displays, vocalizations, and maybe a chase, the cost of "defending" is small yet could be more than the gain from amount of resource needed. A Sharer shares the patch that is not completely needed; she gets to use the ½V she needs no matter what strategy a neighbor uses. An Abandoner abandons use of the patch that is not completely needed, thus losing the ½V that she needs.

The opponent in this game also has 3 strategies: Invade, Share, and Ignore. An Invader invades the patch that the Owner does not entirely need. Such a Neighbor detects the decreased use of the patch by the Owner and responds by trying to acquire the patch for her own territory, but fails half the time if the Owner is an Owner–Defender. A Sharer shares the patch with the Owner. An Ignorer ignores the patch to avoid any confrontation with the Owner.

Sharing is an evolutionarily stable strategy in this scenario (Table 9.6). Abandoners can not invade a community of Sharers nor can Invaders. A Sharer can, however, invade both a community of Defenders, because $E_D(I) < E_D(S)$, and a community of Abandoners, because $E_A(A) < E_A(S)$. The payoff matrix shows that sharing partially needed patches makes sense. Frédérique Dubois and Luc-Alain Giraldeau (2007) modelled more complex sharing options.

Eastern chipmunks maintain territories around their home range centers, where they have food caches, but tolerate sharing of peripheries (Yerger 1953, Broadbooks 1970). Stevens and Stephens (2002) used a game theory approach for modelling sharing of individual food items, which could be applied to sharing individual resource patches. Their analyses started with the premise, however, that the owner of the resource should want the entire resource and not be willing to share.

9.6 War of Attrition Game

Decades ago, Smith (1978, 1981; Chapter 7) showed that North American red squirrels maintain territories that conform to the basic Owner–Neighbor strategy. Not too long after that, Ydenberg and colleagues (Giraldeau and Ydenberg 1987, Ydenberg et al. 1988, Price et al. 1990) tested in the field hypotheses derived from the War of Attrition game (Maynard Smith 1974, Parker and Rubenstein 1981) to gain a more nuanced understanding of territoriality of red squirrels.

The basic premise of the War of Attrition is that 2 contestants face off over a resource that has value V and the one who obtains the resource is the contestant who is willing to out-wait the other. This game is much like the Boxer–Dancer game except that all contestant must be Dancers. Being willing to wait longer might be replaced by being willing to exert extra effort to perform a really convincing dance (display) or otherwise to expend the most energy. In the classic War of Attrition game, each contestant has a time that he is willing to wait but these times are unknown to opponents. So, if contestant A is willing to wait until he accrues the waiting cost of m_A and contestant B is willing to accrue m_B, the winner is the contestant for whom m_X is larger. Assume that $m_A > m_B$, then the payoffs are

$$\text{Payoff for A: } V - m_B$$

$$\text{Payoff for B: } 0 - m_B$$

Note that, although contestant A was willing to accrue a cost of m_A, he actually pays the cost m_B because that was the limit for contestant B, who lost. Choosing and always using the same m_X is not an evolutionarily stable strategy because an intruder who chooses $m_C > m_X$ will win, and such intruders continue to win ad infinitum. One evolutionarily stable strategy is for each contestant to choose m_X from a random distribution, leaving both contestants winning half the time (Maynard Smith 1974). Alternately, contestants could base their decisions about giving up on the value of the resource to each, V_A and V_B, and on the rate at which costs grow, K_A and K_B, or how easily escalation is reached, also called resource holding potential. The contestant with the smaller V_x/K_x should retreat, which is a more realistic evolutionarily stable strategy (Parker and Rubenstein 1981). Given that territories have higher values to owners than to neighbors or strangers who might compete for territories, the Owner–Neighbor strategy is an evolutionarily stable strategy. The possibility of making a mistake when assessing an opponent's V_x/K_x ratio complicates conflicts, causing likely losers sometimes to escalate a conflict, but does not change the evolutionarily stable strategy status of either strategy (Parker and Rubenstein 1981). Ydenberg et al. (1988) showed that familiarity with neighbors minimizes mistakes (for diverse birds, mammals, amphibians, fishes, and insects) and stabilizes the Owner–Neighbor strategy. Owners and Neighbors do not need to escalate interactions.

Working with territorial, arboreal, Neotropical ants, Adams (1990) noted that territorial disputes were usually resolved using the Owner–Neighbor strategy but, when colonies met in neutral areas with new resources, the colony that could raise the largest, most aggressive team of workers the fastest (that is, K is low because raising a big team is easy) claimed the new resource. Adams could manipulate outcomes, however, by manipulating colony size, which directly affected how fast a colony could raise a large force (manipulating K), allowing him to test the predictions of the War of Attrition game.

North American red squirrels maintain individual territories using, at least, calls and chases (Kemp and Keith 1970, Smith 1978, 1981, Boutin and Schweiger 1988, Martinig et al. 2022, McAdam et al. 2022). Using playback calls for red squirrels, Price et al. (1990) rejected the statistical null hypotheses related to 4 biological hypotheses generated by the War of Attrition and related to how V varied with individual squirrels of different classes: 1) Squirrels respond more energetically to the calls of strangers than to the calls of neighbors (V higher for residents than for strangers and, therefore, low cost for residents to have high intensity of their responses); 2) Breeding females respond more energetically to the calls of strangers than do non-breeding females or males (V highest for breeding females); 3) Females breeding early respond more energetically to calls of strangers than do females breeding late (V higher for an early breeder because she will retain her territory while a late breeder will likely bequeath her territory to a daughter); and 4) All classes of squirrels respond similarly to the calls of neighbors (relative V is known for all neighboring squirrels and all residents have adopted the Owner–Neighbor strategy). Given the extensive research that had been done on red squirrels prior to the research of Price et al., they should have used the Hypotheses 1–4 as their biological null hypotheses because squirrels' responses were already well known and those were the hypotheses that they ended up accepting. They might not have been able to reject all biological alternate hypotheses for the biological null hypotheses, stimulating further research. Developing good alternative hypotheses would have required imaginative and creative thinking.

Most modellers, from Maynard Smith (1974) on, note that, although V and K are likely to be correlated, they do not have to be. One can imagine that, while a territory Owner should be in good physical condition, a potential territory Thief might be in poor condition and, therefore, value the territory even more than does the Owner, $V_O < V_T$. A Thief in poor condition, though, would be unable to escalate if need be, $K_O < K_T$.

Enquist and Leimar (1983, 1987) and Englund and Olsson (1990) modelled situations where contestants assess their opponents' willingness to fight as an interaction advances, thereby gaining information about their own versus their opponents' relative abilities (relative *K*s). Englund and Olsson's models predicted that an animal for whom *V* is higher should be more likely to win and that incorporating an opponent's willingness to fight into decisions is an evolutionarily stable strategy. Enquist and Leimar, in contrast, showed that, though netspinning caddis flies did assess and incorporate relative body sizes and resource ownership into their conflicts, they did not assess and incorporate resource value.

Thus, field research has evaluated 3 important variables considered in the Asymmetric War of Attrition game: *V*, *K*, and making mistakes in evaluating an opponent's self-evaluation of V/K. Field research has also presented evidence that this game can provide insight into territorial behavior. Animals' cognitive maps provide the most critical information that territory Owners require to be able to establish *V* and to influence where to set *K*.

Mesterton-Gibbons and Sherratt (2016) modelled the situation of each of two animals believing that he owns a given territory and found that the Asymmetric War of Attrition appeared to match the conditions of the situation, basing their model on damselflies contesting territory ownership (Waage 1998). The contestant who was willing to fight longer gained the territory.

Walton and Nolan (1986), however, documented different results when they used the Asymmetric War of Attrition to derive hypotheses regarding the behavior of male prairie warblers establishing breeding territories after spring migration. Birds were categorized using 2 criteria. First, a bird that set up a territory became a "Resident," whereas a bird that challenged a Resident for his territory was a "Non-resident." Second, a bird who returned to the territory that he had used in years past was an "Owner" whereas one who established a territory new for him was a "Pretender." Residents and Non-residents knew their own status and the status of opponents within this criterion. Owners knew their status and that of Pretenders but Pretenders did not know whether an opponent was an Owner or another Pretender. The consequence was that imperfect information about Owner–Pretender status often generated conflicts. Owner–Residents always won quickly but Owner–Non-residents had some protracted contests, especially when the Pretender–Resident had already attracted a mate and started nesting. V_{owner} would have been large at the end of the previous reproductive season a year earlier and larger than $V_{Pretend}$ in the present year. Owners, however, will have lost information about the territory in the intervening seasons (Chapter 5). Thus, who had the higher *V*, Owners or Pretenders, could favor either contestant. Likewise, the values of K_{owner} than $K_{Pretend}$ could favor either contestant or neither. Thus, the prairie warblers often escalated to extended conflicts. This is a case where the information in 2 animals' cognitive maps conflicted.

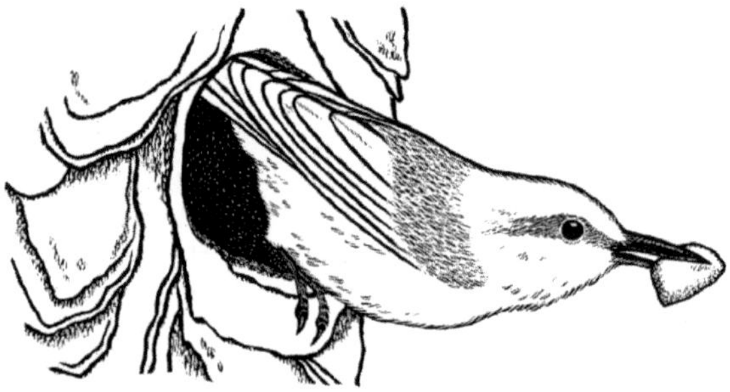

9.7 Territory Boundaries and Landmarks

9.7.1 Boundaries

Maynard Smith (1982) presented a 1-dimensional model, called the Bargaining Game, for establishing boundaries between territories. In the model, each of 2 territories has a core area that extends a distance *d* from the territory center and that supports the owner but is not rich enough to allow reproduction. Figure 9.2 A shows 2 territories with cores; each core has a representative chipmunk's burrow. If the owner were to expand her territory to a distance l_1 she can reproduce. Distance l_1 extends toward the other owner's core. Adams (2001) reviewed territory sizes and shapes and documented that territory

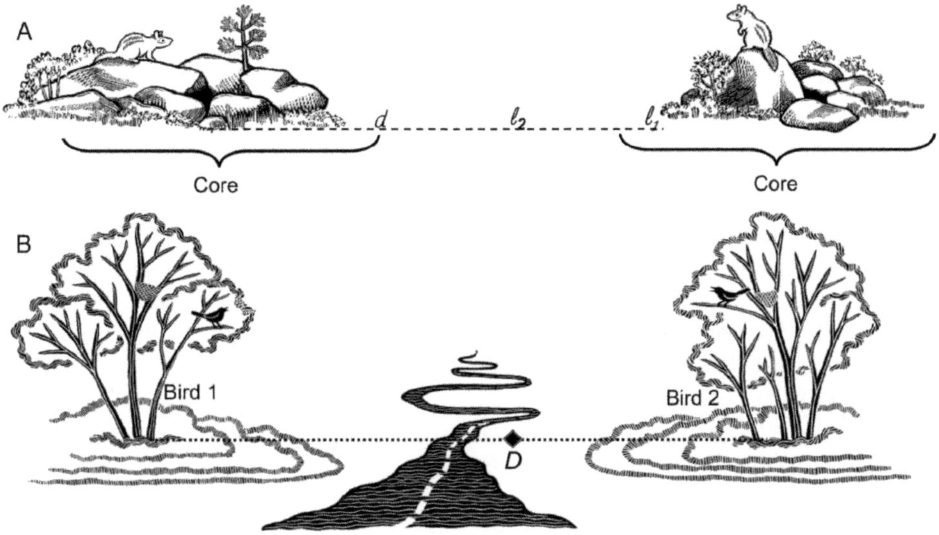

Figure 9.2 A. Two territory owners confront over desires to expand territories, shown as chipmunks' territories. The core for each supports survival but not reproduction. If one owner attempts to expand her territory so that she can reproduce, she can adopt Boxer, Honest, or Bluffer strategies, which lead to different outcomes for l_1 and l_2. l_1 extends somewhat into the other chipmunk's core; l_2 does not. If expansion to l_1 is required for reproduction, then one or the other chipmunk must abandon her territory. If expansion only to l_2 is required for reproduction, then both chipmunks can retain their territories. The Honest strategy is an evolutionary stable strategy for l_1, while Honest is excluded for l_2 and Boxer and Bluffer remain and no evolutionary stable strategy exists. B Two territory owners, shown as generic birds, must establish a boundary between their territories. Territory holding potential decreases at increasing distance from a bird's nest. Bird 1 is larger than Bird 2 and has larger holding potential, leading to *D* being the point where the 2 birds have equal holding potential. *D* falls on Bird 2's side of the stream. Alternately, the 2 birds could adopt the stream as the boundary.

defense decreases with distance from a nest site or other type of core. Maynard Smith modelled 3 strategies: Boxer (which he called Hawk), Bluffer, and Honest. A Boxer displays vigorously all the way to distance l_1 and escalates to a fight if challenged. Each owner pays the cost of fighting, *C*. An Honest owner decreases her display intensity as distance from her core increases from *d* to l_1, consistent with Adams's review; if the other territory owner escalates to a fight, the Honest owner retreats to distance *d*, which may or not be consistent with real animals. A Bluffer displays vigorously all the way to distance l_1 but retreats if challenged.

Honest is an evolutionarily stable strategy if l_1 is inside the neighbor's core (Figure 9.2 A). In this situation, only 1 owner can retain her territory and the other must leave; assume that each owner wins half the time. No escalation occurs.

If, however, the limit needed for reproduction does not extend into the other owner's core but is at l_2 in Figure 9.2 A, then no evolutionarily stable strategy exists but Honest owners are excluded. Maynard Smith noted that, if the 2 owners have incomplete information, confrontations could escalate sometimes when, with complete information, they would not.

Pereira et al. (2003) provided another approach to the Bargaining Game that allows the two territory owners to have territories that have different distributions of habitat and, therefore, different values.

Using extensive data on counter-singing by territorial banded wrens, Vehrencamp et al. (2014) tested Maynard Smith's model against a more general, mutual assessment model of conflict over a resource (mutual assessment occurs when each contestant can assess the opponent's abilities). The data supported the wrens using the Bargaining Game.

9.7.2 Landmarks

The topography and physical structures of a landscape affect the animals that live on it. Topography and vegetation affect territory boundaries for generations of birds (Reid and Weatherhead 1988); reef structure influences territories of resident fish (Ehrlich 1975); topography affects costs of locomotion for terrestrial animals, such as black bears using high ridges as home range boundaries (Powell and Mitchell 1998); and topography can limit or enhance access to resources. Consequently, that animals sometimes establish territory boundaries where they find conspicuous landmarks should not be surprising (e.g. Kodric-Brown and Brown 1978, Kruuk 1972b, Macdonald 1979, Reid and Weatherhead 1988; even humans: Riris et al. 2024). Eason et al. (1999) demonstrated that territorial wasps could be enticed to establish boundaries along horizontal wooden dowels and that defending such landmark boundaries was less costly than defending boundaries without a landmark. When convict cichlids used landmarks for territory boundaries, territories were more elongate than territories without landmarks. Although the cost of maintaining territories with known neighbors was unaffected, territories with exaggerated perimeters led to extra maintenance costs from territory intrusions by strangers (Suriyampola and Eason 2015).

Possible hypotheses for using landmarks are: 1) that landmarks can make detecting intruders easier (lowering defense costs); 2) that topographic features can block vision so that neighbors are unlikely to cross a boundary; 3) that the costs of defense rise abruptly when passing the topographic features; and 4) that landmarks make identification of boundaries easy, thereby minimizing maintenance costs (Craig 1979, Reid and Weatherhead 1988, Eason et al. 1999). Truth for any of these hypotheses would lead to animals incorporating the landmarks as important contents in their cognitive maps.

Mesterton-Gibbons and Adams (2003) explored why animals might use landmarks as territory boundaries by building a simple game theoretic model for how the value of a resource and resource holding potential might change with distance from a territory boundary. Assume that an animal's territorial holding potential, p, increases at rate r_S with the animal's strength or size, S, which represents K in the War of Attrition, and decreases at rate r_D with distance, D, from the territory center, which represents the value, V, of the territory. Assume also that each animal might have a physical structure worth defending, such as a bird's nest, a chipmunk's burrow, a red squirrel's seed cache, or a wolf pack's rendezvous site. A linear equation for p is easy and simple, though any equation that decreases with distance will provide similar results. A linear equation may be appropriate for the many birds that have a center-edge effect (Giraldeau and Ydenberg 1987), while an equation that loses little value as distance from territory center increases until falling at the territory edge would be more appropriate for mammals.

So,

$$p = p_0 + r_S S - r_D D,$$

where p_0 equals an individual's basic holding potential because it owns a territory. Mesterton-Gibbons and Adams (2003) assumed that r_S and r_D have values such that even the weakest territory owner can defend his nest and such that the boundary between 2 adjacent territories falls between the 2 nests.

Letting the sizes of animals 1 and 2 be S_1 and S_2 (let S_1 be the larger) and the distance between the 2 nests be L, then solving for the distance corresponding to equal p for both animals yields

$$D = \frac{1}{2}\left(r_S(S_1 - S_2)/r_D + L\right).$$

D is the single distance that equalizes the territory-holding power of the 2 animals and represents the best boundary (Figure 9.2 B).

If the 2 animals are not able to evaluate each other's strength accurately, a mistake by either or both animals regarding the other's strength could lead to a disagreement regarding the best position of the boundary. If that disagreement escalates, either or both animals could be injured, leading to a greater cost than the cost of a suboptimal boundary (Mesterton-Gibbons and Adams 2003).

Now, suppose that between the 2 nests an obvious landmark exists, such as a waterway, a steep mountain ridge, a gravel road that does not hinder travel, or a narrow passage between 2 lakes. Agreeing to use the landmark as the territory boundary avoids the potential for injury, as long as each animal accepts the potential loss of territory area that could come with accepting the landmark as the boundary. So, accepting the landmark as the boundary depends on each animal's ability to judge the other animal's strength correctly. Mesterton-Gibbons and Adams proved that a set of conditions does exist that establishes an evolutionarily stable strategy and those conditions depend on the cost of fighting and the marginal value of adding more area to a territory. These conditions can exist even when a landmark does not make adding territory awkward, as would be the case if an animal tried to expand his territory to include areas across a river.

Mesterton-Gibbons's and Adams's conclusions lead to the following hypotheses regarding accepting a landmark as a territory boundary. 1) The landmark is near the optimal boundary, especially if the optimal boundary is easily perceived by both animals. 2) The cost of fighting is high. 3) The marginal value of increased territory size is low. 4) A territory owner's ability to apply pressure against his neighbor rises slowly with his strength. And 5) Fighting ability declines rapidly with distance from the nest or the territory center.

Once a landmark is accepted as a territory boundary, it may be accepted by future generations of territory owners. The territory boundaries for wolf packs, for example, can generate olfactory landmarks that may be accepted when territory ownership changes, thereby affecting further the decision of a pack not to invade a neighbor's territory.

9.8 Limitations

Game theory emphasizes equilibria, whereas evolution is an ongoing process. Field biologists often know when they are working in ecological systems that are undergoing change and are, therefore, likely not to be in or near evolutionary equilibria. In those situations, game theoretic models can be more effective for modelling direction of selection than for modelling final equilibrium conditions.

Game theoretic modellers often assume that animals use all information available to them all the time, which can lead to complex models and to models that are more precise than are the behaviors of most animals, whose behaviors are flexible. Real organisms' cognitive maps include myriad information about their environments and use that information in complex ways that are often not understood by modellers, whose models, therefore, do not catch the real complexities of behaviors despite their models' complexities (Houston and McNamara 2005). In addition, small changes in assumptions often cause large differences in model outcomes (Hinsch et al. 2013) and modelling isolated interactions without the context of animals' whole lives can lead to inconsistent outcomes (Houston and McNamara 2005). For example, because territoriality requires that a resource, often food, be limiting, and because reproduction also depends on food supply, modelling territoriality and reproduction without including food availability in the model (e.g. López-Sepulcre and Kokko 2005, Morrell and Kokko 2005, Giuggioli et al. 2011) lacks *the* pertinent biological context (see Hirsch et al. 2013).

With these caveats in mind, game theoretic approaches can nonetheless generate good hypotheses not necessarily obvious using other approaches and, therefore, can increase understanding of animal behavior and of animals' cognitive maps and can even be used to understand cooperation among plants (Wuest et al. 2022). The complex mathematics involved in many recent game theoretic models should not inhibit biologists from testing the resulting hypotheses and should not inhibit biologists (even those with strong mathematical backgrounds) from developing their own simple models that have reasonable biological accuracy and that lead to good hypotheses to test in the field. Creativity and imagination are more important than mathematical complexity. Biology is most productive when biologists with diverse perspectives contribute to areas of inquiry (Messeri and Crockett 2024).

CHAPTER 10

Familiarity Distributions I

Infrequently Asked Questions

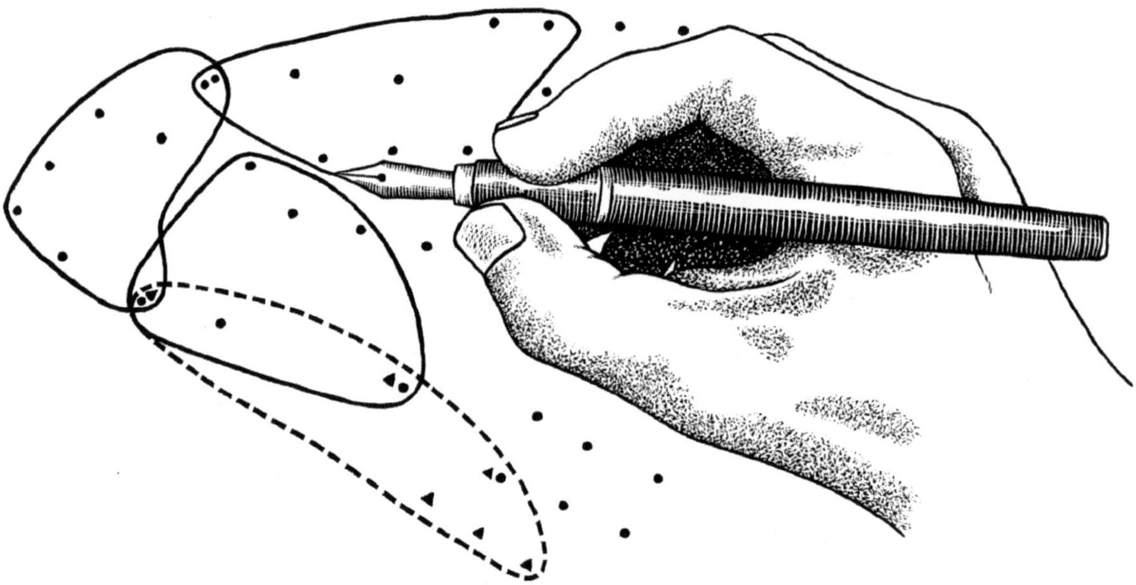

In 1966, Bob Cochrane and a colleague drove madly through central Illinois toward Wisconsin, mostly using township and local roads, sometimes driving through farmers' fields, a yagi antenna mounted on the top of their truck, trying to keep in contact with a thrush wearing a tiny transmitter with a short transmission range. This was the first known tracking at a fine scale of a bird's migration route. At least, that is how I remember the story told to me by Bob in his electronics lab in 1972. At that time, telemetry was beginning to change the type, the scope, and the timescale of the location data collected from animals. Bob was creative with electronics and curious about its potential. The potential is still growing.

Home Ranges of Animals. Roger A. Powell, Oxford University Press. © Roger A. Powell (2025). DOI: 10.1093/9780191949142.003.0010

10.1 Getting Started

One of the very first steps in descriptive analyses of location data, for me at least, is to map the location estimates for my study animals. I map the location estimates on topographic maps and I map them on maps of habitat quality. I remove obvious false data, such as a location in the Pacific Ocean attributed by the Argos system to a fisher from the Sierra Nevada of California, USA. The Argos system did correctly document an exploratory jaunt by a fisher from the northern Sierra Nevada down into California's Central Valley and then back into the mountains. The very high frequency (VHF) and global positioning system (GPS) telemetry that my research team also used then could not have documented such a jaunt. So, I am careful before calling a location false.

Next, I build familiarity distributions and map those. Given that I have argued that home ranges are not maps but are concepts within animals' minds, their cognitive maps, why do I estimate familiarity distributions and map them? Maps help me to visualize how my study animals have moved through and made use of their environments and give me a feel for how their environments can influence their behavior. Maps of familiarity distributions, to be useful, juxtapose animals' movements with those of competitors, with those of potential mates, with accessible resources, and with other characteristics of the local environment. Maps as visualizations allow biologists, especially biologists who are visual thinkers, to concentrate on their data and to gain insights not to be gained from other approaches. I gained such insight when I saw the notch in female bear 260's familiarity distribution that showed how she walked around Little Pisgah Ridge (Chapter 1). Maps are important stepping stones toward analyses that explore home range behaviors more deeply and maps lead biologists to understand more about animals' cognitive maps, which are their real home ranges.

This chapter begins with a bit of history of technology, including data error, and history of methods of mapping location data. Short discussions follow of what familiarity distributions are and are not, what location data represent, and whether building familiarity distributions is necessary. I return to autocorrelation, shifting from an orientation around movements (Chapter 4) to familiarity distributions and then to discussion of excluding data and building 100% vs 95% familiarity distributions.

Animals remember locations within their environments, the resources available at those locations, and the behaviors required to obtain resources at those locations under different conditions, and these different memories are tightly associated in their hippocampus complexes as part of their cognitive maps (Nieh et al. 2021). The consequence is that to use location information only to study and to understand animals' familiarity distributions is incomplete; more information is required to understand why an animal lives where she lives and why she travelled where she travelled.

A familiarity distribution should quantify the animal's probability of having been in different places and it should facilitate quantifying and understanding the importance of different places to the animal at different times and under different conditions. The best familiarity distributions can be combined with analyses of why an animal went where she did when she did, with information on how she perceives her environment, and with analyses of what in her environment was important to her. Using her familiarity distribution in this way can help a biologist to gain insight into her cognitive map, her home range (Peters 1978, Powell 2012a).

A familiarity distribution (descriptive utilization distribution) is sometimes confused with a "utility distribution," a concept borrowed from economics that assigns a value (the "utility") of including a place within an animal's home range (Ellner and Real 1989). Utility distributions need not be probability density functions, though they usually are. To build a utility distribution for an animal's use of its environment, one designs research to show how animals deal with trade-offs between resources. For example, what other resources would an animal be willing to sacrifice to be able to have access to a particular resource? To build a utility distribution across the animal's entire home range, a biologist needs to know what all the trade-offs are that the animal would be willing, or not willing, to make for the resources available at each place. Thus, a utility distribution is a distribution of relative importance, which is what biologists *really* want to know about animals' home ranges. To my knowledge, however, no one has built a utility distribution for an animal. Belovsky (1978, 1984a,b, Belovsky and Jordan 1978) demonstrated how one can quantify trade-offs among several resources.

10.2 History of Data Technology and Quantifying Animals' Locations

The analytical tools used to quantify where animals have been located have come to be called "home range estimators," confounding what animals consider their home ranges to be with what biologists have called their home ranges (Chapter 3). This confusion was inevitable, given that the tools to study and to understand animals' concepts of their home ranges only started to become available in the 1970s, well after biologists had begun quantifying animals' locations. These estimators really summarize an animal's locations and provide insight into the animal's knowledge of her environment. Estimators might better be described as "estimators of environmental familiarity."

The earliest data related to animals' locations, movements, and home ranges came from direct observations and snow tracking, often related to trapping mammals for fur (e.g. Coues 1877, Thompson 1898, Anthony 1928, deVos 1951, 1952, Krohn and Hoving 2010). These data were incomplete and, therefore, the first estimators for how animals used their environments were outlines of where individual animals were located during a period of observation. By the 1930s and 1940s, live-trapping had become the most common tool for studying small mammals and the most common way to describe how a small mammal used her environment was to draw lines to connect the outside capture sites (Dalke and Sime 1938, Schwartz 1941, Lay 1942, Storer et al. 1944). The lines drawn produced an irregular polygon, termed the "minimum convex polygon" by Mohr (1947).

Burt (1943: p. 351) noted that polygons were not necessarily neatly symmetrical in shape: "Many home ranges probably are somewhat ameboid in outline, and to connect the outlying points gives a false impression of the actual area covered." At this same time, biologists also recognized that animals may have travelled beyond trap locations or observations and, therefore, biologists drew home range boundaries to include some area beyond the boundary traps, settling on a boundary strip extending half way to the next trap or next potential observation point. Arguing that minimum convex polygons can include areas obviously not used by study animals, Burt (1940, 1943) and others (Baker 1946, Evans and Holdenried 1943, Stickel 1946) drew their lines around capture points so as to exclude traps where animals were not captured, causing polygons not always to be strictly convex (Figure 10.1). A problem with this approach was, of course, that no 2 biologists would necessarily draw the same lines for a given data set.[1]

[1] Note that in the 1940s, the maps that biologists published were what they literally drew on paper to show their animals' locations. The hand-drawn figures in publications through the middle of the 20th century are engaging.

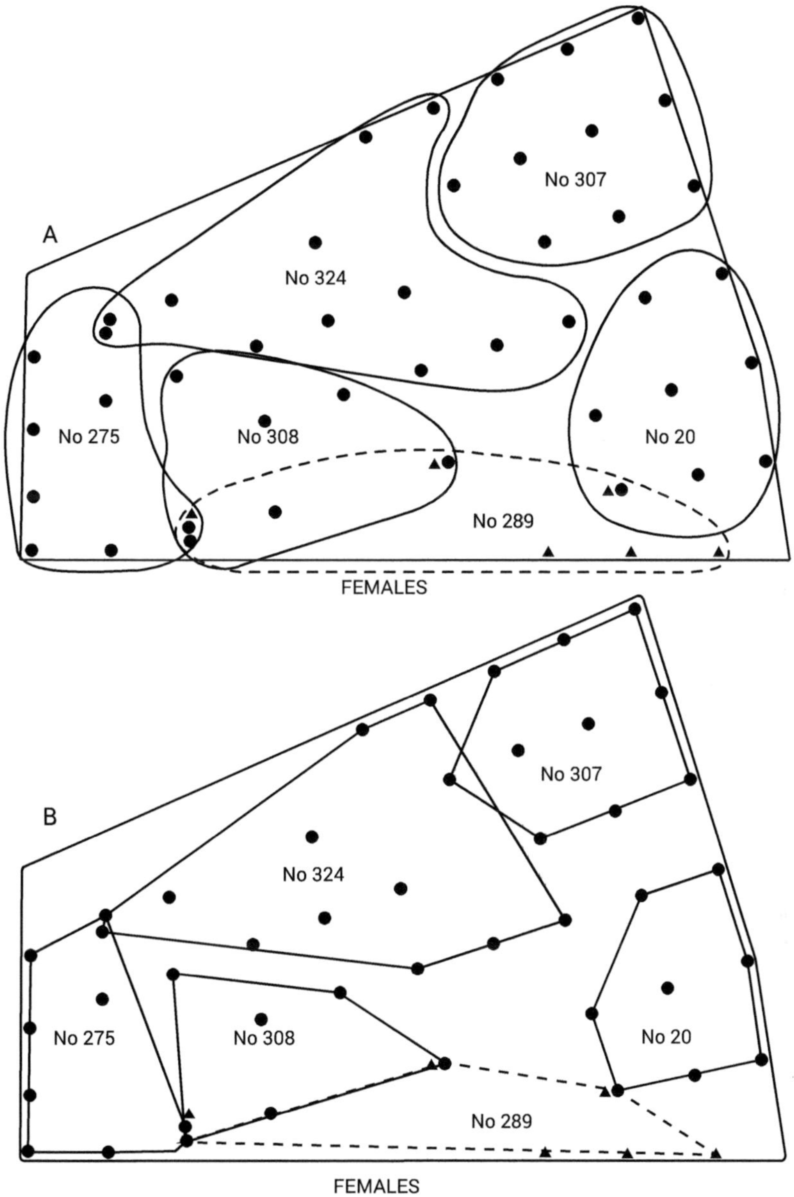

Figure 10.1 Biologists in the 1940s argued that minimum convex polygons (MCPs) included areas obviously not used by their study animals and, therefore, drew lines around capture points so as to exclude traps where animals were not captured, creating ameboid-shaped curves. A. Figure adapted from Burt (1940) with ameboid curves. B. The data from A redrawn as minimum convex polygons.

Mohr's (1947) paper mostly, though not entirely, standardized the use of minimum convex polygons. By the late 1940s, home ranges had become largely quantitative descriptions and maps that seldom included an attempt to understand what animals considered their home ranges to be or what in the home ranges was important. Nonetheless, as early as 1946 Linsdale noted that a home range boundary drawn on a map represents a biologist's abstraction and may have no meaning at all to the animal himself, whose home range it was. In 1949, Leopold wrote that what an animal does in his home range is much more important than the home range size. And in that same year Don Hayne (1949: p. 1) wrote that "understanding of the biological significance of an animal's home range must include some knowledge of the intensity of use, by the animal, of various parts of the area."[2] A biologist today who quantifies only the areas of home ranges needs to understand that limiting home range study to area only is over 75 years out of date. Two home ranges of the same area (size) may have that area for very different reasons. Use of the environment inside the 2 home ranges is usually completely different, making inferences about areas alone misleading. Chapter 12 provides alternatives to using raw areas of familiarity distributions in analyses.

Hayne (1949, 1950)[3] introduced the idea of using numbers of captures per trap as an index of intensity of use, noting that evenly spaced traps actually have different densities at different distances from an animal's center of activity, biasing this index to underestimate the intensity of use of peripheral parts of a home range. He highlighted how trap spacing affects estimates of area used and estimates of intensity of use. Live-trapping data today are still subject to his cautions. Metzgar (1972) and Metzgar and Sheldon (1974) also showed theoretically that estimated shapes and sizes of outlined areas depend on sample size, which has been documented many times since.

Stumpf and Mohr (1962, Mohr and Stumpf 1964, 1966) argued for building composite outline areas by combining data from several individuals at a study site. Their goal was to increase sample sizes so that biologists could gain a better understanding of the general characteristics of areas used by animals within a population. Combining data, of course, eliminates any possibility of understanding how individual animals' home ranges relate to the distributions of resources within them and eliminates the possibility of quantifying variation.

In the end, problems with minimum convex polygons are myriad (van Winkle 1975, Worton 1987, Powell 1987, White and Garrott 1990, Horner and Powell 1990, Powell et al. 1997, Seaman 1993, Stahlecker and Smith 1993). To construct a minimum convex polygon, a biologist discards the vast majority of the data (interior data) he or she worked so hard to collect and keeps only the extreme data points, which are the least trustworthy measures of where the study animal travelled. Building minimum convex polygons also requires large sample sizes (100 or more animal location estimates to approach being repeatable; Bekoff and Mech 1984, Powell 1987, White and Garrott 1990). For very large data sets, minimum convex polygons can be drawn that contain the smallest areas with given percentages of observations (e.g. 25% of observations, 50%, etc.) which estimate contours of intensity of use. These contour polygons do provide information on how heavily individual animals have used the areas within the different contours but the contours have the same problems as do the outermost minimum convex polygon: They include areas not used as intensively as appropriate for their contours. Minimum convex polygons can, however, be used to

[2] One might consider Hayne's statement to be the first concept of a familiarity distribution.

[3] Don Hayne (1949, 1950, 1978) is an unsung hero for biologists who wish to obtain legitimately the most that they can from their data. He was a champion of using biological logic in choice of statistical approaches and logic in the construction of a professional paper. In 1978, Hayne politely highlighted the problems of what we now know as "pseudoreplication," 6 years before Stuart Hurlbert (1984) shoved the concept into biologists' faces and made us pay attention. Hayne's papers of 1949 and 1950 on home ranges gently highlighted major problems with the approaches that biologists were using to quantify home ranges at that time. He noted the importance of direct observations over live-trapping and over track and sign for obtaining movement and home range data. He introduced the idea of using numbers of captures per trap as an index of intensity of use, also noting its drawbacks, and using those numbers as a help for drawing home range outlines. Hayne highlighted the effect of trap spacing on estimates of home range areas and on estimates of intensity of use. He concluded his 1949 paper by calling for biologists to learn what animals' true home ranges are (how the animals perceive their home ranges) and how those perceptions differ from what biologists might measure. Because diverse animals use their home ranges differently, for biologists to gain insights into what home ranges mean to their animals they must develop good hypotheses, develop good research design, and collect data appropriate for testing the hypotheses. Research methods must be objective, repeatable, and biologically appropriate. In the 1980s, Hayne highlighted the logical structure of a scientific paper, taking readers through the past (Introduction) to the near past (Methods) to the present (Results) and on to the future: a Discussion section that highlights new hypotheses to test. His insights are pertinent today for live-trapping and telemetry data, for future data collection with new technologies, and for writing.

define availability for 3rd order resource selection (Horne et al. 2009).

In his review of the literature on mammal movements and home ranges, Sanderson (1966) argued that home ranges will not have standard shapes because the specific needs of individual animals differ. Biologists should, therefore, seek to understand the specific needs of particular animals and the distributions of resources across landscapes and through time rather than to quantify distances moved or the shapes and sizes of areas outlined. Thus, the literature provides a repeat (remember Leopold's and Hayne's comments of 1949) of the idea that knowing what goes on within a home range is more important than being able to measure its area and perimeter.

The first statistical models, beyond minimum convex polygons, applied to home ranges were univariate and circular, describing density of use extending along radii from the center of activity. Authors assumed that a complete data set of an animal's locations would conform to a 2-dimensional normal distribution spreading from the animal's center of activity (Currie and Bellis 1969, Dice and Clark 1953, Hayne 1949, Harrison 1958, White 1964), though Hayne (1949) clearly understood that the change in use spreading from the centers of activity would be idiosyncratic and would depend on resources within a home range. Bivariate models followed (set on an x–y plane; Calhoun and Casby 1958, Jennrich and Turner 1969, Koeppl et al. 1975, 1977a, Mazurkiewicz 1969, 1970, Van Winkle et al. 1973), still mostly assuming normal distributions but now allowing estimates of how animals distributed use across an environment to be elliptical, which is minutely more realistic than circular. These models allowed biologists to view home range estimates as distributions of estimated time spent in different places superimposed on a 2-dimensional map of an environment, that is, home range estimates as "utilization distributions" (Van Winkle 1975), more correctly considered "familiarity distributions."

Almost all familiarity distributions at that time were anchored to the assumptions that an animal's use of its environment spread from its center of activity in a bivariate normal manner and that the "center of activity" for a home range was the arithmetic mean of all locations for an animal (the mean x value and the mean y value; Hayne 1949). Distances moved from the center of activity and how the center of activity related (or failed to relate) to peripheral areas became important subjects of research (Dice and Clark 1953, Harrison 1958, Sanderson and Sanderson 1964, White 1964, Sanderson 1966, Van Winkle et al. 1973). Using the arithmetic mean as the "center of activity" has several problems: 1) if "true home ranges" are not convex in shape (not roughly round but instead crescent-shaped, for example), the arithmetic mean can fall outside of the true home range; 2) the arithmetic mean may not be located near any biologically important part of the familiarity distribution; and 3) if sample sizes (numbers of locations) are small, adding new data can move the arithmetic mean significantly (Hayne 1949, Sanderson 1966, Neft 1966, Dixon and Chapman 1980).

With the advent of radio telemetry in the 1960s, biologists became able to collect more extensive data, both in space and in time, and for more diverse animals than possible by repeated live-trapping. Telemetry data expanded the potential for research on birds (Cochrane 1972) to grizzly bears (Craighead et al. 1995). Each telemetry location was actually a location estimate, however, requiring estimates of location error (Saltz and White 1990, Schmutz and White 1990, White and Garrott 1990, Zimmerman and Powell 1995) and requiring estimates of the effects of location error on familiarity distributions. With wide-ranging data sets and data points not tied to evenly spaced traps, biologists looked for better ways to smooth location data, dropping their assumption of bivariate normal distributions. Thus, telemetry both allowed and forced biologists to develop new estimators.

Because statisticians often used Fourier series to smooth data, Tarter and Kronmal (1970) and Anderson (1982) built familiarity distributions based on the bivariate Fourier series. Anderson's estimator treats each animal location estimate as a spike in a 3rd dimension above an x–y plane, smoothing the spikes into a surface that forms a familiarity distribution. I developed a similar method using spline smoothing techniques (Powell 1987). Both of these estimators show multiple centers of activity accurately and show that centers of activity may be far from the arithmetic mean of the x and y data. The spline estimator had troubles because it generated negative values for probability of use. In addition, both estimators behaved poorly at peripheries of familiarity distributions, because peripheral location data are sparse and do not meet assumptions needed to make the transformations. To address the problem with peripheries, Anderson recommended using animals' 50% familiarity distributions (the smallest area encompassing a 50% probability of use). This recommendation stretched Burt's (1943) basic concept of a home range to its limit by assuming that an animal is on one or another "occasional sally" 50% of the time. These estimators never gained much use.

Human population densities fall in an inverse harmonic mean fashion from centers of urban areas out through rural areas. (The harmonic mean is the reciprocal of the arithmetic mean of the reciprocals of a set of observations.) Dixon and Chapman (1980) proposed that density of animal locations would fall in an inverse harmonic mean fashion from an animal's center of activity. Consequently, and because the harmonic mean center of animal locations "indicates the true center of activity of a home range" (really, an assumption), Dixon and Chapman proposed a familiarity distribution based on harmonic means. This approach also resolved problems of using the arithmetic mean as an animal's center of activity. The harmonic mean distance is calculated from each animal location estimate to each point on a superimposed grid and these harmonic means were used to develop a familiarity distribution (Ackerman et al. 1988). A fatal flaw of the estimator is that each familiarity distribution is unique to the position and spacing of the superimposed grid. Poor choice of grid placement produced familiarity distributions that had sharp peaks at each animal location, or that were overly smoothed, or that had contour lines that did not match the spatial distribution of animal locations (Spencer and Barrett 1984). The estimator had significant bias when the grid was re-oriented subjectively to plot each data set because when a biologist chose contour lines that appeared to be the best fit, distributions could be overly or under-smoothed. Location estimates that fell close to grid cell corners caused significant bias related to differences between the center of the grid pattern and the location closest to the center of activity (Spencer and Barrett 1984).

Nonetheless, we are in luck. Myriad better options have been developed to build familiarity distributions. In 1989, Worton showed that biologists could use kernel smoothing (Silverman 1990) to build familiarity distributions from animal location data. Erran Seaman and I had been working along those same lines with biological statisticians at the University of North Carolina (Seaman 1993, Seaman and Powell 1990, 1996). One can imagine how a kernel estimator works by thinking of location data points as dots on a map board on a table. One-by-one, each data point gets covered by a small hill (all hills are of the same volume and start as the same shape) made of clay or silly putty so that, for points close together, new hills mold themselves over the hills already in place, making hills higher and forming ridges, or even plateaus. The resultant surface becomes a familiarity distribution, showing an estimate of the probability of an animal having been at any point on the map board. Figure 10.2 shows how this process works for a 1-dimension location data set. Kernel density estimators are free from problems of being tied to a grid size or grid placement (Silverman 1990). In general, kernel estimators are easy to use and produce logical results.

Kernel density estimators started being used widely in the 1990s (Laver and Kelly 2005) but are not without their problems. Because kernels used in canned software are round, for example, location estimates at the edges of inhospitable habitat will lead to familiarity distributions that extend into such habitat. Fortunately, variations on the basic concept of kernel estimators are easy to program, leading to diverse software packages available that address different problems. Better yet, anyone with some programming abilities can program an estimator that solves problems pertinent to one's own data and research objectives, such as how to deal with lake shores for terrestrial or aquatic animals. One can weight location estimates by, for example, animal travel speed, by time between location estimates, by time of day, by quality of nearby resources, by locations in inhospitable habitat, and by other behaviors of a target animal (Matthiopoulos 2003b, Katajosto and Moilanen 2006, Fieberg 2007a, Horne et al. 2008, Powell 2012a; Chapter 12).

Local convex hull estimators are an alternative to the widely used kernel estimators, especially

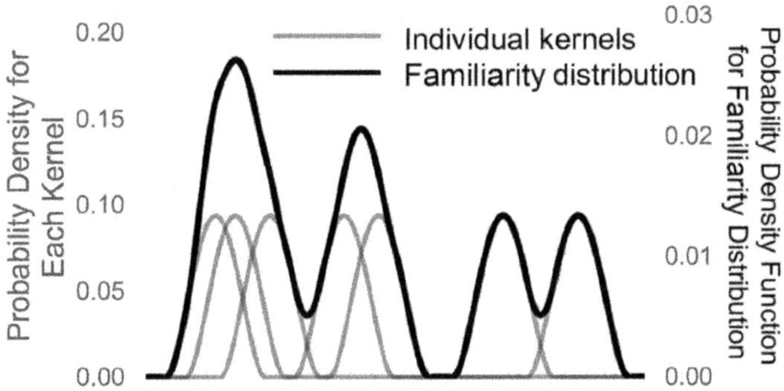

Figure 10.2 Example of kernels arranged along a 1-dimensional axis and the resultant familiarity distribution.

when an animal's use of his environment has sharp boundaries (Getz and Wilmers 2004, Getz et al. 2007). As originally presented, convex hull kernels were constructed by building a minimum convex polygon around the k–1 nearest neighbors for each location (Getz and Wilmers 2004). In more recent developments, convex hull kernels are constructed from all points within a fixed radius of each location, or from all points with a fixed sum of distances, or by weighting pairs of locations by intervening time (Getz et al. 2007). Local convex hull estimators can outperform traditional kernel estimators when habitat boundaries that affect movements are distinct (Chirima and Owen-Smith 2017) but not when true use of an environment varies in a smooth manner (Lichti and Swihart 2011, Noonan et al. 2019b). Familiarity distributions made using any of the local convex hull estimators are based on series of minimum convex polygons, thereby including the problems of minimum convex polygons (such as an animal not using large areas within the convex outline of a polygon).

Both kernel and local convex hull estimators can handle global positioning system telemetry data. My kernel program can calculate the 100% familiarity distribution for a data set with > 15,000 locations in a minute or 2. Global positioning system data combined with accelerometer, magnetometer, and video data can provide complete path trajectories (Chapter 4), which provide more detailed insight into animals' cognitive maps but will have little effect on estimators, which smooth on a scale larger than the refinements of path reconstruction. Camera data, hair snare data, scat identification data, and other data can be added appropriately to telemetry data just as capture data are added to telemetry data now to make data sets as complete as possible.

10.3 Telemetry and Other Data Error

Soon after the advent of very high frequency telemetry, biologists realized that they must take location error seriously. Early users of telemetry used telemetry often to track animals to their locations, as biologists do sometimes today, but quickly triangulation became common. To triangulate an animal's location, a biologist finds the direction of the strongest signal from 2 or more locations and plots the intersection of the lines to estimate the animal's location. Determining the direction of the strongest signal always has error and seldom do 3 or more triangulation lines all intersect in a single point. Without knowing the accuracy and precision of location estimates, a biologist can not know how to treat location data in analyses. When building a familiarity distribution using a kernel estimator, for example, choosing a degree of smoothing that is smaller than location error for the location estimates will quantify inaccurately what is actually known about an animal's moves in an environment. Similarly, location error affects the outlines used to build local convex hull familiarity distributions. Often, however, using triangulation with very high frequency telemetry yields error shapes that are not round. Depending on how far from round are the error shapes, using kernels that are not round can make the resulting familiarity distribution a bit more representative. White and Garrott (1990) provided a good overview of very high frequency telemetry error and how to quantify it.

In the 1980s, time-of-arrival telemetry systems, antenna grid systems, and others were developed. A time-of-arrival system measures the time that a signal takes going from a transmitter to a receiver; the transmitter is mounted on an animal and the receiver on an antenna on a tower or on a satellite. A signal is broadcast from the transmitter with a time signature so that the time needed to reach the receiver can be calculated. A given signal can theoretically reach a receiver on a satellite from anywhere on a sphere whose radius is the distance needed for a signal to use the time calculated. That sphere intersects the earth in a circle. If the signal was received by 3 or more receivers, the 3 circles will (ideally) meet at a single point. Time needed for a signal to travel from the transmitter to the receiver, however, can be increased by signals bouncing, by signals being blocked and weakened from particular directions, and so forth, introducing error into location estimates.

Global positioning system locations also have error, though much smaller error than that for very high frequency telemetry using triangulation. Error reported by manufacturers, however, can be considerably smaller than error for locations in the field. My colleagues and I documented error of about 30 m for global positioning system units predicted by the manufacturer to have error of 10 m (Powell et al. 2022). Consequently, documenting error for global positioning system packages is important for some research. Where global positioning system telemetry is combined with animal mounted video, very precise measures of telemetry error are possible if the video shows environmental details that can be ground truthed. Likewise, animals with global positioning system units who have their photos taken by remotely triggered cameras document telemetry error precisely.

A good way to calculate and to report location error is to use a 2-dimensional distribution around true locations. To calculate this distribution, transform all known locations (for example, locations of transmitters placed in the field to estimate telemetry error) to the origin (0,0) and transform the telemetry location estimates appropriately. Ideally, location error will be circular around the origin and leptokurtic (but without infinite tails; Zimmerman and Powell 1995). If this is the case, then the mean 2-dimensional error will fall on the true location.

A good way to report telemetry error is as percentages of estimated test locations within distances from true locations, as shown here:

Proportion of test locations	Meters from true location	or	Proportion of test locations	Meters from true location
25%	x		x%	25 m
50%	xx		xx%	50 m
75%	xxx		xxx%	75 m
90%	xxxx		xxxx%	100 m

Any telemetry system has biases that can lead to errors with analyses. For example, signal strength varies with habitat, biasing numbers of locations by habitat, and, thereby, requiring compensation during analyses (Tran et al. 2024).

Other modes of data collection also have error that should be considered. Trail cameras, live-traps, and DNA hair snares often use bait to attract animals to sites that the animal might not otherwise visit. Independent data are needed to verify that an animal otherwise travels close to the data sites. Trail cameras and collar-mounted video cameras use extremely wide-angle lenses and, therefore, distort apparent distances and body sizes of animals. Objects photographed appear to be farther from the camera and smaller than is the case. A biologist who wishes to measure distances from a camera needs objective ways to measure distance and animal body size, such as using marked stakes at designated distances from trail cameras and meter-stick-type measures on baited trees. In addition, cameras (digital and film, still and video) "see" colors differently than do humans. Any analyses that depend on colors must be adjusted for the gamuts of the cameras being used. On the plus side, of course, cameras can be modified to respond to ultraviolet and infrared frequencies, facilitating seeing what many birds and insects can see but people can not, and allowing analyses related to heat radiated in the infrared range.

Analyses of DNA collected from hair snares and scats also has error derived from analyses of the

DNA itself: human error during DNA analyses, and sampling error, the latter caused by not collecting and analyzing all hair samples captured by the collecting device and also by selective sampling of animals in the population under study (Langer 2006).

10.4 Before Building a Familiarity Distribution

Understanding several aspects of home ranges and familiarity distributions before choosing a "home range estimator" and attempting to build familiarity distributions is essential. For starters, a biologist can build a familiarity distribution in 3 different ways that usually produce somewhat different results. Van Winkle (1975) defined a familiarity distribution with respect to the relative frequency distribution of the data points. All estimators start by building such a frequency distribution, either using location data directly or by smoothing the location data. Transforming the volume of this distribution to be 1.0 makes the distribution be a probability distribution (first way). Contour lines can be drawn around pertinent probability values. If the values for the probability distribution at each grid point (all familiarity distributions as calculated on a grid of x and y values) are transformed linearly so that the value at the grid cell with the highest probability density value becomes 1.0, then the distribution values range from 0 to 1.0. A biologist can plot contour lines at standard values, such as 0.05 (far from the centrum), 0.50, and 0.95 (close to the centrum). This is a relative probability distribution (second way).

Alternately, one can calculate proportions of the volume of the density distribution. The volume at the grid cell with the highest density value is the density value at that point. At the point with the next highest density value, the cumulative volume of the grid cell with the first point is the sum of their density values. At the grid cell with the 3rd highest value, the cumulative volume is the sum of the 3 values, and so forth. The final value for the volume must be 1.0. The 95% familiarity distribution is drawn around the points with values less than or equal to 0.95 and other percent home ranges are drawn similarly. This is the volume probability distribution (third way). In this case, 95% of the density and 95% of the location data lie within this 95% contour.

The latter 2 methods produce almost equal results when the location data are randomly distributed, which is almost never the case (Figure 10.3). Probability distributions (the first way) must be used to quantify overlap of familiarity distributions. Biologists should report which method was used to make familiarity distributions used in their analyses.

In addition, no "home range estimator" actually estimates a home range. As explained in Chapter 3 on cognitive maps, a home range is far, far more than a familiarity distribution. Such a familiarity distribution describes only a few aspects of what an animal has done during a period of study. Familiarity distributions can not predict what animals will do at other times under different conditions (different seasons, different food availability, different food requirements, different predator population sizes, and so forth). Most familiarity distributions built by "home range estimators" estimate the smoothed densities of animals' location estimates across a study area for a specific period of study in the past. Assuming that each location represents an amount of time, familiarity distributions estimate how much time animals spent in different places during a period of study. Unfortunately, not all location estimates represent the same amount of time. Biologists can deal with this problem by weighting location estimates by, say, speed of travel or by adapting kernel sizes or shapes to activity. Another interpretation is that a probability familiarity distribution estimates the probability of having been able to find an animal in any given place on the study area during the period of study. Thus, a familiarity distribution is purely descriptive and simply building a familiarity distribution does not constitute doing science; nonetheless, it can be one small, and possibly very important, step of doing science.

A familiarity distribution can not predict where an animal will go or what he will do or will have done during any time period outside the study period but, again, it can provide insight. Fleming and colleagues (Fleming et al. 2015) claimed that familiarity distributions built using their autocorrelated kernel density estimator could "anticipate" future use of areas. Their claim, however, represents a misunderstanding. A familiarity distribution can predict areas that its animal will use in the future

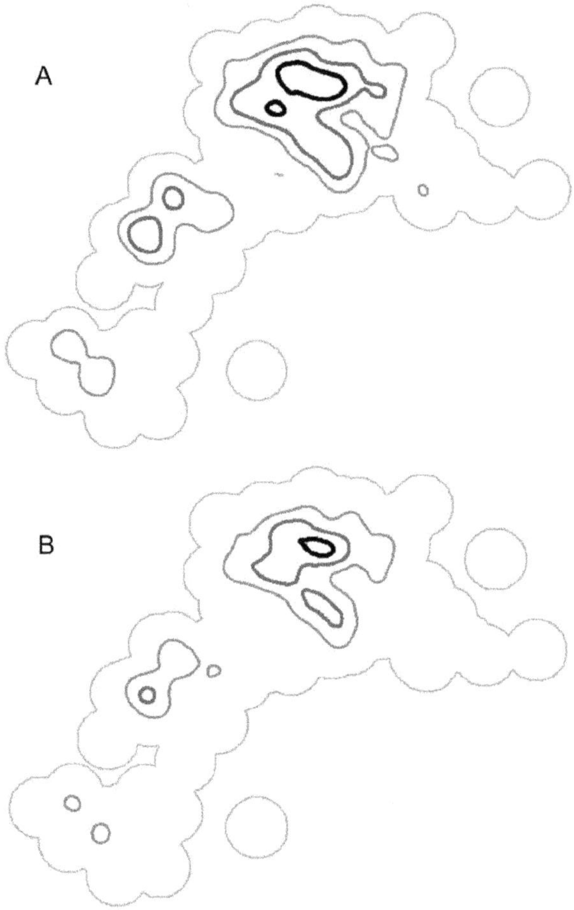

Figure 10.3 Biologists can build familiarity distributions in 3 ways. A. Familiarity distribution built as the proportions of the volume of the probability density distribution (usually called the % home range). B. Familiarity distribution built as the proportions of the maximum probability density. The former (A) emphasizes areas of high densities compared to the latter (B). The familiarity distributions are for female black bear 61 in 1985 in the Blue Ridge Mountains of North Carolina, USA. Contours are 100% (light grey), 90%, 30%, and 10% (black). A familiarity distribution can be built from the probabilities for an animal having been found in any cell of his home range.

only insofar as the animal behaves in the future as it did during the time period for which the familiarity distribution was built or if the band width chosen for the estimator is so large as it includes large areas beyond the location data, even future location data. A familiarity distribution can, however, make general though vague predictions for the future that are better than having no information at all. If environmental conditions change, or the animal's status changes, or any other critical changes occur, the animal may well go to places not previously documented and may spend different amounts of time in places where he was known to have gone in the past.

10.4.1 What Does Each Location Data Point for an Animal Represent?

Certainly, each location data point represents a place where an animal was located, or represents an estimate of that place. The location estimate provides the potential to juxtapose animal behavior data with environmental information. When location data are collected randomly, evenly, or in some other representative manner, then the numbers of location estimates in different areas estimate the proportions of time spent in those areas as long as sample size is large. If the research hypotheses being tested relate only to time spent in different areas, then using a familiarity distribution developed simply from the unbiased, raw location data will suffice.

Time and distance and speed, though related through the defining equation for speed (speed = distance/time), provide different insights into behavior. Distances between telemetry locations are almost always biased low because distance measures straight lines between successive locations whereas animals wander to a greater or lesser degree around the lines (Chapter 4, section 4.2). Speeds calculated for those distances will also be biased low. Location estimates close in time yield distance and speed estimates that are less biased than those spaced far in time because animals have less chance to wander over short time periods. Analyses using distance and speed must be robust to biases, such as limiting data used by time between locations. Tracks in snow provide precise measures of distance but not of time or speed. Speed can sometimes be indexed from stride length or gait, shown by tracks, but one must be careful to understand the assumptions of such indices. Speed can be measured with data added from telemetry or from direct observations (Powell 1979b). Noonan et al. (2019a) showed how to estimate true distances and speeds from distance and time data plus location error.

A familiarity distribution does not show what places were most important to a study animal unless one knows that time spent in a place indexes the importance of that place, which is only sometimes true. Some resources, a desert waterhole, for example, are critically important but need to be visited infrequently or are needed only for short periods (e.g. Noy-Meir 1973, Tsalyuk et al. 2019, Bartlam-Brooks et al. 2020, Rozen-Rechels et al. 2020, Laverty and Berger 2020). If an animal must harvest several foods to balance his diet, some critical foods may be harvested quickly (food patches visited for short time periods but very important; Belovsky 1978, 1984a). Another resource, a resting site, for example, may be used for a long period but be no more valuable than dozens of other potential rest sites (sites used for long time periods but specific sites not important). Birds spend long times incubating eggs and canids spend long times at reproductive dens. In these latter situations, time does index importance. Once eggs hatch, however, parent birds spend much less time on their nests, though the offspring in those nests are more important than the eggs had been. Consequently, unless a biologist knows that time indexes importance, knowing how much time an animal spent in a place can provide some insight into importance but is far from perfect.

Testing well-designed hypotheses is more likely to provide important insight than will quantifying numbers of locations in different places, and is more likely to increase understanding of the biology of study animals. Testing such hypotheses, however, requires information in addition to location data and in addition to distributions of resources. Alternatively, analyses may need to be limited to locations in the vicinity of resources being considered, may need to include foraging rates for different required nutrients, and may need to include information on whether an animal can forage for more than 1 required nutrient at a time.

A way of testing for the waterhole effect is to index patch importance using information from the literature. One index of patch importance is an index of food in the patch based on food density, availability, caloric and nutritional benefits, and based on the patch's physical conditions that facilitate food harvest. Another such index is the ability of a patch to foil predators and to provide safe travel. Still another is providing rest or reproductive sites. A test for the waterhole effect is whether the index values for importance of patches predict the numbers of locations in those patches in the data. If index values do predict numbers of locations, then a familiarity distribution captures at least a component of importance. If the index can not predict numbers of locations, then the waterhole effect applies to target animals in the environment, or the importance index was based on faulty information, or both. Proximity loggers can supplement location telemetry data to establish numbers of locations in potentially important resource locations (Abelia 2024).

A waterhole is an extreme example of importance not being indexed by relative numbers of locations. Less extreme examples exist. Habitat patches that appear very similar to biologists may have decidedly different danger from predators. Many herbivores must spend minimum amounts of time in habitats that provide limiting nutrients that are not required in great quantities and also spend large amounts of time in habitats that are both abundant and important. One can test for this latter condition using the tools provided by Aho and Bowyer (2015). Their index of used and available resources identifies patches with abundant resources that are important. To identify such less extreme mismatches of importance to time requires more or better data than needed for waterhole mismatches.

10.4.2 Familiarity Distributions and Data Smoothing

Most familiarity distributions smooth across animal location estimates to show patterns of central tendency. Statisticians have a long history of seeking and using patterns of central tendency to make deductions. In an unbiased sample from a distribution, the extreme values tend to be unreliable as indices of the patterns shown by the whole distribution. Had the largest or smallest value of the sample been different, the midpoint of the range of a sample might be very different, even if the sample size is large. In contrast, changing only the largest value of a large sample has little effect on the mean (median or mode) and the bigger the sample size, the smaller is that small effect, making the mean a better measure of the central tendency of the sample than the midpoint. Similarly, using the range of a sample to measure variation within the sample again depends heavily on the values of the sample extremes, while sample variance, standard deviation, 50% confidence limits, and other measures of central tendency summarize the variation without being overly affected by sample extremes.[4] By smoothing across distributions of animal location points, familiarity distributions allow biologists to see central tendencies in animals' locations in their environments. Burt (1943) addressed this concept by stating that occasional sallies are to be excluded from what biologists consider an animal's home range to be. If sample size is large and representative, and the estimator used to build a familiarity distribution is robust to extremes, excluding occasional sallies is not necessary and may exclude important information.

Do biologists need to smooth their location data? Not always. Before smoothing, a biologist must know, or assume (with crossed fingers), that a sample of locations is representative relative to the hypotheses being tested. Smoothing can only work if the scale for smoothing dictated by technology matches the scale for environmental patchiness (habitat, physical terrain) and how that dovetails with how a study animal senses that patchiness. For estimators whose familiarity distributions have been compared to known use, accuracy of smoothing (quantified using some statistic such as mean squared error) increases with increasing sample size (Noel 1993, Seaman and Powell 1996, Getz and Wilmers 2004). But, because animals of different species have different movement behaviors, and because different hypotheses can tolerate different levels of accuracy and precision, no universal, minimum sample size exists for smoothing with any "home range estimator." When location data sets are extremely large, smoothing may not be necessary to test many hypotheses. Lynn Rogers and colleagues outfitted black bears in northern Minnesota with global positioning system telemetry collars that located themselves every 10 minutes, yielding data sets well in excess of 16,000 locations for some bears in some years and, thereby, precluding the need to smooth for research on habitat selection (Powell et al. 2022). At the other extreme, data sets with < 20–30 locations per study animal are best used to test hypotheses that require no more than the actual data points for each animal; the insights that come from the emergent properties of home ranges will likely be missing or not accurate because the sample of locations is so small.

10.4.3 Autocorrelation

Data used to build a familiarity distribution are most often sequential locations of an individual animal and these data points are often not statistically independent samples from the true distribution of the animal's wanderings. If the second of 2

[4] Field biologists, however, seldom need to use standard error, which quantifies one's confidence in the mean but does not quantify directly variation within a data set. Standard deviation quantifies variation within a data set, which is almost always the information that biologists need. Understanding evolution depends on understanding variation, not on having confidence in a mean.

successive location estimates for an animal is constrained by the time since the first location, then the 2 locations are statistically autocorrelated in time (Chapter 3, section 3.2). Therefore, how much new statistical information (not necessarily biological information) is provided by each new animal location depends on the time since and the distance from the previous location. The expected time between locations beyond which the constraint is eliminated can be calculated (Swihart and Slade 1985).

Autocorrelation must be considered when filling gaps in movement paths (Chapter 3). And autocorrelation can be used to establish site fidelity, which is crucial documention that an animal has established a home range (Swihard and Slade 1985).

Nonetheless, the importance of autocorrelation for home range studies is not so clear. Animal location data are *always* biologically autocorrelated, even when they are not statistically autocorrelated, because animals use their knowledge of their home ranges, use their cognitive maps, to plan movements. An animal's every movement is planned within its cognitive map and within the context of everything the animal knows about its environment. No movement is independent of the preceding movements and many are not independent of movements yet to come that have already been planned. Thus, I and others have made disparaging remarks regarding the importance placed on statistical autocorrelation by some biologists (1987, 2000, 2012a, Kie et al. 2010).

Consider this example. Every move I make within my house is related to my knowledge of every place in the house. If I have not been in the basement for 2 weeks, the reason for not going there relates to other activities planned that outranked activities planned for the basement. Even though analyses of my movements might show that my visits to the basement are statistically independent, they depend biologically on everything else that is in my house and everything else that I have done and plan to do. If I stay in the kitchen all day, the reason relates to my plans to cook meals for all the days of the coming week, which I know I can not do in my study or in the bedroom or in the basement. My visits to the bathroom are not occasional sallies and they are not caused by my cooking all day but, instead, by my drinking tea all day. My trips to the bathroom might appear minor in an activity budget and an energy budget but they are critically important to me (literally a waterhole problem).

Thinking of home ranges as cognitive maps forces a biologist to keep the extended context in mind all of the time. Understanding the spatial aspects of home ranges means understanding how and why animals plan their foraging routes, learn new escape routes, and monitor potential mates, and all depend biologically on past movements that are not statistically autocorrelated with planned movements. Where and when an animal does those things is *always* biologically autocorrelated with where it does *everything else*. An animal makes daily decisions to return to certain food patches but *not* to return to other patches that have not yet renewed, showing that the statistically independent locations and statistically autocorrelated locations can result from a common decision rule used by the animal: Return only to optimally renewed patches (Chapter 5). Setting a time limit after which all locations that are considered to be independent (Swihart and Slade 1986) has poor biological support. For a biologist to advocate that statistically autocorrelated location data must be avoided for unbiased analyses is like a starving wolf obsessing over a flea.

Luckily, lack of statistical independence is not a great problem for most analyses (Lair 1987, Powell 1987, Andersen and Rongstad 1989, Gese et al. 1990, Reynolds and Laundrè 1990, Fieberg 2007b; but see Noonan et al. 2021). At the occasional times when statistical autocorrelation appears biologically important, biologists must be aware of statistical autocorrelation within their data sets. They must know under what circumstances autocorrelation must be handled by data analyses. They must accept it and deal with it in biologically logical ways.

In addition, important biological information can exist within statistically autocorrelated data that can not exist when all data points are independent. For example, statistically autocorrelated data provide information on how long an animal stays in a local area during each visit, that is, bout length. A large data set of independent locations can estimate the total amounts of time that an animal spends in different local areas but provides no information on behavior bout lengths while in those areas; a statistically autocorrelated data set provides data on bout length. A black bear decides how long to forage in each berry patch. In a year with abundant berries, the bear may decide to stay in a patch for a long time but, when berries are sparse, she will move to another, perhaps distant, patch after a short time. The consequence is that foraging conditions and this bear's decision rules (abundant berries, stay long; sparse berries, leave quickly) affect statistical autocorrelation of her location data, which informs a biologist about foraging conditions. The bear's cognitive map guides her planning of foraging bouts and the results of foraging bouts update the cognitive map.

Similarly, large data sets with independent location points provide little if any information on travel routes, though travel routes are critical for planning future moves.

If a biologist wishes to have a statistically unbiased estimate of how much time a given animal spends in patches of different resources, then estimating and incorporating autocorrelation in analyses may be appropriate. Alternately, a biologist can weight each location estimate by the time since the previous and until the following location estimate as a proportion of the time for statistical independence; locations close together in time get weighted more lightly than locations long apart (Katajisto and Moilanen 2006, Johnson et al. 2008). Fleming and colleagues (Fleming et al. 2015, Fleming and Calabrese 2017) developed a way to reduce statistical bias from autocorrelation when using certain kernel estimators. Their weighting approach can be used to build familiarity distributions.

10.4.4 95% Familiarity Distributions and Excluding Data

Most familiarity distributions built from sufficient data describe approximately how animals have distributed their time across their environments and, in one manner, provide a description of animals' "normal activities" (Burt 1943). A probability level criterion might conceivably be used to eliminate "occasional sallies." Most biologists use 95% as their arbitrary but repeatable probability level: the smallest area with a cumulative probability of use equal to 0.95 is often defined as an animal's home range. Nonetheless, no biological logic supports this choice of 0.95. Eliminating from a familiarity distribution some small percentage of locations that cover the area where a study animal was least likely to visit (his occasional sallies) will eliminate areas visited quickly and then abandoned. Similarly, a statistical argument exists for excluding some small percentage of peripheral location data points because extremes are not reliable, that is, they tend not to be repeatable. The extremes may be due to a sampling regime that just happens to record an animal's visit to one peripheral location but not to another. Nonetheless, constructing familiarity distributions using a large data set and a reliable estimator will yield low probability values for peripheries, solving the problem of extreme values.

Removing extreme data is not the same as removing obviously false data. False data can be identified using criteria independent of a data set, such as removing a location that could not have been reached by a study animal (such as a fish "located" on land).

If a biologist determines that he must remove extremes, he must inspect his data closely and have an objective definition of "extreme" that does not specify that precisely 5% of data points be excluded. Of course, inspecting one's data closely is *always* a very good thing. In the end, though, even this approach can eliminate critically important, distant parts of a home range. Distant data points can be caused by animals' movement decisions that biologists do not understand, such as animals visiting known, distant places that have dependable resources during times of scarcity. Consequently, distant locations can be critically important for study animals. Animals retain such places in their cognitive maps and may update information about those places regularly from a distance.

Using 95% familiarity distributions may be widely accepted because it appears consistent with the use of 0.05 as the (also) arbitrary choice for the limiting p value used to evaluate statistical significance. Nonetheless, the 95% familiarity distribution developed from 100% of one's data will seldom match exactly the 100% familiarity distribution developed from 95% of the data after weeding out extreme points. Along those same lines, the 95%

familiarity distribution built using the 3 methods of building familiarity distributions yield different 95% distributions; the 100% distributions will also differ but have the same boundaries.

A kernel estimator designed to use a bivariate normal kernel (which extends to infinity) produces familiarity distributions of infinite area, making the 100% familiarity distribution undefined. To be practical, all publicly available kernel estimators using bivariate normal kernels (which is most of them) must be programmed to select an arbitrary truncation for the kernel but, to my knowledge, that limit is not publicly available. How big an effect the truncated distributions have is unknown. One could argue that no matter what arbitrary rule is used to truncate the periphery of a familiarity distribution, those areas represent a small proportion of an animal's time and, therefore, make little difference, an assumption that is not quite always true.

For large data sets developed using representative sampling, leaving extreme data points in the data sets and calculating the 100% familiarity distribution provides a total perspective and has few effects on analyses. A biologist may then choose, or choose not, to concentrate on how an animal uses the environment within, say, 95% (or some other arbitrary percent) of that use. If a biologically pertinent percent exists for a given data set or for a given analysis, then using that percent makes sense.

10.5 Area of a Familiarity Distribution

Familiarity distributions depend on sample size. Dividing a data set of animal locations randomly

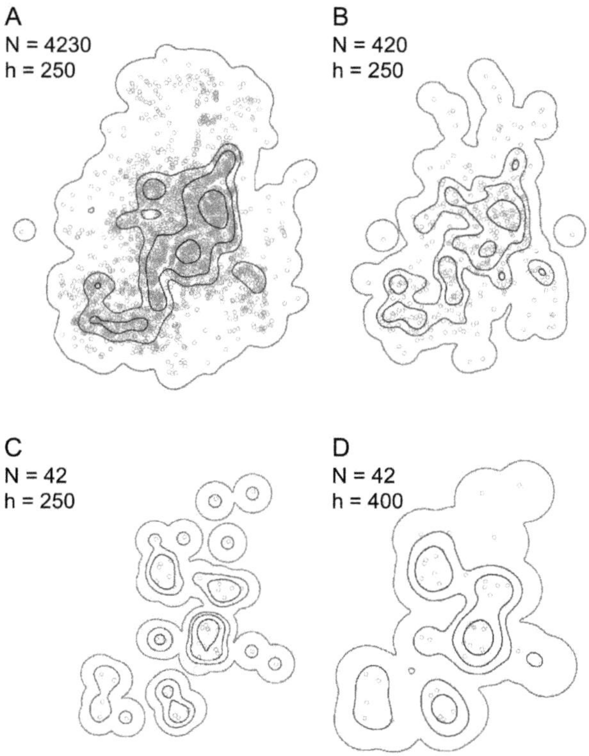

Figure 10.4 Examples of 100% familiarity distributions for a mule deer from New Mexico in 2020. The complete data set has 4230 global positioning system locations. A. Using all data. B. Using 10% of the data chosen randomly, simulating a large data set collected using very high frequency telemetry. C. Using 1% of the data chosen randomly, simulating a much smaller data set collected using very high frequency telemetry. D. Same as C but with band width h = 400 m.

into 2 unequal subsets, or even equal subsets, will produce 2 different, though similar, familiarity distributions (Figure 10.4; Fieberg 2007a, Kochanny et al. 2009, Fieberg and Börger 2012). The 2 familiarity distributions almost always have different areas and areas different from a familiarity distribution made from the entire data set.

Familiarity distributions, and their areas, generated by different estimators and by variants of a given estimator usually differ as well. Each estimator has parameters that must be chosen to match the biology of the target animal, the data, and the hypotheses being tested. For kernel estimators, for example, kernel shape and extent of smoothing (band width h) must be selected, while the convex hull estimator requires selection of the number of nearest neighbors to be associated with each location estimate or the limiting distance from the location.

The edges of animals' home ranges can be diffuse (Gautestad and Mysterud 1993, 1995), making the area of a home range undefined. Nonetheless, an island with a fractal, and therefore potentially infinite, coastline still has a finite limit to its area. The same is true of familiarity distributions: their areas can always be estimated. Whether area is an important measure of a familiarity distribution or a misleading measure is a different question and often depends on how the familiarity distribution correlates with what an animal considers the patches in her realized home range to be. Mike Mitchell and I (Mitchell and Powell 2008) showed that an animal can use, and consider as its realized home range, only relatively small patches within a large area. A familiarity distribution that smooths too much will miss such details of the true home range. A familiarity distribution that incorporates perceived resources not visited and that excludes areas of non-habitat that are avoided does provide information on landscape quality and patchiness, and may provide information about individual choices of home ranges and about population viability, growth, and density. Building a mechanistic distribution may be a better idea. Areas that monitored animals can not visit, such as the middle of a large lake, should have probabilities of location set to 0 (Benhamou 2011).

To compare areas of familiarity distributions among different studies, one must know, at least: 1) sample sizes used to build all familiarity distributions to be compared; 2) how data were excluded, if at all; 3) what estimator was used to build each familiarity distribution; 4) what parameter values were used for the estimator; 5) what percent familiarity distribution was built and what reported; and 6) how autocorrelation was handled when building each familiarity distribution. A familiarity distribution with clumping implies that landscape patchiness is critically important for understanding landscape quality. This information can be used in decisions about population management or landscape management.

CHAPTER 11

Building Familiarity Distributions

This chapter covers and compares using the most common statistical approaches for building familiarity distributions, starting with some simple approaches and then covering kernel, bridge, and local convex hull estimators.

Figure 11.1 juxtaposes familiarity distributions built using grid cell, fixed kernel, Brownian bridge, and local convex hull estimators for a representative bear tracked with very high frequency telemetry and the familiarity distributions for a single mule deer tracked with global positioning system telemetry.

11.1 Simple Familiarity Distributions

A minimum convex polygon provides a simple representation of where an animal lives if plotted on a topographic map or a map of resources. Chapter 10 highlighted the myriad problems of minimum convex polygons. Such polygons can sometimes be

Home Ranges of Animals. Roger A. Powell, Oxford University Press. © Roger A. Powell (2025). DOI: 10.1093/9780191949142.003.0011

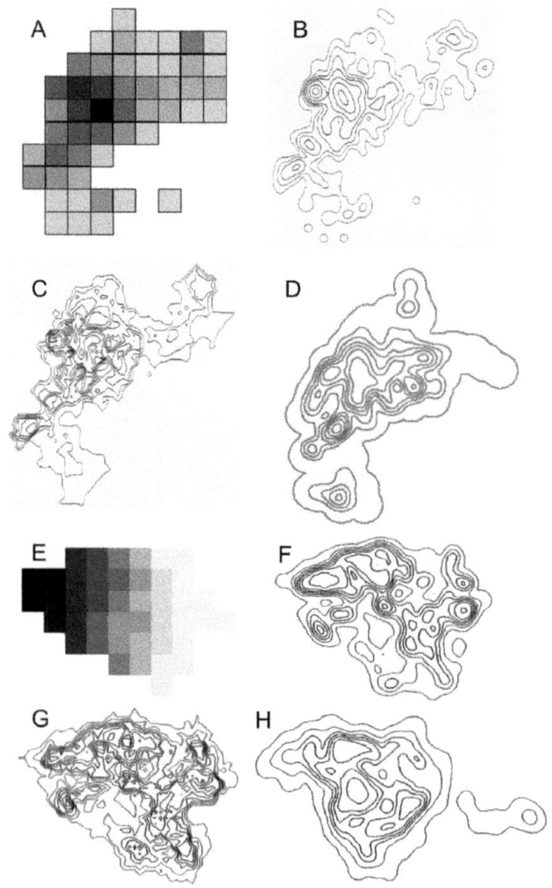

Figure 11.1 Comparisons of 4 familiarity distributions for a black bear and a mule deer. A. Grid cell. B. Fixed kernel. C. Local convex hull. D. bridge familiarity distributions for female black bear 61, 1984, Southern Appalachians, North Carolina, USA. E. Grid cell. F. Fixed kernel. G Local convex hull. H. Bridge familiarity distributions for mule deer in New Mexico, USA. For A and E, grid cell = 500 × 500m. For B, D, F, and H kernel = K2, *h* = 250, 100%. For C and G, *k* = 25. A, B, D, E, F, and H are built from cumulative proportions, as is done generally to build descriptive per cent familiarity distributions.

handy, though, to gain a quick, rough idea of where animals' home ranges lie and what resources are in the local environment. For me, though, a minimum convex polygon gives less insight than does a plot of the raw location data because the latter shows the distribution of the locations. A biologist can build a familiarity distribution by building a series of minimum convex polygons. First, build a minimum convex polygon using all location data; then build the smallest polygon possible using 95% (or another percentage) of the locations; then build another polygon using 90% of the data, and so on. Building a familiarity distribution from minimum convex polygons is awkward and computer running time increases rapidly with sample size and the number of contours. Familiarity distributions built from most other approaches produce equivalent or better information and are easier to build.

The simplest familiarity distribution to build superimposes a grid on a map of a study area and weights each cell in the grid by the number (n) or by the proportion of location estimates for a study animal within that cell (e.g. Horner 1986, Doncaster and Macdonald 1991, Zoellick and Smith 1992; Figure 11.1 A, E). This approach may seem outdated, given the many sophisticated home range estimators available; nonetheless, its simplicity and its direct use of data make it a good choice for many analyses (Figure 11.1 E). Calculating the proportion of locations in each cell produces a likelihood, or probability density, distribution that shows the probability that at any time and day during the study period an animal will have been found in that particular cell. The density estimator does no smoothing. To make a volume familiarity distribution, sort cells by decreasing density and calculate the per cent of all cells with densities higher than the value of each cell. These values make the volume 100% familiarity distribution.

This estimator should be considered for global positioning system telemetry data collected at short time intervals (multiple locations per hour) because it avoids having to make many assumptions required by estimators that are used commonly at present, such as how the data will be smoothed. If grid cells are small, this approach will create travel paths that show rest sites and foraging patches (Figure 11.2). The density estimator is good for habitat analyses. Grid cells need to be large enough to handle telemetry error.

Gautestad and colleagues (1998, Gautestad and Mysterud 1993, 1995, Gautestad et al. 1998) argued that $n^{1/2}$ indexes density of locations better than does n. Using $n^{1/2}$ emphasizes lightly used cells (Figure 11.3). Note that Figure 11.1 A, which I built from cumulative proportions (which can yield 100%, 95%, 75%, or other percentage familiarity distributions) differs distinctly from Figure 11.3 A

Figure 11.2 Grid cell plot of over 7000 GPS locations collected every 10 minutes for female black bear Dot in northern Minnesota, USA, in 2013. Each cell is 100 × 100 m. Many lightly shaded cells show travel routes. Black fill equates to Dot being found in a cell over 30 times. Clumps of such cells show foraging patches.

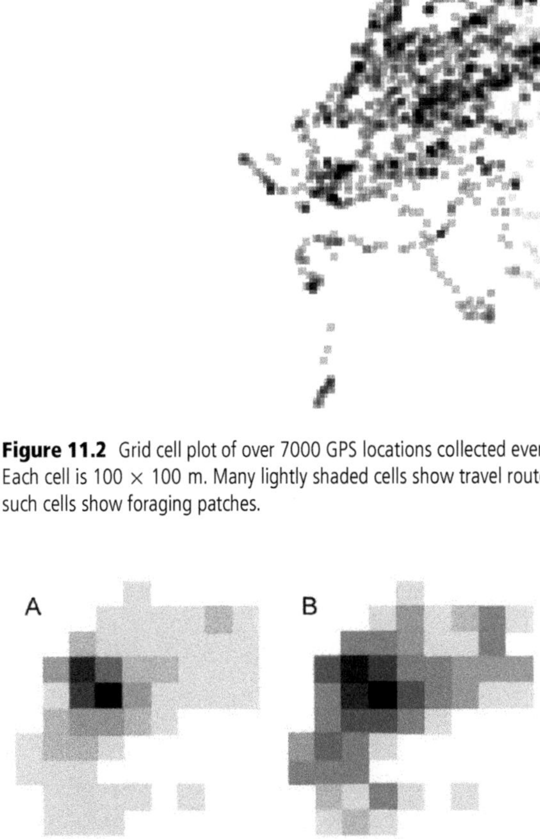

Figure 11.3 Grid cell plots of locations of female black bear 61 in 1984, in the Southern Appalachians of North Carolina, USA. Cell size is 500 × 500 m. Sum over all cells in each figure = 1. A. Darkness of the fill in each cell is proportional to, n, the number of locations/cell. B. Darkness of the fill in each cell is proportional to $n^{1/2}$. Contrast A and B to Figure 11.1 A, which was built from cumulative numbers of locations/cell. Figure 11.1 A resembles 11.3 B here more closely that it does 11.3 A.

(which was built on n) but is similar to Figure 11.3 B (which was built on $n^{1/2}$). The comparison shows that familiarity distributions built from cumulative densities emphasize differences among lightly used cells, as does the $n^{1/2}$ approach, compared to probability density distributions, which show the estimated probability of an animal being in each cell during the period of study. Building familiarity distributions using cumulative proportions has become the norm but distorts the actual probability of an animal having been located in each cell by emphasizing peripheries.

Doncaster and Macdonald (1991) estimated the familiarity distributions of red foxes as a retrospective count of the grid cells known to have been visited during specific time periods. This approach is equivalent to treating the cells as marked individuals for a mark–recapture study and estimating the true density and the familiarity distribution sizes using a "minimum number known alive" approach (Krebs 1966). At any time, a fox's familiarity distribution included those cells that had been visited before that time and that would be visited again in the future. This approach allowed Doncaster and Macdonald to follow foxes' familiarity distributions as they drifted across the landscape. The authors used survival estimators to calculate the rates at which cells were lost from familiarity distributions and new cells added (Doncaster and Macdonald 1991). With this approach, potential "occasional sallies" are easily identified as those cells visited only once, as long as the sample size for locations is large. Identifying these locations as occasional sallies, however, does not mean that they are not important. They may have never been visited again or they may have become important central cells after the research ended.

Choosing cell size is an application of hierarchy theory and can be a problem for analyses using grids (Lloyd 1967, Vandermeer 1981). For data on animal locations, grid cell size should incorporate, in some objective way, information about location error, information about the radius of attraction for trapping data, information about the radius of an animal's perception, the scale of habitat data, and knowledge of the appropriate scale for the hypotheses being tested. For comparisons among animals in a population, cell size must be equal for all animals. To compare results for animals of different species, cell size relative to the mean size of a circle or polygon circumscribing the data for each species must be equal or otherwise scaled among the populations compared. Because cell size is related to the scale of the behaviors being studied, changing cell size can change results of analyses (Lloyd 1967, Vandermeer 1981). The smallest cell size is dictated by location error but the ecological processes of the hypotheses being tested otherwise dictate the appropriate cell size.

11.2 Kernel Estimators

Although kernels (the hills in my explanation in Ch 10.1) can be of any shape and do not have to be symmetrical, all canned kernel estimators appear to use 1 of 2 symmetrical shapes: bivariate normal (most software), or biweight (also called K_2; Silverman's [1990] preference because it is more efficient than the bivariate normal kernel and is mathematically more flexible). Unfortunately, kernel shapes are not always detailed in the information readily available for software. The 2 kernel shapes produce slightly different familiarity distributions and, consequently, can sometimes lead to slightly different results of analyses. The breadth of a kernel is quantified by its band width, designated as h. The statistic used to measure h differs, however, for the 2 kernel shapes because band width is incorporated differently into the individual equations for the 2 kernel shapes. For a bivariate normal kernel, h is the standard deviation for the distribution. For the biweight kernel, K_2, h is the radius at the bottom of the "hill." Thus, the entirety of kernel K_2 is included within its h but that is not the case for a bivariate normal kernel, for which a large portion of each kernel lies outside its h. To be able to replicate analyses or to compare home range statistics across different studies, biologists must know both what kernel was used in each study and must know h. Knowing how software truncates bivariate normal kernels would also be useful. When all kernels used to build a familiarity distribution for an individual animal are of the same shape and have the same h, then the estimator is called a "fixed kernel estimator."

The equation for the circular, bivariate normal kernel, $(\phi(x,y))$, is

$$\phi(x,y) = \frac{1}{2\pi h^2} \exp\left\{-\frac{1}{2}\left[\left(\frac{x-x'}{h}\right)^2 + \left(\frac{y-y'}{h}\right)^2\right]\right\} \quad \text{Eq. 11.1}$$

where x' and y' are the values for x and y at a location data point and x and y represent all other points on the $x - y$ plane. The equation for K_2 for location x is

$$K_2(x) = (3/\pi)\left(1 - (x'x)^2\right)^2 \quad \text{for } x'x < 1 \quad \text{and}$$
$$K_2(x) = 0 \quad \text{for } x'x \geq 1, \quad \text{Eq. 11.2}$$

where $x'x$ is the distance from the evaluation point to the observation point (the location estimate) divided by band width, h. To be consistent with equation 11.1, the equation for location (x,y) can be rewritten as

$$K_2(x,y)$$
$$= (3/\pi)\left(1 - \sqrt{\left(\left((x-x')^2 + (y-y')^2\right)/h\right)^2}\right)^2$$
$$\text{for } \left(\left(\left(x-x'\right)^2 + \left(y-y'\right)^2\right)^{1/2}\right)/h \leq 1$$

and

$$K_2(x,y) = 0$$
$$\text{for } \left(\left(\left(x-x'\right)^2 + \left(y-y'\right)^2\right)^{1/2}\right)/h \geq 1.$$

Figure 11.4 shows normal, leptokurtotic, and 2 kernels in 1 dimension. All 4 kernels shown have area = 1 below the curves. For comparison, $h = 1$ for kernels K_2 and K_4 and h for the bivariate normal kernel was chosen so that the 2 curves have similar maximum values. A bivariate normal kernel has infinite tails, leading to the potential to overestimate use of peripheral areas of an animal's home range. I assume that software packages that use bivariate normal kernels truncate the distribution at some distance from the location data point. Alternately, software could truncate density values at some distance from the center of the familiarity distribution. Kernel K_2 has a normal-ish shape but does not have infinite tails; kernel $K_2 = 0$ for distances beyond h. For the K_4 kernel, the exponents in equation 11.2 are changed from 2 to 4. (I used K_4 in Chapter 6, section 6.7.) Using a bivariate normal kernel appropriately truncated, a biologist can choose h to be equal to or smaller than the standard deviation of the location error distribution and generate a curve that is taller and narrower than the normal distribution using h = standard deviation, such as the tallest curve in the Figure 11.4.

Choosing band width is a critical yet difficult aspect of developing a kernel estimator for calculating animals' familiarity distributions (Silverman 1990). The optimal K_2 band width to minimize statistical error is known for independent data (not autocorrelated) that have an approximate bivariate normal distribution, often called the "reference bandwidth, h_{ref}" (Worton 1989). For a data set of d dimensions, h_{ref} for each dimension $i(i = 1, 2, \ldots, d)$ is

$$h_i = A\sigma_i n^{-1/(4-d)},$$

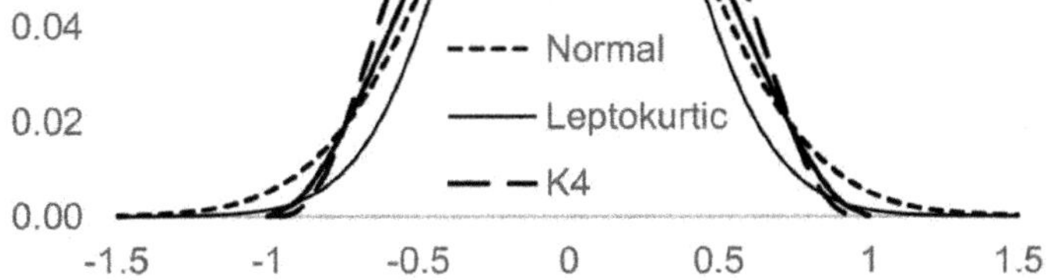

Figure 11.4 Four 1-dimensional kernels, showing the differences. All are drawn with the same area under the curve. Few software packages make clear which kernel is used, though the normal kernel appears to be used most often. It is also used in many theoretical analyses of how kernel estimators function. h for the normal kernel equals the standard deviation of the function as shown in the upper right. Unfortunately, the normal distribution has infinite tails, requiring software writers to truncate the distribution, usually at an unspecified place. K_2 and $K_4 = 0$ at h and $-h$. K_2 has an approximately normal distribution but de-emphasizes the peak and the extremes slightly. K_4 demonstrates how the kernel changes shape with a different exponent than is used in equation 11.2. Good field biologists using triangulation with very high frequency data produce error distributions that are somewhat leptokurtic but without infinite tails. The $K_{0.75}$ approximates the leptokurtic but equals 0 at $-h$ and h.

where A is a constant particular for the chosen kernel being used and σ is the standard deviation for the data in dimension i (Silverman 1986). For 2 dimensions, this equation collapses to

$$h_{ref} = \hat{\sigma} n^{-(1/2)},$$

where $\hat{\sigma} = \left(1/2\left(\sigma_{x^2} + \sigma_{y^2}\right)\right)^{1/2}$. Many software packages calculate h_{ref}, so biologists seldom need to calculate it themselves should their location data approximate a bivariate normal distribution.

Nevertheless, animal location data seldom, if ever, approximate bivariate normal distributions. For distributions that are not normal, band widths that minimize statistical error can be chosen using least squares cross validation (Seaman and Powell 1996), maximum likelihood cross validation (Horne and Garton 2006b), and mean integrated squared error (Fleming et al. 2015). These approaches choose an h that optimizes the fit between the location estimates and the familiarity distribution that is constructed, given certain characteristics of the data. Fleming et al. (2015) proved that autocorrelated data for which every location has the same potential error (assumed to be bivariate normal) causes least squares cross validation to underestimate the optimal h and that the underestimate increases with greater autocorrelation. The result is that, when building a familiarity distribution from autocorrelated data, the h chosen by least squares cross validation under-smooths data compared to the h chosen by mean integrated squared error. Given that the error for good, triangulated data resembles a leptokurtic distribution without infinite tails, underestimating the statistically optimal band width will be a good thing under some circumstances. In addition, different software packages use different algorithms for least squares cross validation and, therefore, can choose vastly different band widths for a single data set. Thus, to evaluate whether the h chosen is appropriate for the hypotheses being tested and the objectives of the research when using least squares cross validation to choose h, one needs to know the kernel used (not always available) and also the algorithm used for cross validation (never available to my knowledge), and the degree of autocorrelation of the data.

Mean integrated squared error appears at present to be the best statistical method for choosing h when: 1) using bivariate normal kernels; 2) using data with bivariate normal location error that is the same for all locations; and 3) using data that are autocorrelated with all data points having the same amount of autocorrelation (Fleming et al. 2015). Those restrictions do not match many data sets for locations of wild animals. In the end, the effects of violated assumptions may make the optimal h chosen by any statistical approach not to be the statistically optimal h and not to be the biologically optimal h, either.

Needless to say, and perhaps thank goodness, data, research goals, and hypotheses being tested most often dictate that h be chosen not solely on statistical grounds. The goal for choosing h is not to have the statistically optimal h but to choose h that provides the most productive biological insights for a given study. For different analyses of the same data, using different hs may be important. One rule of thumb is that h should never be chosen such that the kernel is narrower and taller than the distribution of location error. To match location error, one can try values of h in equations 11.1 and 11.2 to find the value of h that leads to the best match to the distribution of location error. Setting $K_2 h = 0.75$ produces a kernel with a shape similar to that of the leptokurtic kernel in Figure 11.4 but without infinite tails, matching triangulation error for biologists who are well-practiced with very high frequency systems. If no match is very good, then choose an h that generates a kernel that is broader than the distribution of location error.

For location data with small error, such as for global positioning system data, for snow-tracking data mapped precisely, and for video-collar data, biological criteria or research-specific criteria must be used to choose h for the hypotheses being tested. Choosing kernel band width also involves a trade-off for smoothing (Fieberg 2007b). Narrow kernels reveal small-scale details in a data set and, consequently, tend also to highlight measurement and sampling error (location error or trap placement, for example). Wide kernels smooth sampling error but also hide local detail. All statistical approaches for choosing h tend to choose large band widths with extensive smoothing for small data sets and smaller band widths with less smoothing for large data sets (Figure 10.4). When using large data sets with small location error, choosing a small h can produce highly disjunct familiarity distributions that may appear to be under-smoothed but may be smoothed correctly for quantifying habitat selection.

To understand the overall environment with which a target animal is familiar, a large h may be appropriate. If the animal is a resident, she has developed her cognitive map based on extensive experience in the local environment. She will know environmental patches not visited recently. Using an h large enough to leave no or few fairly disjunct areas of a 100% familiarity distribution probably resembles the area she considers when planning moves.

To test some hypotheses, the best band width should be chosen for each hypothesis to build each study animal's familiarity distribution. If a biologist is evaluating individual differences and behaviors, then using a different h for each individual animal might be appropriate. For other hypotheses, all familiarity distributions may need to be estimated using the same band width. If a biologist is interested in population-wide characteristics, then using the same h for all animals makes the most sense. Using the same band width for all can minimize over-smoothing of small data sets and under-smoothing of large data sets. Animals within a population are generally faced with similar conditions and, therefore, using the same h for all is a conservative move. Finally, if a biologist must compare familiarity distributions for animals of different species, or animals of the same species from different populations that experience different conditions, then choosing h proportional to mean daily movements for animals in each population (or another measure of movement) makes sense.

Other objective methods for choosing band width based on the biology of animals include using the speed that animals can move (Katajisto and Moilanen 2006, Johnson et al. 2008); using a biologist's goals, such as making 100% familiarity distributions contiguous (Kie et al. 2002, Berger and Gese 2007, Jacques et al. 2009, reviewed by Kie et al. 2010); using characteristics of the technology used to collect data, such as location error; using the scale of resource patchiness, including food resources and escape resources; and using study site topography.

Kernels do not have to be hill-shaped and symmetrical or all of the same volume. One can think of a grid cell familiarity distribution, as a kernel estimator with a square, columnar kernel, $h = 1/2$ the grid cell dimension, and each location estimate rounded to the (x, y) location as the center of its grid cell.

I recommend that biologists program their own familiarity distribution estimators (both familiarity and mechanistic). I use the Statistical Analysis System (SAS) to program kernel, bridge, and local convex hull estimators; some of my recent graduate students have used Python. The Continuous Time Movement Model (ctmm) system uses R and has open code (https://ctmm-initiative.github.io/ctmmwebdoc/).

When modelling one's own kernel estimator, a biologist can choose kernel shape and h that match research goals best. For one study, I made kernels that were distorted toward the wind direction, thus emphasizing the area an animal can smell at the time its location was estimated (Chapter 12). I have used larger hs for travelling animals than for foraging animals than for resting animals but all with the same volume. One can also weight data points by time between adjacent data points (for autocorrelated data), by activity, by energy expenditure or food acquisition, by aspects of the environment and resources, and by other variables pertinent to the hypotheses one wishes to test (not all of the same volume; Chapter 12; Powell and Mitchell 2012). Finally, once a biologist has generated a density distribution, densities for grid cells can be set to 0 for locations that can not be reached or that can not be used, such as lakes for obligate terrestrial animals and land for fish (Benhamou 2011, Powell and Mitchell 2012).

The University of Maryland website for the Continuous Time Movement Model system states that "ctmm is an R package for analyzing animal tracking data as a continuous-time stochastic process." The package provides FFT variogram and Lomb–Scargle periodogram analyses, and Kalman-filter maximum-likelihood estimation; evaluates and chooses the bridge that works best to complete a movement path; and builds kernel familiarity distributions for autocorrelated data (Calabrese et al. 2016). The program even has a website to analyze data and to produce the familiarity distributions online (https://ctmm.shinyapps.io/ctmmweb/). The package does allow a number of analyses related to animal movement and to building familiarity distributions but, as I have noted earlier in this chapter, I favor analyses that emphasize biology and biological assumptions above statistics and statistical assumptions, given that statistical assumptions are almost always violated. As do most kernel estimators, those built using the ctmm packages can extend into areas never visited by target animals and that are unavailable to them.

11.3 Adaptive Kernel Estimators

Adaptive kernel estimators, in contrast to fixed kernel estimators, vary h with the local density of location data, making kernels narrower but higher (all still with the same volume) where locations are densely packed, such as near activity centers, and wider and lower where locations are sparse, such as peripheral locations (Silverman 1986, Seaman and Powell 1996). Silverman (1986) suggested the following technique for calculating adaptive kernels. First, choose a pilot band width, call it h_p. A measure of location error could be one option, or using h chosen for an associated fixed kernel estimator is a good choice. Next, using h_p, make a density distribution and note the density at every data point (location estimate) and then calculate the geometric mean of those values (g). The local band width factor for location estimate i, λ_i is

$$\lambda_i = (f(X_i)/g)^{-1/\alpha},$$

where $f(X_i)$ is the value of the pilot density distribution at location estimate i, and α is a sensitivity parameter with a suggested value of 2. The value of adaptive h used at each location estimate i (h_i) becomes

$$h_i = \lambda_i h_p.$$

Ideally, α is chosen based on biology of the research animals, on research objectives, on research technology, and such that h_i is greater than location error for all data points. In addition, for large global positioning system data sets, g approaches 0, requiring subsampling of data sets to calculate appropriate values of λ_i.

As appealing as adaptive kernel estimators may appear, they generally overestimate the use of peripheral areas (Seaman and Powell 1996). With some calculation, a biologist might find a value for α that works well for a given data set; the value chosen for α must be reported along with the kernel used and h.

Alternately, a biologist can develop her own approach for adaptive kernels.

11.4 Bridge Estimators

Bridge estimators are most useful when research goals include identifying specific travel routes, such as road crossings and migration routes (Horne et al. 2007, 2019, Sawyer et al. 2019, Kranstauber et al. 2014, Kranstauber 2019). Variances must be estimated for different movement behaviors to bridge gaps in movement paths (Chapter 4; Kranstauber et al. 2012). Given that animal movements are usually confined to their home ranges, understanding constraints on movement paths provides insights into cognitive maps (Chapter section 4.4; Miller et al. 2011) and understanding cognitive maps, in turn, contributes to understanding constraints on movement paths.

Most estimators assume that any 2 sequential location data points are spatial only and that the time sequence information is irrelevant. Bridge estimators do not ignore the time sequence information available with most data on animal locations (Chapter 4). All bridge estimators add ridge-shaped kernels between consecutive location estimates, while the heights, widths, and shapes of the ridges depend on the time and distance between locations, on the bout behavior, and on travel capabilities of the target animals (Chapter 4, Chapter 10; Bullard 1999; Horne et al. 2007, Powell 2012a, Kranstauber et al. 2012). The ridge-shaped kernels are incorporated into kernel estimators along with the kernels established at location estimates (Figure 11.1 D, H).

Kays et al. (in press) built movement paths using global positioning system data for kinkajous on Barro Colorado Island in Panama using the Occurrence Distribution Model of ctmm and they built descriptive bridge familiarity distributions for autocorrelated data using the ctmm model, which can choose an over-large h. They found that the fine-scale movement paths exhibited remarkably refined territory edges where paths of adjacent kinkajous abutted each other with no overlap. The bridge familiarity distributions, in contrast, showed considerable overlap of familiarity distributions for adjacent kinkajous. Unfortunately, Kays et al. did not specify how h was chosen or how movement variance was calculated for the bridge calculations. Reducing h and the variances used to calculate bridges will cause a bridge familiarity distribution to converge on the straight-line movement path estimates. Therefore, h and movement variance chosen on biological information may have produced different bridge familiarity distributions. Nonetheless, plots of estimated movement paths provided consequential insight into interactions among neighboring kinkajous.

When locations are mostly independent statistically, as is often the case of very high frequency telemetry data, a bridge estimator adds nothing to a familiarity distribution built using a conventional kernel estimator that it modifies. A bridge estimator also adds nothing to the conventional kernel estimator when global positioning system telemetry locations are collected at such short intervals that

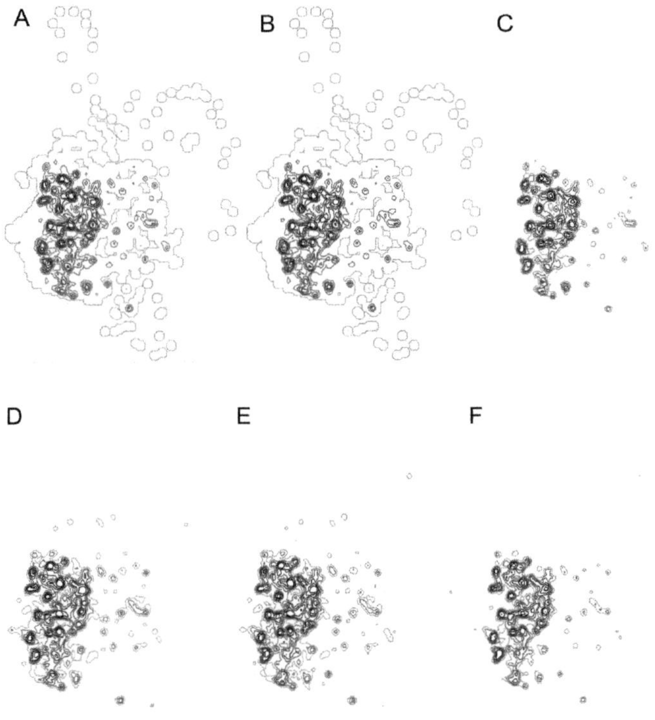

Figure 11.5 Six familiarity distributions for female black bear N001 who lived within the city limits of Asheville, North Carolina, USA, and was followed using global positioning system telemetry in 2014. N001's global positioning system unit located itself 4 times/minute when N001 was active. For all familiarity distributions, kernel = $K2$ and $h = 100\ m$. A and D. 100% familiarity distribution. B and E. 100% familiarity distribution. C and F. 95% familiarity distribution. Panels A, B, and C used all the data. Panels D, E, and F used filtered data with 1 location per 2 hours.

target animals do not travel far enough between locations to warrant building bridges (Horne et al. 2007, Huck et al. 2008, Powell 2012a).

For comparison, I developed 6 volume familiarity distributions for female black bear N001, who lived within the city limits of Asheville, North Carolina, USA (Gould 2020). I assumed that Gould's bears in Asheville had the same time limit for autocorrelation as the bears that I had studied on neighboring land on the Pisgah National Forest (Chapter 4). During 5 months of activity in 2014, N001 was located 4× per hour when she was active. Using all the data, I developed a 100% fixed kernel familiarity distribution, a 100% bridge familiarity distribution, and a 95% bridge familiarity distribution (kernel = K_2, $h = 100$ m; Figure 11.5). I developed the same types of familiarity distributions using 1 location each 2 hours (mimicking very high frequency data or less frequent global positioning system data). The familiarity distributions were for activity because rest sites were undersampled. Using all the data, the 100% kernel and 100% Brownian bridge familiarity distributions are nearly identical (Figure 11.5 A vs B) because locations were so frequent that few bridges were built. The 100% and 95% bridge familiarity distributions differ considerably, however, (Figure 11.5 B vs C) because the 95% familiarity distribution eliminated most bridges and other areas with low density of use. In contrast, the 100% bridge familiarity distributions for the 2-hr data (Figure 11.5 D–F) show slightly more detail, thanks to the bridges, than does the 100% fixed kernel familiarity distribution. The 95% Brownian bridge familiarity distribution for the 2-hr data, however, resembles the 100% kernel slightly more closely than it does the 100% bridge, again because cells with low density of use have been removed. As Horne et al. (2007, 2019) noted, however, bridge estimators are more likely to connect well-spaced areas of heavy use than are kernel estimators.

11.5 Local Convex Hull Estimators

Local convex hull estimators (often abbreviated as LoCoH) are an alternative to the widely used kernel estimators, especially when the environment in which an animal lives has sharp boundaries, such as the shoreline of a body of water for terrestrial and aquatic animals, large highways that animals avoid, or clear-cut forest stands for animals restricted to clear-cut habitats or to the adjacent standing forest (Chirima and Owen-Smith 2017, Getz and Wilmers 2004, Getz et al. 2007, Lyons et al. 2013). As originally presented, convex hulls are the minimal convex polygons around the $k-1$ nearest neighbor locations of each location estimate for an animal (Getz and Wilmers 2004; Figure 11.1). Each minimum convex polygon has a density of locations proportional to the inverse of its area. The hulls are then compiled and transformed to generate a density familiarity distribution and a volume familiarity distribution, if desired, for the animal during the period of data collection (Figure 11.1 C, G). Contours of these familiarity distributions often follow the edges of the minimum convex polygons.

Subsequent variations of the local convex hull estimator include hulls constructed from all points within a fixed radius r of each location estimate; from all points whose cumulative distance to the focal point, ordered closest to furthest, is less than or equal to a; and by weighting distances for these various methods by the time s between each location and the focal location (Getz et al. 2007, Lyons et al. 2013). Using simulations to test the models against known distributions, the s and a variations improve on the r variation, which works better than the original k model.

The values of parameters k, r, a, and s can be chosen using various approaches similar to cross validation algorithms for minimizing statistical error (Dougherty et al. 2018). The parameters k, r, and a should be chosen so that all minimum convex polygons are greater in size than location error. Lyons et al. (2013) recommended choosing parameters that minimize "spurious holes" in the familiarity distribution, thus tending to reduce Type I errors. Spurious holes are empty spaces in the familiarity distribution created by having used small parameter values that produce what Lyons et al. call a "Swiss cheese" pattern. Topography or landscape features (e.g., steep ridges, lakes) sometimes do cause animals to avoid certain areas and parameter values should be chosen so that familiarity distributions do include such "true holes," thus reducing Type II errors. Getz et al. (2007) recommended identifying landscape features that could naturally produce holes in the local convex hull output and using those features to help choose *k*, *r*, or *a*. Plotting the areas of the familiarity distributions against the values of *k*, *r*, or *a* used to build the familiarity distributions often shows a plateau after all spurious holes have been covered. The area increases again when real holes begin to be covered. Thus, identifying a plateau and adopting a midpoint value for *k*, *r*, or *a* appears to be the best way to choose parameter values. This is the only approach to choosing parameter values that is related to biology. Like a minimum convex polygon plot of all location data, the local convex hull estimators are geometric, although they are related to intensity of use (Signer and Fieberg 2021).

Local convex hull estimators have been used to build familiarity distributions for diverse animals (e.g. Chirima and Owen-Smith 2017, Zehnder et al. 2018, Cross et al. 2019, Franckowiak et al. 2019). They can outperform standard kernel estimators when habitat boundaries that affect movements are distinct, such as highways and waterways (Chirima and Owen-Smith 2017), but not necessarily when a kernel estimator sets density = 0 for unavailable locations (not available in standard estimators but an easy option to incorporate *post hoc* or to incorporate into an estimator that a biologist programs). Local convex hull estimators generate familiarity distributions similar to those of kernel estimators when true use of an environment varies in a smooth manner (Figure 11.1; Lichti and Swihart 2011, Zehnder et al. 2018, Cronk and Pillay 2021). Although local convex hull estimators sometimes delineate distinct travel patterns when animal locations are relatively close together in time, the precision does not include probability distributions around estimated travel paths, as generated by bridge estimators. Local convex hull estimators can conceivably be modified to provide insight into an animal's perceptions of his surroundings, by choosing *k*, *r*, or *a* and orienting the minimum convex polygons based on the animal's sensory perceptions, such as wind direction. A familiarity distribution built in this fashion can be tested against a null hypothesis based on choosing *k*, *r*, or *a* on, for example, location error.

Logical, biological ways to choose *k*, *r*, or *a* are not as complicated as choosing *h* for kernel estimators. The *s* variant uses time sequence information in data sets, as do bridge estimators, but not standard kernel estimators. The question arises: How do familiarity distributions for these estimators compare?

11.6 A Comparison

Chirima and Owen-Smith (2017) compared the abilities of the grid cell, fixed, and adaptive kernel estimators, and local convex hull *k* and *a* estimators to build 99% familiarity distributions to match the recorded species ranges of impala, wildebeest, waterbuck, and sable antelope in Kruger National Park, South Africa. For the grid cell estimator, they chose 5 × 5 km cells. They used least squares cross validation available from ArcMap 10.1 to choose *h* and used ArcMap to run the fixed and adaptive kernel estimators. They set $h = 2.2$ km for the fixed kern00651 estimator and 2.22 (pilot value) for the adaptive kernel. I assume, but do not know, that the kernel was bivariate normal. They used ArcMap to run the local convex hull *k* and *a* variants. They chose *k* and *a* by finding values when the plateau began to form, showing when spurious holes were eliminated and true holes had not been.

Chirima and Owen-Smith concluded that all methods provided similar results but differed in different details. The local convex hull *a* appeared to provide the most accurate results. The 2 kernel estimators provided almost identical results but both tended not to show gaps in the actual ranges. The kernel estimators provided almost identical results to those of local convex hull *a* when *h* was chosen on criteria other than least squares cross validation and these criteria yielded smaller values for *h*. Finally, the authors concluded that all these estimators estimated species ranges well.

CHAPTER 12

Familiarity Distributions III

Overlap, Cores, Activity and Sensory Kernels, Areas

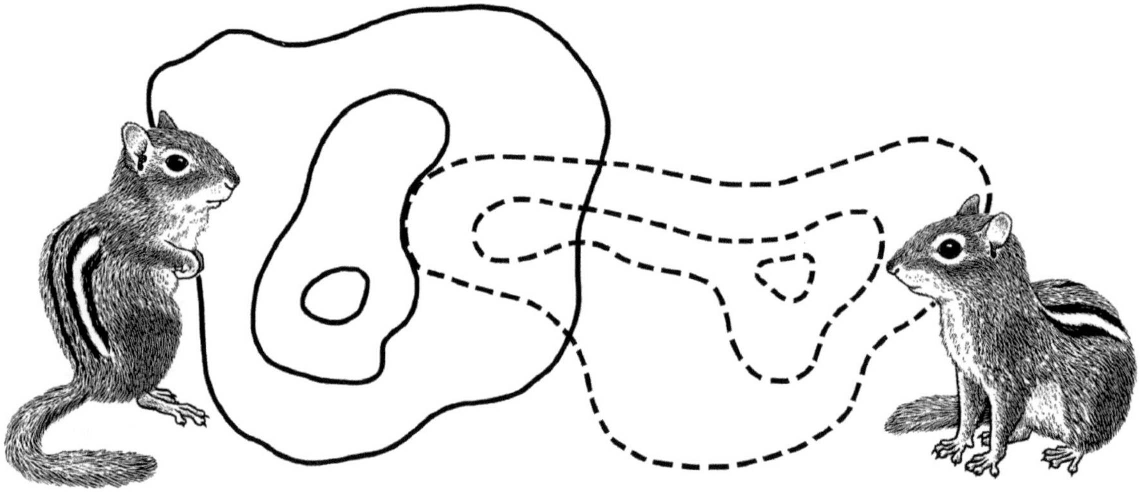

Closing the door behind me, I checked my watch and started my walk from home to the university. I walked south to the end of the road and then cut laterally (southwest) across the prairie dog town to reach 15th Street. I continued south on 15th till I could cut over to the grassy courtyard at the university. There, I walked along paved paths, circumnavigating the north and west sides of the courtyard to reach the Biological Sciences building. Checking my watch again, I walked upstairs to my office and wrote down the times and my exact route. I was documenting my own home range.

12.1 Overview

This chapter explains ways to quantify both static and dynamic overlap of familiarity distributions and then how to identify cores for familiarity distributions. The chapter then discusses weighting locations and building behavior-specific familiarity distributions. This chapter presents examples that illustrate how understanding animals' perceptions of their environments and understanding how they build cognitive maps can be incorporated into familiarity distributions. Examples include varying kernel band widths; weighting locations by visual, smell, and hearing abilities, and using currencies other than numbers of locations; and shaping kernels to fit wind direction and speed and to fit topography. Next, the chapter uses analyses of familiarity distributions for my daily life to highlight shortcomings of familiarity distributions as estimators of my real life. The chapter ends by showing ways to combine evenness of location distributions with familiarity distribution area to gain a better understanding of the amount of a landscape that an animal actually uses.

Home Ranges of Animals. Roger A. Powell, Oxford University Press. © Roger A. Powell (2025). DOI: 10.1093/9780191949142.003.0012

12.2 Overlap of Familiarity Distributions

Non-territorial home ranges of conspecifics often overlap, sometimes extensively. Relatives may often use areas of overlap simultaneously, while non-relatives may avoid simultaneous use. Animals that raise young without help from a mate may avoid areas of home range overlap with competitors or conspecifics with the potential to commit infanticide. The conditions that lead to, and how animals in a population respond to, home range overlap is critical for understanding cognitive maps.

For members of some species, territorial behavior has been documented objectively, especially for the many birds, lizards, and mammals that can be observed extensively. For many mammals, however, territory defense or responses to scent marks or vocalizations are difficult to document and apparent lack of overlap of familiarity distributions is the only evidence of territoriality. Overlap of familiarity distributions can be compared statistically among, for example, populations that appear to differ in territorial behavior but for whom territorial behavior has not been manipulated experimentally. When a biologist quantifies overlap incorporating probability of use of different parts of familiarity distributions, she has more information than simply documenting area of overlap. Indeed, documenting only area of overlap can lead to misinterpretation of results (Rennie 2015).

Overlap of familiarity distributions can be static or dynamic (Doncaster 1990). Measuring static overlap quantifies the probabilities of 2 (or more) animals being in the area of overlap of their familiarity distributions at the same time. The overlap relates to the potential for interference competition. Dynamic overlap involves inter-dependent movements of the 2 (or more) individuals whose familiarity distributions overlap. This overlap has the potential for both exploitative and interference competition.

12.2.1 Static Interactions

The area of overlap of 2 familiarity distributions usually varies in probability of use for the 2 individuals; the overlap may be large for both individuals but be used little by one or the other, or maybe both. In contrast, a small but critical area of overlap may be used intensively by both. Indeed, the biology is most interesting when the animals respond both to resources and to each other.

A couple of indices work well to quantify pairwise overlap of familiarity distributions for each of 2 animals. The first, I_p, can be related to interference competition and is

$$I_p = \Sigma p_{ki} \cdot \Sigma p_{kj} \qquad k \in O \qquad \text{Eq. 12.1}$$

where p_{ki} and p_{kj} are the independent probabilities of animals i and j being in cell k of the study site and the sums are over all cells k in the area of overlap, O, of the animals' familiarity distributions. I_p ranges from 0 to 1. This index is the simple probability that the 2 animals will be in the area of overlap at the same time were they to move independently of each other. Given that most animals do not move without considering the movements of other animals, static interactions should be studied in conjunction with dynamic interactions. If the product of the proportions of locations for each of the 2 animals in the area of overlap at the same time is greater than I_p, then the animals may have been attracted to each other; if the product is less than I_p, then they may have avoided each other. One can not test for attraction or avoidance, however, with a sample size of 1 pair. For a population of animals, one can test for mean I_p not equalling the products of the proportions. For the black bears I studied in the Blue Ridge Mountains, I_p for the adult females did not differ from the proportion of the products but the distribution of the products was distinctly bimodal around I_p. This result suggested that some pairs of females distinctly avoided each other while other pairs were distinctly attracted to each other (Powell et al. 1997). This analysis yielded results consistent with the analysis of mechanistic distributions presented in Chapter 6.

I_p, is a probabilistic measure of overlap that rises in the distinctly non-linear fashion of a square function as both $\sum p_{ki}$ and $\sum p_{kj}$ increase. One can force this index into a linear index by taking the square root of I_p (Bhattacharyya 1943, Fieberg and Kochanny 2005, Winner et al. 2018) but, by doing so, the index is no longer a probability, even though the statistic may appear to be better behaved. And, by not being a probability, the linear index lacks the biological interpretation of I_p; it is just an index of similarity of 2 animals' familiarity distributions. Winner et al. (2018) provided formulas for confidence intervals for this linear index.

Spearman's coefficient of rank correlation is a robust index that can be used broadly to index overlap of familiarity distributions (Doncaster 1990). A biologist can calculate Spearman's r for the probability density values in the cells in a grid used for the familiarity distributions of 2 animals, or for the frequencies of use of cells. The index behaves well and non-linear responses of the index outside of the area of overlap (where one individual has probability of use equal to 0) do not affect the overall usefulness of the index.

Because an animal's familiarity distribution can overlap with the familiarity distributions of several other animals, not all pair-wise index values of overlap within a study site are strictly independent. Similarly, for studies that follow some individuals for more than one year, index values for different years may not be independent. Statistical tests must be controlled for both individuals in each pair-wise overlap when testing for differences among sites, among years, or among populations of different species.

These indices of overlap, and other similar indices (Hurlbert 1978) can be used to compare overlap between sites, to compare changes in overlap with changes in food or other limiting resources, or to deduce territorial behavior when active defense, scent marks, or calls have not been documented.

12.2.2 Dynamic Interactions

Several approaches can be used to quantify and to test whether 2 individuals affect the behaviors of each other. The indices test predominantly for attraction or avoidance and are important for testing hypotheses related to pair bonds, to the existence of extended family groups, or to other social behaviors.

Lloyd's (1967) "mean crowding," $\overset{*}{m}$, is an excellent population measure of overlap, probably still the best available (Hurlbert 1984). $\overset{*}{m}$ measures the density of conspecifics that each individual perceives to be nearby. If individuals distribute themselves patchily by living in family groups or by foraging in proximity to each other, $\overset{*}{m}$ is larger than mean density, $\bar{m}$; if individuals spread out, then $\overset{*}{m}$ is smaller than $\bar{m}$. Lloyd originally applied mean crowding to slow-moving litter invertebrates that he could count without them moving significant distances during his period of counting. With a few modest adjustments, however, $\overset{*}{m}$ can be applied more widely. For each individual i of N individuals, how many other conspecifics, x_i, are within its area of movement yields

$$\overset{*}{m} = \Sigma_i x_i / N,$$

or the mean number per individual of other individuals who are close. $\overset{*}{m}$ can also be calculated using quadrants, yielding the mean number per individual of other individuals in the same quadrant. The size of the quadrants should be based on the biology of the animals, such as perceptual range and movement distances. A strength of Lloyd's mean crowding is that it measures overlap for an entire population at once and allows one to calculate standard measures of central tendency (mean, median, standard deviation).

From a sample of individuals wearing telemetry collars, a biologist can choose a time period to represent "at the same time" and a distance between individuals that constitutes "near" enough to affect a target animal's behavior. For my research on black bears in the Blue Ridge Mountains of North Carolina, USA, during 1981–2002 (Powell et al. 1997, Reynolds-Hogland et al. 2007), we used very high frequency telemetry and a truck-mounted antenna to locate up to 13 black bears within 2 hours, which we would repeat 3, 5, or 11 times (for 8, 12, or 24 hr total, and occasionally 48 hr). The data set has 10 periods with 11–13 female bears tracked at the same time, which constituted most of the adult females within the central part of our study area. I set "near" as 300 m because bears often hear intrusions in their forests as far as several hundred meters away. They often smell foods from > 1 km. Mean crowding for this sample was $\overset{*}{m} = 0.36$ (sd $= 0.26$), meaning that over a third of the bears had a neighbor within 300 m while the 11–13 collared bears were tracked simultaneously.

The best estimate for the number of adult female black bears that the entire 250 km^2 study area supported was 24 (Mitchell and Powell 2012), which converts to a mean of 0.39 bears in an area of 300 m radius, which is a measure of density. For this population, $\dot{m} \approx \bar{m}((0.36 \approx 0.39)$, confirming that the bears were neither patchily nor evenly distributed on the landscape.

In contrast, adult female black bears in northeastern Minnesota, USA, do maintain territories (Powell et al. 1997, 2022). These bears, tracked using global positioning system telemetry during 2010–2014, provided 28 periods with 9–10 tracked at the same time, which constituted all of the adult females in the study area. With "near" set at 300 m, the mean $\overset{*}{m}$ for this sample was 0.09 (sd = 0.15). Mean crowding differed significantly between the 2 populations. "Near" had to be set at 1300 m for the Minnesota population before $\overset{*}{m}$ approached 0.36. These results confirm that the female bears in the 2 populations maintained different social organizations, territorial vs not territorial (Chapter 7; Powell et al. 1997, 2022) and is consistent with the analysis using I_p above in section 12.2.1.

χ^2 and G tests can be used to explore whether 2 animals located at the same time (approximately) tend to be found together. Paired and unpaired distances can be near (animals associated) vs far (not associated), choosing "near" based on the biology of the target animals or on the limits of technology used in the research (e.g. Doncaster 1990, Horner and Powell 1990). Minta (1992) defined "near" as two animals being within an area of overlap at the same time. The N paired locations (those estimated at the same time for animals 1 and 2) estimated how often 2 animals were close together. The distances from animal 1 to the $N - 1$ locations of animal 2 that were not taken at the same time yield a sample of $N^2 - N$ distances that can be assigned as near and far; these near and far assignments estimate how often the 2 animals would be near each other if the movements of each were unaffected by the other. Minta (1992) noted that χ^2 statistics behave better than G with small sample sizes. Minta (1992) developed further tests for attraction to or avoidance of an area of overlap that allow biologists to test for more specific use of areas of overlap and that accommodate more diverse data. He showed how to test whether one animal of a pair is attracted but the other not, how to test for attraction when animals are not always located simultaneously, and how to test for attraction when home ranges are not known but the area of overlap is.

Li et al. (2013) developed a method of testing for attraction and avoidance that uses a permutation test (measuring the distance between all pairs of locations of 2 individuals). They showed that traditional measures of distance between 2 individuals and the frequency of the 2 individuals meeting does document attraction or avoidance at large and small values but that a large area of uncertainty exists at intermediate measures that the test of Li et al. can differentiate. Although the test is computationally expensive with large data sets, Li et al. presented 2 short cuts.

Finally, social network analyses have become a productive approach for understanding how members of groups interact and how those interactions depend on resources and landscapes (Dalziel et al. 2008, Sih et al. 2009, Godfrey et al. 2013, Pinter-Wollman et al. 2014, Firth et al. 2017, Casey et al. 2020, Jones et al. 2020, Pacheco 2020). The vast bulk of social network analyses deal with direct interactions of individuals within groups and deal only occasionally with indirect interactions (Brent 2015). Nonetheless, network analyses have the potential to clarify how, for example, the interactions between non-group-living[1] male and female mammals are affected during the breeding season by their interactions before the breeding season and, in turn, how the before-breeding interactions affect interactions afterward. Thus, network analyses have the potential to show the importance of indirect connections between non-group-living

[1] The term "asocial" is often used to describe mammals who do not live in groups. The term "asocial", however, is misleading because all mammals maintain knowledge about conspecifics who live in close vicinity. Individual mammals keep track of locations of neighbors, identities of neighbors (at least individual identities, sex, age, breeding condition, and aspects of health), sharing of resources with neighbors, and so on. Consequently, individuals who do not live in groups have active social lives from a distance, communicating with neighbors over flexible distances and periods of time.

individuals. Unfortunately, results of network analyses depend heavily on particular techniques for data collection (Castles et al. 2014). Network theory has been used to understand relationships within and among meerkat groups and how meerkat interactions affect the spread of tuberculosis (Madden et al. 2009, Drewe et al. 2009, Drewe 2010).

Quantifying interactions between individuals can be plagued by 3 types of bias that lead to underestimation of interactions (Creel et al. 2013). The first is Step bias, caused by large distances and long times between successive telemetry locations and exacerbated by non-linear travel paths between telemetry locations and by rapid travel. The second is Scatter bias, caused when not all members of 2 target groups are outfitted with telemetry, leading to some interactions not being documented. Extrapolating from interactions between monitored individuals can lead to under- or overestimates of interactions among all group members. The third is Strength bias, caused when interactions between a group and single individuals are not counted because interactions between the group and unmonitored individuals can not be documented. Creel et al. noted that these biases are particularly important when quantifying responses of groups of prey, such as large ungulates, to cursorial predators, such as large canids. Quantifying overlap of familiarity distributions will be affected by these 3 biases.

12.2.3 Testing for Territoriality

The best test for territoriality is to quantify territorial behaviors such as territory advertisements and overt territory defense. Another good test comes from obtaining fine-scale movement paths and looking for patterns of overlap or lack of overlap (Kays in press). For many animals, however, fine-scale location data and observations of territory advertisements and defense are difficult or impossible to obtain. Even so, patterns of overlap of familiarity distributions can be indexed using $\overset{*}{m}$, I_p, or Spearman's r to deduce territorial behavior. $\overset{*}{m}$ can be calculated to quantify how much groups overlap, not just individuals. I_p and Spearman's r can be used to test if 2 distributions differ for 2 groups but need additional information to be able to determine territoriality. I_p is little affected by cell size as long as cell size is relatively small compared to the size of a familiarity distribution, because probabilities of being in cells are summed over the area of overlap. I_p, however, can be used only to test for differences between or among populations or classes of individuals. Spearman's r is sensitive to cell size and, therefore, has the potential to give different results with different cell sizes. Cell size must match the grain or scale of the hypotheses being tested and must be the same for all data being compared. Spearman's r can, however, identify spatial attraction or avoidance on a scale much finer than that of individual home ranges (Doncaster 1990).

12.3 Cores for Familiarity Distributions

Within an animal's home range, some areas are able to contribute more to the animal's fitness than can other areas. Foods and other resources are patchily distributed (e.g. Curio 1976, Goss-Custard 1977, Frafjord and Prestrud 1992, Powell et al. 1997) and, therefore, those parts of a home range with high densities of critical resources ought, logically, to contribute more to its owner's fitness than do areas with few resources. Trying to identify as the core that part of an animal's familiarity distribution that is most disproportionately important has a long history (e.g. Burt 1943, Kaufmann 1962, Ewer 1968, Samuel et al. 1985, Seaman and Powell 1990, Wilson et al. 2010). Identifying cores, if they exist, contributes to the understanding of how the qualities of resource patches are incorporated into animals' cognitive maps and how differences among individuals within a population affect their choices to use their resources. Stiegler et al.

(2022) reported, for example, that shy European hares had smaller 50% familiarity distributions than did bold individuals but had larger peripheries, suggesting that shy hares may have had smaller cores.

Understanding cores of familiarity distributions has 2 parts (Powell 2000). First, a core is used disproportionately heavily or is used in some other way that shows that it is important and used preferentially. This first part leads to the second: a core, therefore, can not be defined by an ad hoc level of use (i.e. a 25% familiarity distribution; or all areas used more than average) but must be determined by the animals themselves. The literature is filled with definitions of cores that are ad hoc or subjective. Were an animal to use her home range randomly or evenly, the home range would have a core by such definitions even though no areas were used preferentially. Hefty and Stewart (2018) and Tórrez-Herrera et al. (2020) documented that the golden-mantled ground squirrels and the capuchins that they studied did not avoid the peripheries of their familiarity distributions and even performed some behaviors more often in the peripheries. Calculating a core by using an arbitrary level of use would misrepresent the use of their environments by these mammals. Using an ad hoc definition can lead to spurious results and lead a biologist to identify random use of an environment as being significant. To have a biological meaning, a core must have a definition that has a functional, biological interpretation related to preferential use.

Using the notation of Hooten et al. (2017), a familiarity distribution H (for Home range) documented to have clumped locations can be partitioned into 2 non-overlapping subsets C and $C^{'}$, core and periphery, while $H \equiv C \cup C^{'}$. C and $C^{'}$ may each have disjunct subregions (the familiarity distribution had disjunct pieces). One also assumes that a specific isopleth of a familiarity distribution, identified by the symbol ϕ, that exhibits clumping or clustering can be found that divides the familiarity distribution into 2 levels of density (f_C and $f_{C'}$) that correspond to core and non-core, or to core and periphery. If isopleth ϕ is known, then C and $C^{'}$ can be identified as can n_c and $n_{c'}$, the numbers of cells in the core and the periphery.

Thus, to document a core, animal location data must show significant clumping, not even use or random use of the environment. Two objective and intuitive approaches (not ad hoc) have been developed to identify cores that meet this requirement. One is more graphical using log-survival curves to identify cores, is based on observable behavior, and requires only simple math and statistics. The other uses maximum likelihood to identify a core.

Clumping—One can calculate $\dot{m}^{'}$, Moran's *I*, or Ripley's (1976) *K* and *L* statistics (Wilson et al. 2010, Hooten et al. 2017) to document clumped location data.

During the period when Don Hayne was publishing about animals' home ranges not being used evenly, Patrick Moran (1950) published an index, *I*, of spatial distributions that ranged from −1 for even distributions through 0 for random distributions to 1 for tightly clumped distributions.

One can calculate Moran's *I* either from a grid cell familiarity distribution, or from an animal's probability distribution for locations placed on a grid:

$$I = \frac{N}{W} \frac{\sum_i \sum_j w_{ij}(x_i - \bar{x})(x_j - \bar{x})}{\sum_i (x_i - \bar{x})^2},$$

where:

- N is the number of cells indexed by i and j in the grid used to build the familiarity distribution,
- x_i is the number of locations within cell i,
- x_j is the number of locations within cell j, which cell is adjacent to or close to cell i,
- $\bar{x}$ is the mean number of locations in all cells,
- w_{ij} is the weighting given to the product within the double sum, for example, the inverse of the distance from cell i to cell j, and
- W is the sum of all the weights.

x_i can be a measure of something other than the number of locations within a cell. It can be, for example, the quality of habitat within each cell, or the amount of food harvested by an animal in a cell, or some other quantity measuring something about how the animal uses its environment.

Many software packages can calculate Moran's *I*, though knowing how to program it is always good.

Alternately, one can use $\overset{*}{m}$ to calculate the index $\overset{*}{m}'$ that also ranges from −1 for even distributions to +1 (approaching tangentially) for highly clumped distributions:

$$\overset{*}{m}' = \frac{\overset{*}{m} - \bar{m}}{\overset{*}{m} + \bar{m}}.$$

Note that $\overset{*}{m}$ here is not a measure of conspecifics being close but of other cells with animal locations being close.

Finally, K:

$$K(d) = \lambda^{-1},$$

where $\lambda = n/A$, A = the area within an animal's familiarity distribution that is within distance d of a given location for the animal, and n = the total number of locations. $K(d)$ is the mean number of locations within distance d across all locations. For animal location data, one must adjust for locations near the edge of an animal's familiarity distribution. Thus,

$$\widehat{K}(d) = \frac{A\Sigma p_i f(d_{ij})}{n^2},$$

where p_i is the proportion of a circle of radius d around location i that is within the boundaries of the animal's familiarity distribution, and $f(d_{ij})$ is a function that has the value 1 if the distance between locations i and j is $\leq d$ and otherwise 0. The equation for L is

$$\widehat{L}(d) = (\widehat{K}(d)/\pi)^{1/2} - d.$$

Significantly positive values document clustered locations at distance d, while negative values document an even distribution.

Cores—*Log-survival*—Erran Seaman and I (1990, Powell et al. 1997, Powell 2000) developed an objective, intuitive technique that allows animals themselves to define their home range cores when location data are clumped. Our technique is based on log-survival functions, which have been used to identify the start and stop of behavior bouts (Fagen and Young 1978, Slater and Lester 1982) and identifies clearly the areas used most intensively within an animal's familiarity distribution (Seaman and Powell 1990, Powell 2000). Resource patches vary in size and quality, leading to cores that differ appropriately for different animals. Bingham and Noon (1997) and Vander Wal and Rodgers (2012) used the same method, and Harris et al. (1990) may have but their explanation is not clear.

To identify the core of an animal's familiarity distribution using the log-survival approach, start with an animal's grid-based, volume familiarity distribution. If time spent in a location is an index of importance, then the familiarity distribution can be used directly. If time spent in a location is *not* an index of importance, then cells of the familiarity distribution must be weighted by importance; importance must be quantified from information independent from location data. For each cell, identify its volume familiarity value. So, for example, if a cell lies within the 95% isopleth (within the 95% familiarity distribution) but outside the 94% isopleth, then that cell's isopleth is 95%, or 0.95. Also, for each cell, calculate its relative probability of use, i.e. its probability density divided by the maximum probability of use for cells within the familiarity distribution. Thus, the cell with the highest probability of use has the value 1.00 for its relative probability of use and the cell with the least non-zero probability of use has a value of nearly 0. Now plot the values for each cell on an x–y graph with proportion of highest probability on the x-axis and cell isopleth value on the y-axis (Figure 12.1). Both the x- and y-axes go from 0 to 1.

Were an animal to use the environment within its familiarity distribution randomly, all points would cluster around the straight line going from $(x,y) = (0,1)$ to $(1, 0)$ with slope $= -1$ (Figure 12.1 A). If, however, use of the environment by an animal is clumped, then the points will sag below the line of random use; the more the points sag the more clumped are the data (Figure 12.1 B). In contrast, if use of the environment is extremely even (almost all areas used with equal intensity), the points will remain on a plateau with a high, stable y-value from x equals 0 up to x equal almost 1 and then plummet (assuming that a cell or cells with maximum probabilities exist; Figure 12.1 C).

For clumped data, the graph will slump below the line for random use somewhere, indicating the part of the animal's familiarity distribution with

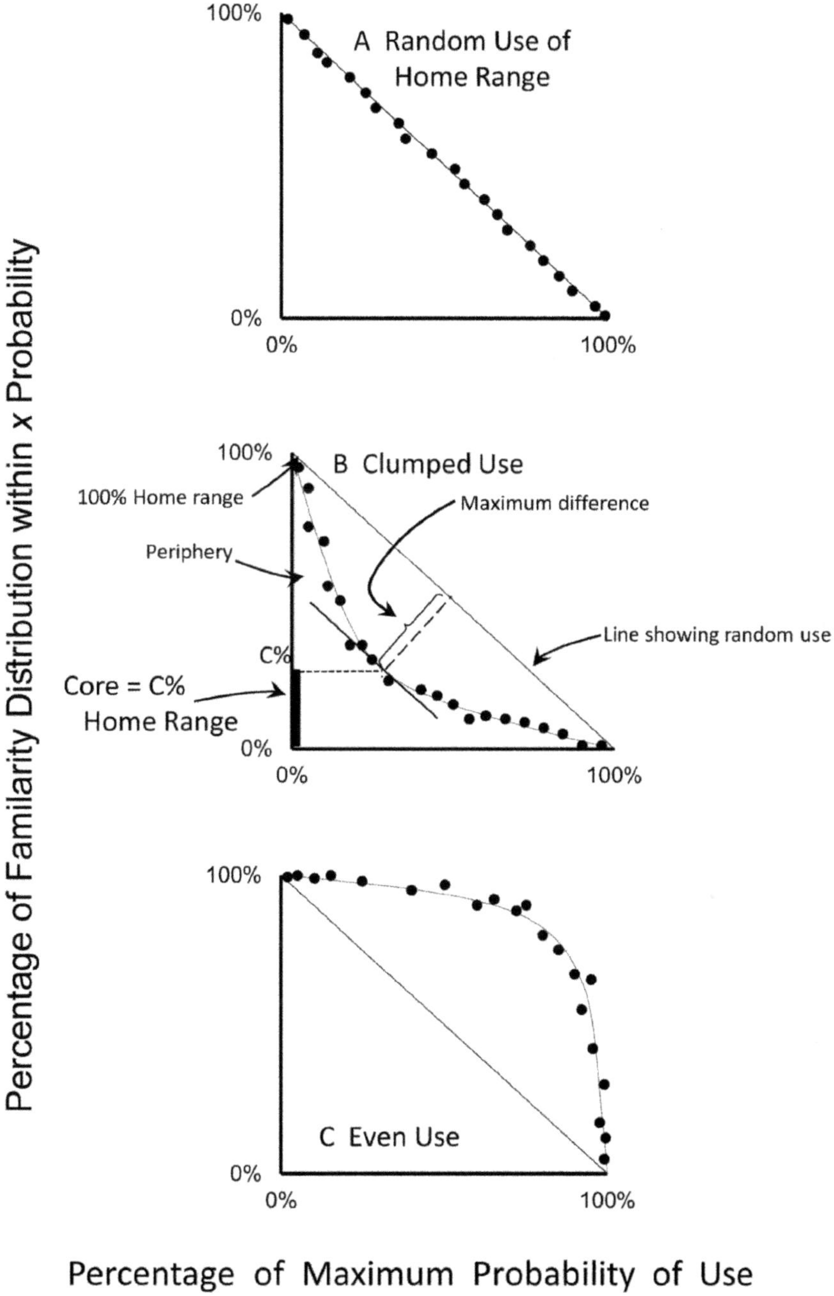

Figure 12.1 For an animal's familiarity distribution, whether the animal's use of its environment is random, clumped, or even affects the relationship between probability of use and the percentage of the home range with that percentage of use or greater. The *x*-axis is the probability of use for areas within an animal's familiarity distribution calculated as the percentage of the maximum probability of use for any cell. The *y*-axis is the percentage familiarity distribution (the 0% familiarity distribution going up to the 100%). When an animal uses its home range randomly, location estimates fall along a line with slope −1. When an animal uses its home range in a clumped fashion, the curve of use dips below the line for random use. The point where the line is furthest from the line of random use designates the core. A. Relationship for random use by an animal of the area within its familiarity distribution. B. For clumped or patchy use of the familiarity distribution. C. For even or over-dispersed use of the familiarity distribution.

clumped locations. Most graphs slump over most of their lengths (Figure 12.1 B), showing extensively clumped use of the environment. Going from low probability of use to high, the slumping curve first moves farther and farther below the random use line, indicating more and more clumping as probability of use climbs. At some place, the curve starts moving back closer to the line of random use, suggesting that use of the environment begins to be less clumped at high probabilities of use. The point where the curve is farthest from the line of random use identifies the threshold for the home range core (Figure 12.1 B; Seaman and Powell 1990 after Fagen and Young 1978; Slater and Lester 1982), this point identifies the isopleth ϕ, the threshold isopleth for the core. This point is also where the tangent of the animal's curve is parallel with the line of random use and slope $= -1$. Those cells of the familiarity distribution mapped onto the steeply descending slope along the y-axis are used least and constitute the periphery of the home range. Those cells of the familiarity distribution mapped onto the gently descending slope along the x-axis are used most and constitute the core of the home range. Plots of actual data that do not yield smooth curves, usually because of small sample sizes, can be fit with smoothed curves.

Figure 12.2 shows log-survival curves for female fisher 20940 in the northern Sierra Nevada, California, in 2015, and female feral pig Calico Kate on Fort Benning, Georgia, in 2006 (data from Sparklin et al. 2009, Ditchkoff et al. 2012, Facka et al. 2016, Powell et al. 2025). Note that Calico Kate's curve starts above the line for random use but then falls below that line for areas used at high probabilities. This pattern shows that Calico Kate used the periphery of her home range more evenly than random while using the core of her home range in a more clumped manner than random. Curves for some individuals rise above the random use line for areas used at high probabilities (high x values). Such curves show that these individuals use their central cores evenly.

Likelihood—Wilson et al. (2010) developed a more statistically heavy approach to estimate core areas. Their model allows a biologist to find the isopleth ϕ that separates maximally clumped location estimates from less clumped and, thereby, to define ϕ as the isopleth that estimates the core. Although Wilson et al. specified using familiarity distributions derived from kernel-derived estimators, the source of the familiarity distribution should not affect the outcome.

To use this method, first, superimpose a grid of m cells over the location estimates and the familiarity distribution. m, the number of cells with familiarity distribution values > 0, should be large; an m that is 5–20 times the number of locations works well. Wilson et al. adopted a multinomial framework to model grid cell counts:

$$y_C \sim \text{MN}\left(y_c \mid n_c, p_c, \phi\right),$$
$$y_{C'} \sim \text{MN}\left(y_{c'} \mid n_{c'}, p_{c'}, \phi\right),$$

where y_c is an $\text{m}_c \times 1$ vector of the cell counts in the core, n_c is the number of location estimates in the core, and $\boldsymbol{p}_c \equiv (1/m_c, \ldots 1/m_c)$ are the grid cell probabilities for the core. The second line above is for peripheral areas. Hooten et al. (2017) note that the core area intensity is $a_c n_c / m_c$, where a_c is the area of the core. Intensity of the periphery is $a_c, n_{c'} / m_{c'}$. The total likelihood for the model is

$$[y|\phi] \equiv \text{MN}\left(y_c|n_c, \boldsymbol{p}_c, \phi\right) \times \text{MN}\left(y_{c'}|n_{c'}, \boldsymbol{p}_{c'}, \phi\right),$$

where y is the $m \times 1$ vector of cell counts. A biologist can use maximum likelihood approaches to find the value of ϕ that maximizes the above equation, or use Bayesian approaches. One can easily program the product of the intensities (core and periphery) to find the value of ϕ that maximizes the product.

Comparisons—Both the log-survival approach and the maximum likelihood approach of Wilson et al. can be used within a resulting core to identify a core within the core. The log-survival method shows where cores sometimes have evenly distributed use across the most intensively used areas.

Figure 12.3 shows the log-survival and the maximum likelihood cores for female fisher 20940 and feral pig Calico Kate, whose log-survival curves are shown in Figure 12.2. For both animals, the log-survival core landed within the maximum likelihood core.

I calculated the cores for home ranges of 10 simulated animals whose locations were updated every 14 minutes (100 times per day) for a year. Each animal was able to perceive resources up to 75 distance units away and resource patches were of different

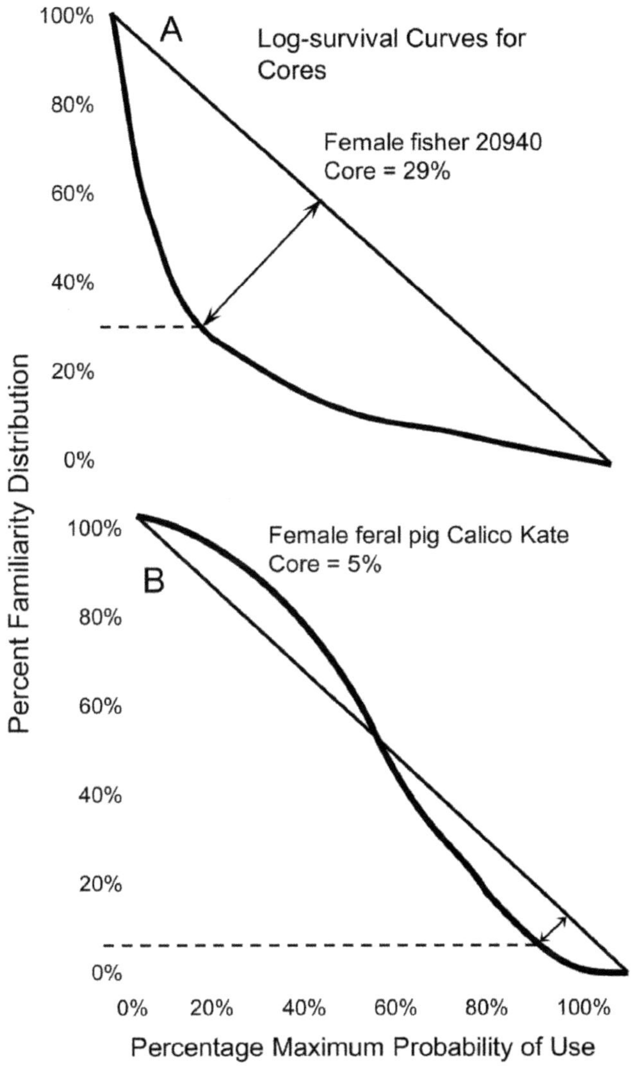

Figure 12.2 Log-survival curves for A. female fisher 20940 in the northern Sierra Nevada, California, in 2015, and B. female feral pig Calico Kate on Fort Benning, Georgia, in 2006. The double arrow lines show where the log survival curves are maximally below the straight line that represents random use. The dashed lines show where those maxima meet the *y*-axis, representing the percent familiarity distribution that identifies the core.

sizes, had clumped distributions, and the abundance of each resource was seasonal with different timing. A forager travelled toward and then foraged at the resource patch with the highest yield discounted by the distance from the forager's last location. The mean log-survival isopleth, ϕ, was the 25% familiarity distribution (sd = 10), while the maximum likelihood method produced a mean isopleth ϕ of 44% (sd = 10). These values for ϕ differed (paired *t*-test, $t = 6.3$, $p < 0.0001$). For each animal, the log-survival ϕ was smaller than the maximum likelihood ϕ.

Data for 96 black bears, studied by my research team in the Blue Ridge Mountains of North Carolina, USA (Powell et al. 1997, 2002; 54 females, 44 males over 19 years for 167 bear-years of data) produced mean log-survival cores with ϕ at the 21% familiarity distribution (sd = 8), while the maximum likelihood cores averaged the 45% familiarity distribution (sd = 7), and these differed

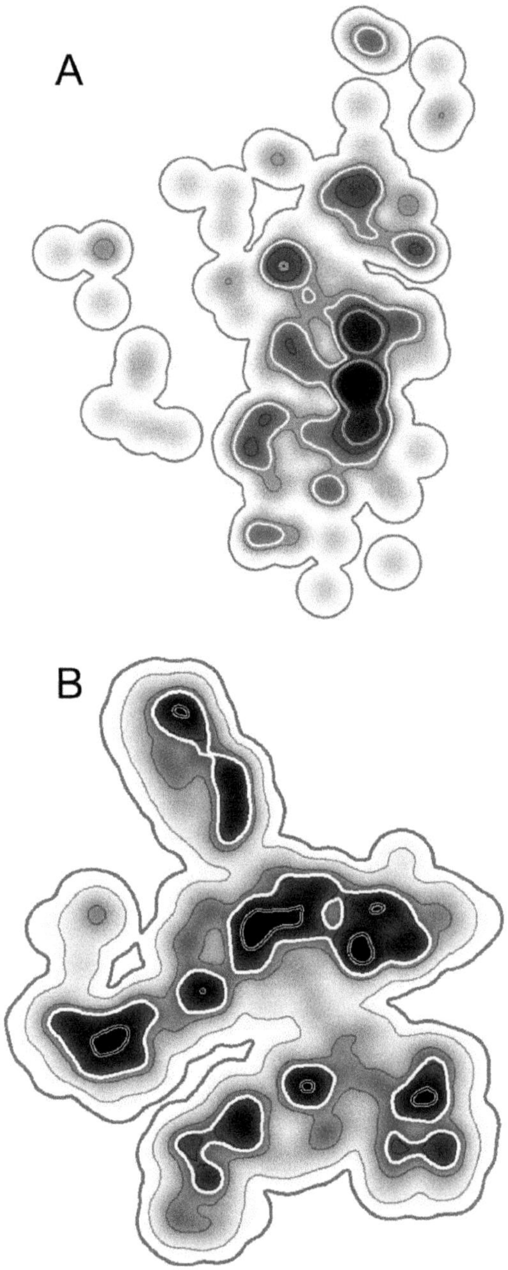

Figure 12.3 Examples of both log-survival and cluster cores. Maximum likelihood cluster cores are shown in wide white curves; log-survival cores in wide grey lines inside the cluster core curves. Outside outline is the 100% fixed kernel familiarity distribution. Shading light to dark shows intensity of use from low to high. Thin grey lines show the 35% and 65% contours. A. Cores for female fisher 20940 in 2015 in northern Sierra Nevada, California. Kernel = K_2, h = 400m. B. Cores for female feral pig Calico Kate in 2006, Fort Benning, Georgia. Kernel = K_2, h = 250m.

(t-test, $t = 27.1$, $p < 0.0001$). Averages for males (log-survival 21.0 ± 7, maximum likelihood 47.8 ± 9) were larger than those for females (log-survival 20.6 ± 8, maximum likelihood 43.5 ± 5; blocked by method, glm, $F = 4.3$, $p = 0.04$; effect size suggests that these sex differences truly are statistically significant but I doubt their biological significance). Blocked by sex, the average isopleths still differed between the 2 methods (glm, $F = 4.67$, $p = 0.03$). Of the 167 bear-years, 26 home ranges (12 female, 14 male) did not exhibit significant clumping and, therefore, did not have home range cores.

That the log-survival and maximum likelihood methods produce different results is not surprising because they define home range cores differently. The maximum likelihood method finds the optimum boundary between location estimates that are slightly clumped and those that are highly clumped. The log-survival method identifies a change in behavior from areas with increasing clumping to areas with decreasing clumping, the latter of which are the most heavily used areas within a familiarity distribution. Given that home ranges and home range cores are behavioral concepts, I prefer a behavioral definition for home range cores; many biologists may prefer a statistical approach. One can use either method to define peripheries of familiarity distributions as well. This maximum likelihood method often identifies the dividing line for cores and peripheries as an isopleth in the 40% to 50% range. Thus, by this definition, peripheries constitute a little over half of an animal's home range. By the log-survival method, peripheries constitute an even larger part of a familiarity distribution. Both methods can be used to divide the "periphery" into 2 or more non-overlapping sets, just as they can be used to divide the entire home range into non-overlapping sets. Biologists must report what method they use.

12.3.1 Disproportional ≠ Preferential

The analyses presented above assume that an animal's ability to use each patch in her familiarity distribution is independent of her ability to use other patches. If the animal is more likely to move through the central part of her familiarity distribution to reach more widely spaced patches, then the animal will spend more in the center patches than might be her preference when more preferential patches exist. This situation can lead to disproportional time in the center and a center bias. Don Hayne (1949) noted this problem and, to date, it has not been addressed.

12.4 Customizing Familiarity Distributions

This section and the following (section 12.5) present unconventional approaches for building familiarity distributions that have the potential to give biologists new hypotheses to test in the field and new insights into animals' home ranges, their cognitive maps. The approaches presented apply to kernel estimators but new approaches can be imagined to use with other estimators, including estimators not yet invented. The only limit to discovering yet more insightful approaches is curiosity, imagination, and creativity. Go for it!

12.4.1 Reassigning Familiarity Distribution Values

Traditional kernels are round (equation 10.1, equation 10.2) but this is not required for building familiarity distributions. Kernels can have any shape that makes biological or technological sense (Figure 12.4). For global positioning system telemetry data, building a familiarity distribution using a circular kernel makes sense. Nonetheless, seldom are animals' distributions of use of their environments so "standard" as to allow biologists to apply black box software packages for any familiarity distribution estimators blindly (Börger et al. 2006). When animals reuse reproductive den sites or resting sites repeatedly, for example, locations at den sites should be removed from data sets before familiarity distributions are estimated, then reinserted with a very small h right at the den sites and all probabilities transformed appropriately.

Local convex hull estimators usually exclude areas obviously not used by the animals being studied. Even so, some choices of k, a, or s can lead to inappropriate spurious holes in familiarity distributions (Chapter 11, section 11.4). Kernel-generated familiarity distributions sometimes incorporate areas obviously not used. Many terrestrial animals

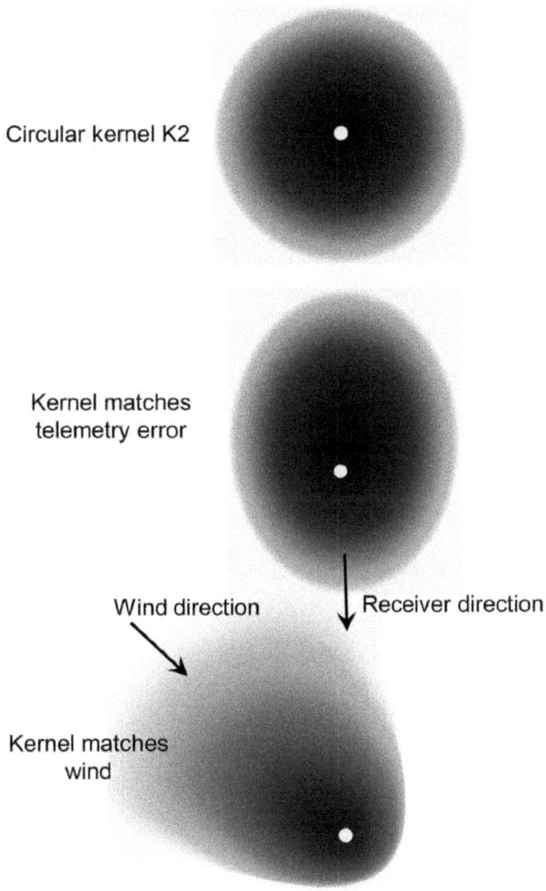

Figure 12.4 Three examples of the many shapes that kernels can have. In each example, the white dot represents the estimated location of a study animal. The top example is a simple, circular K_2 . The middle example is elliptical to match telemetry error typical of triangulating very high frequency telemetry signals. The bottom example matches the shape of scents on the wind.

avoid swimming across lakes, for example, yet parts of lakes will be included in familiarity distributions if a target animal moves along the shoreline, and especially if the animal forages close to the water. Similarly, if roads or other physical objects (cliffs, ridge tops, stark habitat edges, etc.) delineate home range boundaries, probability densities on the wrong side of the boundary can be set to 0 manually, or to an otherwise low probability, for non-habitat, and then probabilities for areas that can be used recalculated. The probability for locations in a lake, for example, can be based on how often target animals swim across lakes of different sizes or on how often target animals go fishing in shallow water.

12.4.2 Behavior-Specific Familiarity Distributions

Animals' foods are usually habitat-specific, leading animals to travel, often at intent speeds, in linear moves between different food patches (Belovsky 1978, Powell 1978, Powell et al. 1997, Berger et al. 2001, Beschta and Ripple 2009 document some of the complexities of travelling to food patches). Using their cognitive maps, animals plan travel to foraging sites. Frugivorous bats and birds often plan travel routes to incorporate more than 1 foraging site (Sarel et al. 2017, Ormond and O'Keefe 2022) and I hypothesize that most mammals and birds plan foraging moves to include more than 1 foraging site. Movement within food patches is often fairly slow and localized.

Here I generate some hypotheses related to behavior categories and use behavior-specific familiarity distributions to test them. Members of some species, such as bears, fishers, and martens, are hypothesized to rest in a safe site fairly close to the last foraging patch. This is a pattern for some ungulates, as well. Members of these species may even choose the order for visiting foraging patches so that the last patch in a bout has resting sites nearby. For essentially continuous data, one can test whether animals rest near terminal foraging sites. Even though most telemetry data have locations estimates that are separated by at least a few to several hours, accelerometer data may allow a biologist to fill in data between locations. For such telemetry data, the hypotheses to test are that the distribution of resting sites overlaps extensively with the distribution of foraging sites, while neither of those distributions overlaps as much with the distribution of travel.

Wolves, in contrast, hunt for large herbivore prey, particularly in winter, while travelling and rest in close proximity to the carcass of their most recent kill until the kill is picked clean. Wolves hunting in summer for beavers or hares, or fishing, forage in localized areas, sometimes as lie-and-wait predators (Gable et al. 2018a, b, 2020). Also in summer, wolf packs rest in rendezvous sites that are used for weeks at a time when pups are small (Mech and Boitani 2003). Thus, for wolves during summer, the hypothesis to test is that the distribution of resting sites differs from the distributions for hunting both large and small prey, which also differ. In winter, the hypothesis is that the distribution of travel overlaps only modestly with the distribution of rest sites.

Large herbivores move slowly while grazing or browsing and are likely to make linear moves between foraging patches (e.g. Seidel and Boyce 2016). To rest, they find sites where they can be vigilant. Distribution of travel, foraging, and resting may have similar, low overlaps. If ungulates rest in sites close to the last foraging site or in anticipation of the next foraging site, then overlap of foraging and resting will be high.

With these ideas and hypotheses in mind, I analyzed data for fishers, wolves and pronghorns and for black bear in 3 populations, all for whom I had access to large data sets. Using telemetry data, my research teams and colleagues have differentiated rest, travel movements, and localized foraging movements for the 6 populations (Horner and Powell 1990, Powell et al. 1997; data from McQuivey 2015, Gable et al. 2018b, 2020, Facka and Powell 2021, Barber-Meyer and Mech unpublished data). I built separate familiarity distributions made from resting, local foraging, and travelling location estimates, and I built 100% familiarity distributions from all the location data from each individual animal (Figure 12.5). For very high frequency telemetry (VHF; black bears in the Blue Ridge Mountains, North Carolina, USA, and most fishers studied in the northern Sierra Nevada, northern California,

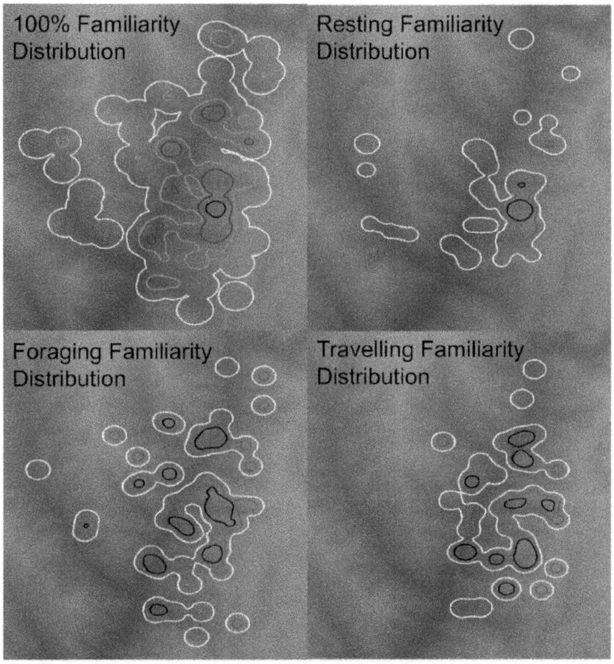

Figure 12.5 The 100%, Resting, Foraging, and Travelling familiarity distributions for female fisher 20940, California, USA, in 2015, built using a fixed kernel estimator with the K_2 kernel and h = 350m. This fisher rested most in one area where she also foraged heavily. She did not emphasize that area as heavily for travelling; she travelled between foraging areas. The background shows relative elevation, low elevations dark, higher elevations have shades of lighter grey.

USA), we used activity monitoring data, signal variability, movement speeds, and location clusters to assign locations to the 3 behaviors (Horner and Powell 1990, Powell et al. 1997, Anderson and Lindzey 2003, Chatterjee et al. 2023). For global positioning system telemetry data (black bears in the city of Asheville, North Carolina, USA, black bears in northern Minnesota, USA, wolves in northern Minnesota (each wolf in a different pack, most data in summer), pronghorns in Arizona, USA, and some fishers), we used location clusters, movement speeds, and accelerometer data. I calculated overlap as I_p.

The mammals of the 6 populations showed considerable diversity of overlap patterns for familiarity distributions, with no consistent patterns among the populations (Figure 12.6). Although most populations exhibited modest to high overlaps of most of the behavioral familiarity distributions with the 100% familiarity distributions, pronghorns and Blue Ridge black bears had considerable variations in overlaps. The Asheville and the Blue Ridge bear populations showed distinctly different patterns of overlap, despite having been adjacent to each other. The data for the 2 populations were separated by 10 years, however, and differed by one population being urban (Asheville) and the other within a national forest (Blue Ridge). Asheville bears rested and foraged in the same places but travelled between separated rest–foraging sites. Blue Ridge bears had modest to low overlap of all activities. Minnesota bears exhibited high overlap of all activities though the small differences did differ statistically.

I hypothesized (above) that for black bears the overlap of familiarity distributions for resting and foraging is high compared to overlaps of both resting and foraging with travel. This was the case for the Asheville bears and somewhat so for the black bears in northern Minnesota (Figure 12.6). In contrast, for the Blue Ridge bears the overlap of resting with foraging was nearly identical to

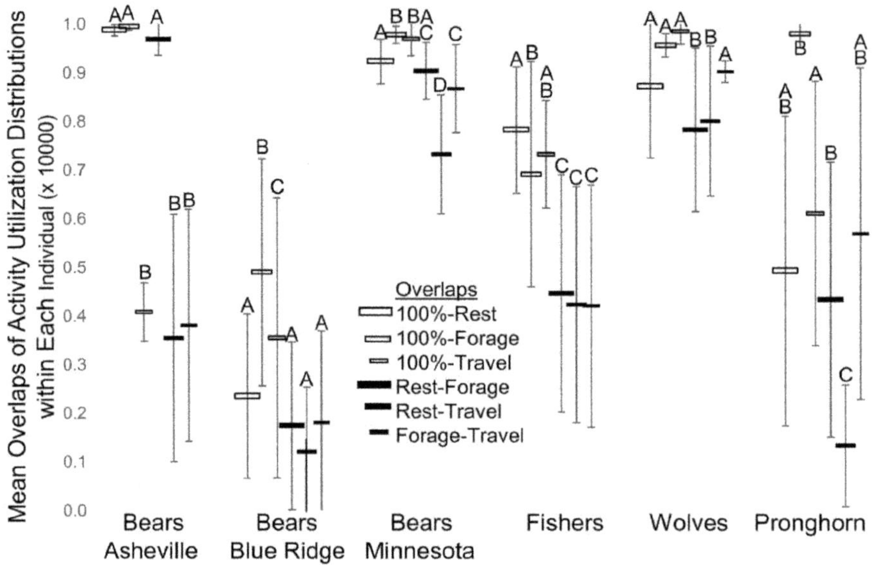

Figure 12.6 Mean (±SD) probability overlap (I_p) of descriptive behavioral familiarity distributions for 3 black bear populations, a fisher population, a wolf population, and a pronghorn population. All wolves were in different packs. Overlap categories with a common letter do not differ significantly. Overlap patterns differ considerably among the 6 populations. Even the 3 bear populations show considerable diversity: all behavioral familiarity distributions overlap for the Minnesota bears while most of the behavioral familiarity distributions for the Blue Ridge bears overlap little. In all populations but the Blue Ridge bear population, the resting familiarity distributions overlap relatively extensively with the 100% familiarity distributions compared to other overlaps. For black bears in Asheville, North Carolina, and black bears in northern Minnesota, the overlap between resting and foraging was more extensive than resting or foraging with travel. For fishers, all behavioral familiarity distributions overlap heavily with the 100% familiarity distribution and overlap significantly less along themselves. Wolf packs appear to rest, forage, and travel in the same places (data are predominantly for summer). Pronghorns show the greatest variances in overlaps; they also constituted the only population in which individual animals did not make independent decisions regarding movements.

that for both resting with travel and foraging with travel; nonetheless all these overlaps were low. This result suggests modest support for the hypothesis that bears often rest near where they forage but, ironically, does not hold for the population that suggested the hypothesis (my impressions from observations).

I hypothesized that the fisher familiarity distributions for resting and foraging overlap more than do resting or foraging with travel. I reject this hypothesis because fishers' familiarity distributions for rest, foraging, and travel all had very similar amounts of moderately low overlap. This result suggests the hypothesis that fishers readily switched among these 3 basic behavior categories, despite the patchiness of the study area or perhaps because of it.

I hypothesized that patterns of wolf travel, rest, and restricted movements differ from those of fishers and black bears, leading to a pattern of high spatial overlaps for all pairs of behaviors. Indeed, overlaps of all pairs of familiarity distributions were high and similar. This result suggests the hypothesis that wolves exhibit behaviors of all 3 categories similarly throughout their territories in summer. The wolves appear to have separated resting a bit from travelling and foraging in summer, which is consistent with them using rendezvous sites.

Lastly, I hypothesized that pronghorns would have similar and low overlaps of their familiarity distributions for all 3 behaviors. The overlaps for pronghorn behaviors exhibited the greatest variation, both within comparisons and among means for comparisons. Most overlaps were moderate, with the overlap for rest and travel being low and the overlap of the 100% familiarity distribution with that for resting being very high. Thus, I reject the hypothesis for similar and low overlaps of behavioral familiarity distributions for pronghorns.

Overlap of the 3 activity familiarity distributions with each other shows how much these activities share the same patches of an animal's environment. These analyses show that animals living under different conditions distribute behaviors differently across their environments, which is not a novel insight. Thus, planning of movements by these animals (their cognitive maps) show few similarities, which is also not surprising. I have just exposed the tip of what we can learn about how animals plan their activities.

12.4.3 Perception of Environments and Sensory Kernels

To build familiarity distributions that provide different insight into study animals' cognitive maps, one can use kernels that represent animals' sensory perceptions. Most terrestrial animals perceive their environments with at least our standard 3 senses: vision, hearing, and smell; some animals use magnetism, electroreception, and other senses that we humans have trouble imagining, let alone perceiving (Schumacker et al. 2017). Though humans have experience with the 3 standard senses, we do not perceive with those senses as do other animals (Figure 12.7; Chapter 3). Birds and insects, for example, perceive visual wavelengths that humans do not and have different visual acuities. Mice and bats hear ultrasound and elephants hear and feel infrasound at frequencies that humans do not hear but may feel. Dogs, and many other mammals, can smell odors that humans can not because our sense of smell is relatively depauperate. When weasels hunt, for example, they rely on hearing more than smell, and smell more than vision (Gillingham 1986); weasels probably use hearing more than smell because hearing has pertinent information about time. Nonetheless, humans have enough experience with the 3 senses to gain an idea of how perception might be modelled.

Estimating perception ranges includes determining signal detectability and response bias (Fletcher et al. 2013). Perception ranges vary with body sizes for mammals (Mech and Zollner 2002, Prevedello et al. 2010) and vary with physical obstructions, such as trees and tree canopies (Aben et al. 2021), as well as with physical sensory capabilities. In fact, though animals remember the layout of their home ranges, they are more likely to move to places that they can perceive, even within a complex environment with obstructions, than to places that they remember but can not perceive immediately (Aben et al. 2021). When perceiving odors on the wind, animals identify odors of interest, note their intensity and patterns of being intermittent or constant, and note directional motion (Schmuker et al. 2016, Kadakia et al. 2022). Intensity and intermittence provide information on distance to the odor source.

To gain perspective on how differences among the senses might affect animals' perceptions of their environments and, therefore, what information becomes incorporated or updated in their cognitive maps, I modified a kernel estimator to develop familiarity distributions of black bears in northern Minnesota (data from Powell et al. 2022) and feral pigs in Georgia (data from Sparklin et al. 2009) by modifying the kernels used for each population to match the shapes of perception under different weather and other ecological conditions (Figure 12.8). I compared the familiarity distributions to fixed-kernel familiarity distributions. Vision is affected by light conditions, including moonlight, by precipitation, and by obstructions; hearing and

Figure 12.7 Different animals have different perspectives on the world. A. A weasel perceives the world through smell and hearing more than through vision. Yet, the visual perspective of a weasel differs from that of a wolf or a human. B. A wolf's visual perspective differs tremendously from a weasel's. C. A person's visual perspective is more like that of a wolf than a weasel. Nonetheless, because a wolf sees the world from a lower perspective, the view still differs significantly from that of a human. (Photos by the author.)

smell are affected by wind and by precipitation, especially heavy rain (Garstang 2004, Fairbanks and Dobson 2007, Hollen et al. 2011, Rockhill et al. 2013, Hall 2016, Clermont et al. 2017, Smielak et al. 2023). Hearing and smell can be improved by light winds but decrease when winds are strong. I varied h by time of day and rain for vision kernels. I developed a kernel that is distorted toward the direction of the wind with the distortion relative to wind speed (Figure 12.4). I then downloaded weather conditions from the nearest airport to each study site. I chose to use data from these 2 study populations to avoid effects of mountainous topography on wind.

Because the kernel used to make the familiarity distribution for vision was round, the 100% vision familiarity distributions resembled the fixed-kernel familiarity distributions; the differences were caused by variations with light and weather on the h used for the vision familiarity distribution (Figure 12.8). Because individuals of the 2 species are most likely to smell important odors from distances further than they can see, and distances larger than the h usually used for building standard, fixed-kernel familiarity distributions, the olfaction familiarity distributions have the least detail but the widest geographical range. I chose h for hearing to be intermediate between the h for vision and that for smell, and varied it with precipitation, leading the hearing familiarity distribution to have more detail than that for smell but less than that for vision. Combining the familiarity distributions for the 3 senses yielded familiarity distributions that resemble the fixed-kernel familiarity distributions but extend beyond the fixed kernel boundaries because of the abilities of most mammals to hear and to smell odors far away (Figure 12.8). The combined familiarity distributions sometimes had more (Figure 12.8 B) and sometimes less (Figure 12.8 A) detail than the 100% familiarity distribution due to details of the data sets for the individual animals. Sensory familiarity distributions for birds, with vision or hearing superior to that of mammals but, perhaps, inferior smell, would look quite different than those for most mammals (Wikelski et al. 2021). A biologist could build visual familiarity distributions for foraging and for flying distances.

Comparing the familiarity distributions for senses combined vs that for a standard fixed kernel provides insight for understanding animals' cognitive maps. Animals are able to update their cognitive maps for sites as far away as shown by their sensory familiarity distributions. With improved information and data on sensory abilities (data mostly not yet available), sensory familiarity distributions and combined familiarity distributions will contribute even better understanding of animals' cognitive maps.

12.5 Using Different Currencies

Another approach for gaining insights into animal's cognitive maps (and how they live their lives) is to build familiarity distributions using biological

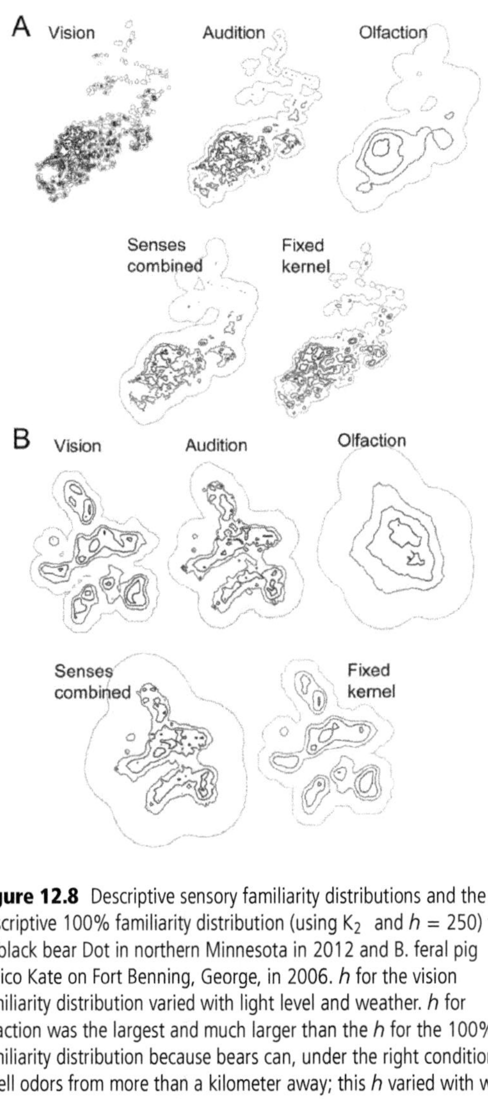

Figure 12.8 Descriptive sensory familiarity distributions and the descriptive 100% familiarity distribution (using K_2 and $h = 250$) for A. black bear Dot in northern Minnesota in 2012 and B. feral pig Calico Kate on Fort Benning, George, in 2006. h for the vision familiarity distribution varied with light level and weather. h for olfaction was the largest and much larger than the h for the 100% familiarity distribution because bears can, under the right conditions, smell odors from more than a kilometer away; this h varied with wind direction and speed and with weather. h for audition was intermediate between h for vision and olfaction and was affected by wind and weather. Low to intermediate wind speeds increased h while high wind speed decreased h; rain decreased h. The vision familiarity distribution shows the greatest detail while that for olfaction shows the least detail but the greatest range. The familiarity distributions made by combining those for the separate senses (giving each equal weight) are similar to but larger than the 100% familiarity distribution simply using locations; Dot's combined familiarity distribution shows somewhat less detail than does the 100% familiarity distribution while Calico Kate's combined familiarity distribution shows somewhat more detail.

factors to weight locations (Powell 2000, 2004, Powell and Mitchell 2012). For example, build a familiarity distribution that weights the kernel for each location by rate of energy expenditure by the animal when the location was recorded.

Ultimately, one would like to build a distribution of contributions to fitness (Powell 2004). Using different currencies provides different ways to address potential importance to an animal of its environment and has the potential to identify critically limiting resources that are seldom used.

12.5.1 My Home Range

Home ranges differ among animals of different species, among individuals within a species, and even within individuals over time. Nonetheless, all animals use their home ranges to provide food, to avoid predators, and to reach other resources. In this respect, humans are no different from other mammals: we go places to gain resources (e.g. to earn money and to buy groceries) and we avoid dangerous places (e.g. walking on a busy highway at rush hour or in a dark alley). We even appear to make many of the same irrational choices that many other animals make, such as misjudging the importance of conditions far in the future because we use a discount rate that is set for too short a time (about 5 years for humans; Real 1996). Consequently, home ranges of people have the potential to provide critical insight into aspects of home ranges of animals because humans can communicate about their home ranges. In addition, humans can gain near complete information about people (i.e. better data).

To understand how different currencies provide insights into animals' home ranges, I quantified my home range using 5 different currencies (Powell and Mitchell 2012). This exercise highlighted several conceptual challenges that have led, in part, to my writing this book. By attempting to quantify my home range, Mike Mitchell and I (Powell and Mitchell 2012) realized that familiarity distributions ***unequivocally*** failed to quantify what my home range was to me. The exercise was a conceptual beginning for re-evaluating what the home ranges of animals really are and how they might best be understood and studied.

During sabbatical at the University of Wyoming in 1990–1991, I recorded my location continuously during 65 random days. I recorded the beginning and ending times for all activities and the exact routes travelled (always within 10 m, most often within 3 m, and sometimes 1 m), the mode of travel, and the irregularities in travel speed, if any. I recorded what I ate, where I ate it, how much money I spent, and for what. For example, on a typical weekday, I noted when I awoke, when and what I ate for breakfast, when I left my house for the university, how I travelled (walk, bike, or drive) and the exact route to the university (including walking on a sidewalk vs center of a street vs through the prairie dog town in a vacant lot), arrival time at the university, movements through the biology building, cups of coffee drunk, lunchtime, and so forth. When I went into the field (Medicine Bow Mountains, approximately 50 km west of town), I recorded times and exact travel routes, return times, and so forth.

My data were continuous and event driven, as are most direct, observational data, because animals' lives are continuous and event driven. In contrast, most data used to build familiarity distributions are discrete and time driven. The only conventional home range estimator that can deal with continuous, event-driven data is the grid cell estimator. Despite their completeness and desirability, continuous data can not be used by other estimators. Therefore, to illustrate my home range, Mike and I used the grid cell estimator and also forced my data into discrete locations and used a kernel estimator to build a familiarity distribution, thereby losing (important) detail. First, Mike and I superimposed a 10 × 10 m grid over a map of the Laramie area and calculated the duration of time that I spent sequentially in each cell, calculated money spent in each cell, calculated energy gained in each cell (Davis 1970), and calculated energy expended (Kleiber 1975, Powell 1979b, Taylor et al. 1970) in each cell (details presented by Powell and Mitchell 2012). We would have lost less detail had we used smaller grid cells, say 3 m, but would have dealt with many more cells. We then converted total use of my Wyoming environment in each currency to a grid cell familiarity distribution that summed to 1.0.

Second, we calculated a 95% fixed-kernel familiarity distribution for each currency, using K_2 and setting $h = 250$ m to smooth over the minute details of my use of my Wyoming environment and to make my familiarity distributions appear typical of those calculated for wild mammals.

Figure 12.9 A shows my typical 95% fixed kernel familiarity distribution, excluding travel outside of the city of Laramie. The prominent features of my estimated familiarity distribution are my house, the biology building at the university, my walking routes to the university, the grocery store and other businesses downtown, and the house of Steve Buskirk, a friend. Figure 12.9 B shows the 100% grid cell familiarity distribution for where I actually spent time, including my diverse travel routes from my house to the biology building at the university.

Comparing Figures 12.9 A and B highlights several points.

First, because my data are complete for all 65 days Mike and I did not need to estimate travel routes between locations because we knew them (no bridges needed). For those 65 days, my use of the Laramie environs is precisely shown by Figure 12.9B.

Second, a conventional familiarity distribution (the kind most biologists appear to want; Laver and Kelly 2008, Kie et al. 2010) grossly mis-estimated my use of the Laramie environs. The 95% fixed-kernel familiarity distribution includes vast areas where I never went and that I did not consider to be part of my home range (for example, other people's houses and yards), but did not include areas where I did go and that I did consider to be part of my home range (for example, all of the route to Steve's house, all of which would have been included in a 100% kernel familiarity distribution and that was included in my 100% grid cell familiarity distribution).

Third, my 65 days are only a sample of my movements during my entire sabbatical, affecting the ways one can use the data and the resulting familiarity distributions. My 65 days were representative of my behaviors and movements and, thus, provide solid insight into my general movements and other behavior patterns. Consequently, 65-day samples may be adequate to test population level hypotheses about where other Laramie citizens spent time. But, for population questions, Mike and I needed samples for many other people in the population (the population of Laramie or the population of

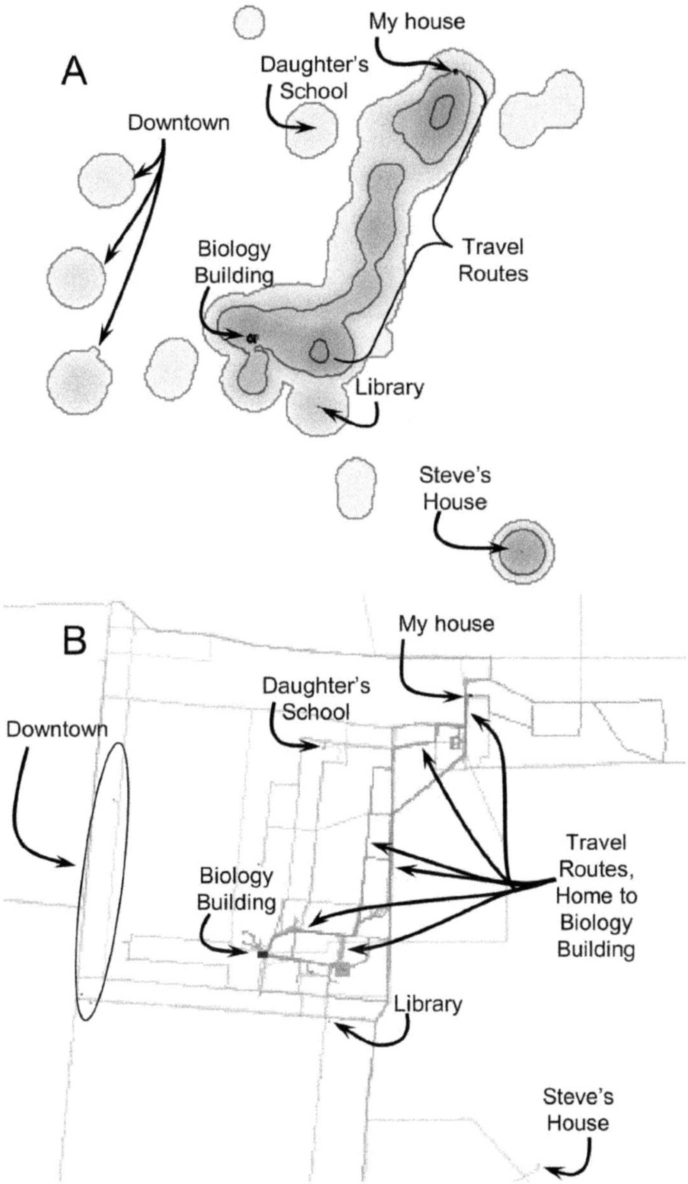

Figure 12.9 A. My 95% fixed kernel familiarity distribution, showing major places where I was estimated to have spent time. Shading shows relative amount of time estimated to be spent in different places. B. My 100% grid cell familiarity distribution, showing the exact places where I spent time. Shading shows relative amount of time actually spent in different places; the oval highlights my visits to downtown Laramie. (Modified from Powell and Mitchell 2012.)

university faculty, for example), data that we did not have.

Places not visited by me in those 65 days fell into 2 categories: 1) places that I visited on other days because those places had resources that I needed on those particular other days; and 2) places that I avoided. During days not sampled, I visited places that I considered to be part of my home range (such as other friends' houses and restaurants downtown). In contrast, I actively avoided houses of unknown people and businesses of no use to me and I did not consider them to be part of my home range. Neither Figure 12.9 A nor B is of any value for differentiating between these 2 categories of non-use.

Obviously, no home range estimator can make up for a small sample size. To understand my complete use of the Laramie environs, Mike and I needed data for all days of my sabbatical. Building a familiarity distribution can hide the nasty fact that a sample size sufficient to answer some questions may not be adequate to answer other questions (also noted by Fieberg and Börger 2012).

Fourth, I noted distinct boundaries to my home range where private property restricted my movements but I did not consider my home range to have distinct boundaries where my familiarity with the landscape was limited by numbers of visits or by knowledge of environments. In the Medicine Bow Mountains, for example, my familiarity with different places varied in a continuous manner with numbers of visits, time spent in those places, and my knowledge gained from maps, and then how I added potential but unexplored routes between places to my cognitive map.

Along those same lines, when Consie, my wife, and I and our dogs went to the Medicine Bow Mountains to cut firewood, I knew from previous visits to the Medicine Bow National Forest where to find good, dry, dead and down lodgepole pines. We planned firewood runs based on weather and on snow conditions being appropriate for our Newfoundland dogs to haul logs from where I cut them to our truck. The firewood sites, knowledge of weather and snow, and understanding what needed to be done to cut and haul firewood were all part of our group home range, part of our cognitive maps, and used by us to plan and then to make firewood runs (Consie's and my cognitive maps and those of the 3 dogs). Consie's and my plans for using our group home range (our shared aspects of our cognitive maps) was exactly parallel to how fruit bats use their home ranges to plan foraging trips to fruit trees, how black bears use their home ranges to plan foraging trips to berry patches in summer and to oak stands in autumn, and how white-tailed deer plan foraging trips to lakes near our present house to forage on aquatic plants to meet their sodium requirements.

Figure 12.10 shows my use of the Laramie environs indexed using the 5 different currencies. No 2 panels in Figure 12.10 are exactly the same and each provides a different picture of what was important to me during my sabbatical year. All of the currencies show parts of my home range and I used that information in my cognitive map to plan travel and to make decisions about when to visit different places. Energy and money could be gained and money spent only in restricted places, while I expended energy and spent time everywhere I went. That energy gained and money spent were spatially restricted might suggest that sites for these activities were limiting. That was true only in part. Although Laramie had more than the 1 grocery store that my family used, the other stores did not have the foods my family wanted at competitive prices. I learned about the other stores on days not sampled or from Consie. Nonetheless, I did not frequent all places where I could have gained energy or spent money. I was selective.

I did not include the groceries I bought at stores as energy gained. Instead, I recorded energy gained from those groceries only as energy from meals at home. In this way, my behavior mimicked a North American red squirrel squirrelling away conifer cones and dried mushrooms in his food cache. Should I have counted, say, half the energy gained at the grocery store and then the other half when I actually ate the purchased food, counted the energy gained when eating, and then subtracted the waste from the previously recorded energy gained recorded for the grocery store? Other approaches should be considered.

I earned money in part everywhere that I was involved in money earning activities, which included university-related activities, which occurred across Laramie and into the surrounding mountains, and freelance activities, which occurred mostly at my house.

Figures 12.9 and 12.10 provide, consequentially, different perspectives on how different aspects of my life contributed to how I used the Laramie environment and how that use contributed (indirectly) to my fitness. Even more consequential is

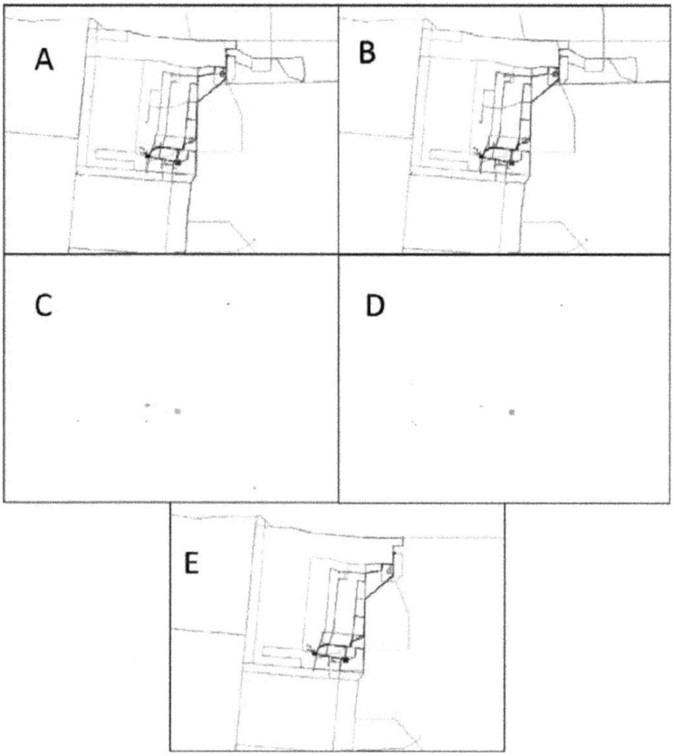

Figure 12.10 My descriptive grid cell familiarity distribution for life in Laramie, Wyoming, presented in different currencies. A. Time spent in different places. B. Energy expended. C. Energy gained. D. Money spent. E. Money earned. Darkness of line represents the proportion of each currency I "used" in each place. Note that I gained energy and spent money in few places and that both were highly concentrated at the small grey rectangles and tiny dot in panels C and D. (Modified from Powell and Mitchell 2012.)

how the maps provide readers with insight into my cognitive map. No single figure provides complete insight into my cognitive map nor do all of them together. For me, my cognitive map was not a Euclidean map as shown in these 2 figures but, instead, an amalgam of the views and smells and sounds and concepts of the costs and contributions related to all the places that I remembered, including my memories of how to travel among the many places. My home range also included my knowledge of how to make my most efficient use of time in the grocery store, of how to minimize gas used to drive to the grocery store if I did or did not also have to drive to the hardware store, and of how much time would be spent at a restaurant and how I could use that information to decide where to go out to eat, and so forth. All that information was part of my home range but not included in the maps in Figures 12.9 and 12.10.

Figures 12.9 and 12.10 may not be able to provide readers with complete insight into my cognitive map *but* readers can use the 2 figures to inquire into their own cognitive maps and how their home ranges include places where food was eaten, money spent, energy expended, and how they plan their use of those places. Understanding their own cognitive maps extends insight into the home ranges of study animals. I encourage readers to consider what currencies they might use for their own home ranges.

I still remember my walking routes to the university, though some of them undoubtedly no longer exist. I have not updated my cognitive map of Laramie for decades.

12.5.2 Fear and Other Currencies

I explored characteristics of my own home range by making familiarity distributions using currencies that made sense for a human. Except for money earned and spent, those currencies also make sense for other animals but many other potential currencies also make sense for other animals, and even humans. Biologists who live at least part time in inner cities, as I did as a graduate student in Chicago, could use the landscape of fear in addition to knowledge of locations for resources and knowledge of potential and used travel routes.

Laundré and Brown introduced the concept of a "landscape of fear" (Laundré et al. 2001, Brown et al. 1999, Brown and Kotler 2004), which can be used as a currency. Laundré et al. used the per cent of time that elk and bison were vigilant as a measure of fear. Although vigilance can be measured, it can not be defined for habitats or for parts of a landscape where target animals do not go. In addition, many small mammals and birds appear to be vigilant at all times, making different degrees of caution impossible to differentiate without much more detailed understanding of those animals' behaviors. Brown and colleagues have used "giving up densities," or the limiting harvest rate at which a forager leaves a resource patch (measured as resource density), as a measure of a forager's level of fear in different habitats (Brown 1999, Brown et al. 1999, Brown and Kotler 2004 and subsequent). This measure is repeatable using experimental, or otherwise measurable, food patches established in different habitats. As with the work of Laundré et al., no information is gained about habitats not sampled or where animals do not go. Given that resources are generally considered to be positive (all the resource currencies that I used for my familiarity distributions were positive), using some inverse of a measure of fear, a "landscape of safety," would allow a biologist to compare a safety familiarity distribution to an energy-gained familiarity distribution, or a rate of energy expended familiarity distribution.

Density of reproductive sites, densities of available foods, densities of conspecifics, and many other currencies have the potential to provide insights into the home ranges of study animals. The better that biologists understand the cognitive maps of their study animals, the more currencies will become obvious.

12.6 Beyond Home Range Area

I have not hesitated to disparage areas of familiarity distributions as meaningful measures, even emphasizing that over 7 decades ago Don Hayne (1949) noted that to study home ranges one must evaluate what happens inside the boundaries. Thus, to quantify areas of familiarity distributions adequately, one must quantify: 1) the area outlined by a familiarity distribution, A_{ud}; and 2) the distribution of importance across that area. If an animal is not subject to the waterhole phenomenon, then the amount of time or energy that it spends in a habitat patch may index adequately the importance of that patch. If this is the case, then an index of the patchiness of locations can provide insight into the distribution of importance of patches. Such an index can be built using Moran's I or $\overset{*}{m}'$ (section 12.2 of this chapter).

One additional calculation can help with an area index: the minimum number of cells needed to contain the number of locations recorded. For a data set with L locations and a maximum M locations in any 1 cell, if all cells were to contain M locations, then the minimum number of cells needed to hold L locations is L/M cells, or A_{min}. The familiarity distribution for this data set will have A_{fd} cells. The range from A_{min} to A_{fd} spans the possible numbers of cells (areas) for the data set. Moran's I or $\overset{*}{m}'$ can index the continuum between these extremes; let $V = (1 - I)/2$ or $V = \left(1 - \overset{*}{m}'\right)/2$. V varies from 0, for a tightly clumped distribution of resource patches, to 1, for an even distribution. Therefore:

$$A_{effective} = A_{min} + V \cdot \left(A_{fd} - A_{min}\right).$$

$A_{effective}$ is the effective area the animal is actually using; that is, $A_{effective}$ is how much area the animal would need were it able to pack all of its resources into the smallest area possible. An animal with a small $A_{effective}$ uses its environment in a highly clumped manner. The larger $A_{effective}$, the more evenly the animal uses its environment to obtain its required resources. Thus, $A_{effective}$ is an index of an animal's use of the environment within her familiarity distribution. If the animal uses the entirety of her familiarity distribution evenly (I or $\overset{*}{m}' = -1$, $V = 1$), then she uses all places within her familiarity distribution with equal intensity and really does use, and probably needs, the entire area: $A_{effective} = A_{fd}$. Any use of the familiarity distribution that is less than even means that the animal actually

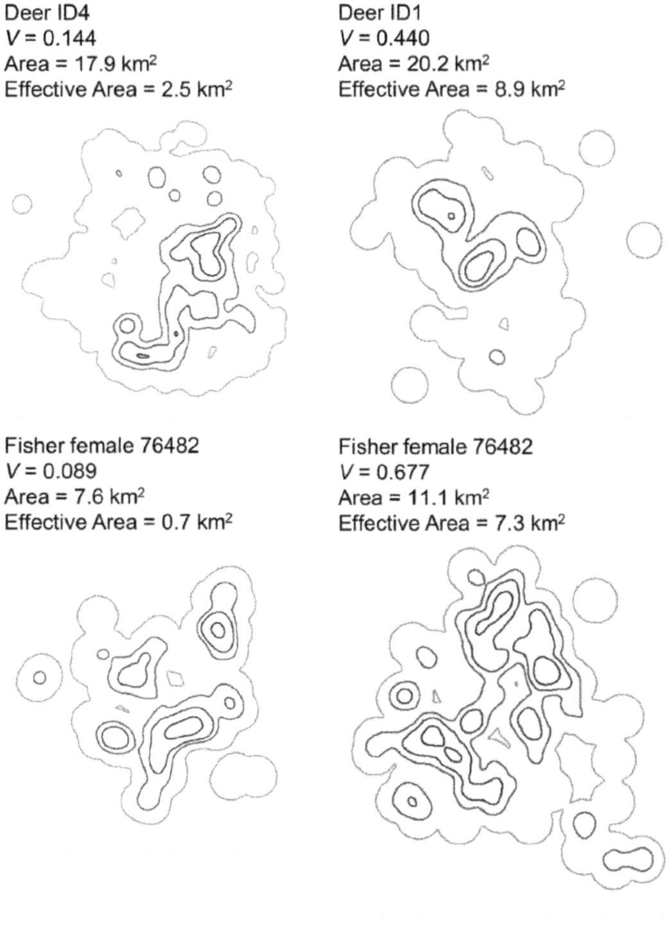

Figure 12.11 Deer ID4 and Fisher 76482 had small values for V, used the areas within their familiarity distributions in highly clumped manners, and had low values for $A_{effective}$. Deer ID1 used his environment in a nearly random fashion, having V slightly less than 0.5. Female fisher 76,482 used her environment in a manner somewhat more even than random, having $V = 0.667$. (Data for mule deer from Sawyer et al. 2019.)

uses, and probably needs, less than the entire area, but must travel between important habitat patches, which are not all of the same value. This is the effect that Mike Mitchell and I modelled (Mitchell and Powell 2008) and that he and I and Aaron Facka have demonstrated for fishers in northern California.

Figure 12.11 shows 4 familiarity distributions: 2 for mule deer and 2 for fishers. Deer ID4 and Fisher 76482 had small values for V (and values for I or $\overset{*}{m}'$ near 1), used the areas within their familiarity distributions in highly clumped manners, and had low values for $A_{effective}$. Deer ID1 used its environment in a nearly random fashion, having V slightly less than 0.5. Female fisher 76482 used her environment in a manner somewhat more even than random.

Consequently, to be able to make sense of the area of an animal's familiarity distributions, A_{fd}, and to understand the animal's use of the environment within that area, both A_{fd} and $A_{effective}$ are critical. An alternative is to report A_{fd} and $\overset{*}{m}'$ or V. Other creative ways of considering areas of familiarity distributions certainly exist and will provide insights. Reporting A_{fd} alone is inadequate.

CHAPTER 13

A Doorway to New Cogitations on Home Ranges

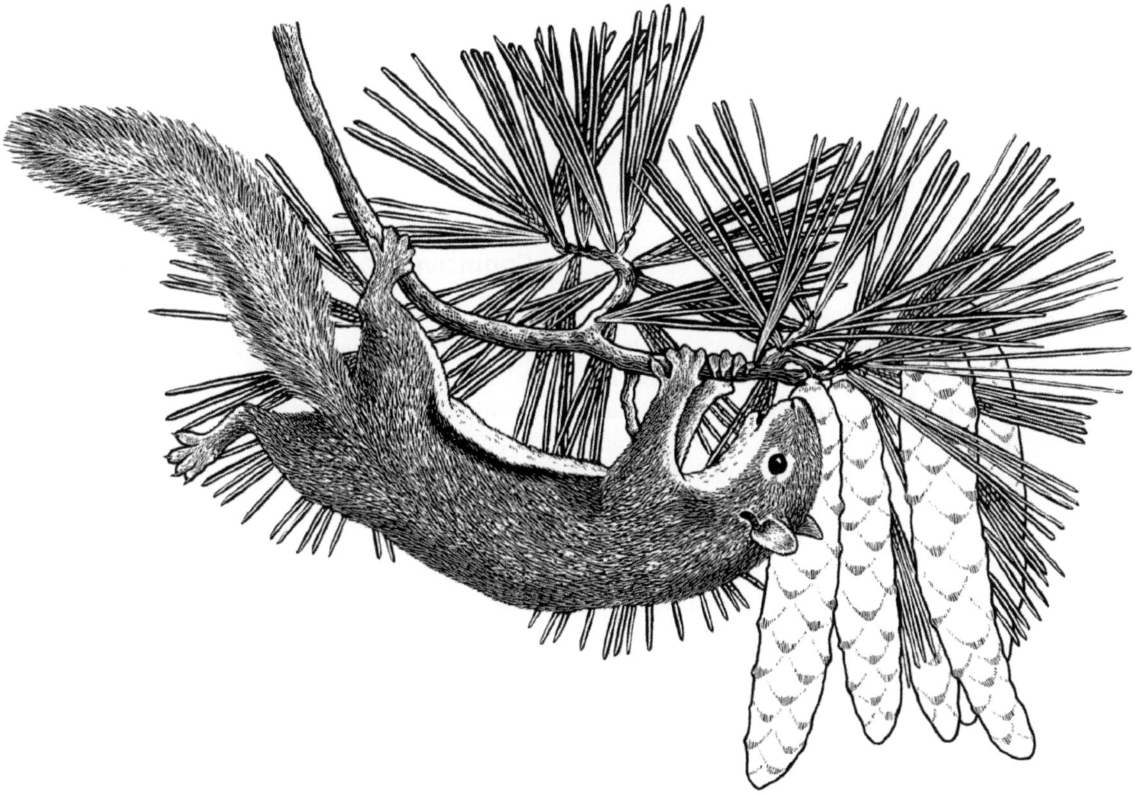

To be honest, all the "stuff" that animals know and the knowledge that they use every day absolutely leaves me in awe. Animals are so, so amazing! All animals. I feel privileged and honored that all the non-human animals, *and* plants and fungi and all other life share their glorious Earth with me. What a treat! I thank them.

For me, studying home ranges is the way to learn the most about animals because home ranges contain so much information and mean so much to animals. What is not exciting about that? By learning as much as I can about animals, I hope to contribute to their continued existence and to the ongoing and future evolution of life on this splendiferous

Home Ranges of Animals. Roger A. Powell, Oxford University Press. © Roger A. Powell (2025). DOI: 10.1093/9780191949142.003.0013

Earth. Selfishly, I want to learn more so as to continue the dopamine rush of living in awe. I hope that this book has helped at least some readers to understand a bit more about how awesome animals are.

13.1 Pulling Some Thoughts Together

This world that leaves me in awe faces a biodiversity crisis that continues to worsen. The diversity of organisms within ecological communities affects community stability (May 1973, 1989, through and beyond Hatton et al. 2024) and the equity and well-being of humans and our communities (Costanza et al. 1997, Hernández-Blanco et al. 2022, Richardson et al. 2023, Sangha et al. 2023, Kubiszewski et al. 2024, Drupp et al. 2024). The biodiversity crisis has the potential to have far-reaching effects, especially combined with the climate crisis. Consequently, the better the science that biologists use to learn the biology of the organisms that live on our planet, the better prepared humans will be to deal with humans' effects on our planet and on its other inhabitants.

Unlike the ships of the Phaeacians, which understood by instinct where their crews wished to go (Homer, *The Odyssey*, Book 13), biologists do not understand instinctively what animals think and why they wish to do what they do. Therefore, biologists use observations of animals, field data, and creative thinking and modelling to develop good hypotheses, and then use good research design to test those hypotheses and to gain insight into animals' cognitive maps (expanding from Leopold 1949). The rewards of such research are new knowledge of how and why animals use their environments related to resources, and of how and why such use relates to sex, maturity, experience, social status, safety, and nutritional and physiological condition, among myriad other things. Studying home ranges is an interdisciplinary endeavor that gains insight from field research, lab research, and diverse approaches to modelling, and that gains insight from at least animal behavior, animal ecology, evolutionary ecology, neurobiology, and behavioral psychology.

The heart of this book is my attempt to help readers become aware that animals' home ranges, maintained as their cognitive maps, are complex concepts within the animals' minds and not simply maps of the animals' environments. Think of what you know about your environment, about how you decide where to go and by what means, about when you decide not to go anywhere but to stay put, and why you make that decision: think of the sale on eggs at the grocery store, of the road construction on your major commuting route, and of how the weather or your neighbor affects your decisions of what to do on Saturday morning. These thoughts are all part of your cognitive map and, therefore, part of your own home range.

The bulk of this book presents different techniques and approaches to building hypotheses that can be tested in the field, techniques and approaches to support studying animals' home ranges as their cognitive maps, and how these techniques and approaches can provide complementarity and synergistic insight into those home ranges.

13.2 The Hippocampus Complex, Cognitive Maps, and Home Ranges

Eichenbaum (2017) summarized the role of the hippocampus complex to be that of maintaining the relational organization and flexibility of an animal's cognitive map. Hippocampal networks "map" spatial and nonspatial memories and map knowledge domains that share an emphasis on relational organization, including multiple navigational strategies. The hippocampus complex organizes experiences in memory, for which spatial mapping and navigation are both a metaphor for and a prominent application of relational memory organization. Thus, the role of the hippocampus complex extends far beyond simple environmental mapping and navigation. Daniel Levitin used the metaphor of the card catalog found in a library before the digital age. The hippocampus complex has the information of where in an animal's brain a certain memory can be found, such as on which shelf in which wing of the library a book can be found (Leviton 2022: 7). The hippocampus complex is the "cognitive map," but it is not a "map" as the word is used in everyday language; instead, it is a center for organizing and planning—the hippocampus complex "maps" ideas and plans pertaining to the memories that it has organized.

Consequently, in Chapter 3 Cognitive Maps, I defined a home range as: "That part of an animal's cognitive map that the animal keeps up to date via all of the animal's senses; includes places not visited recently but that are held in memory for future, specific use."

Yes, an animal's cognitive map does map his home range (in the sense of making what we generally do call a "map") but, far more importantly, his cognitive map is in charge of organizing all his memories about places in his home range: how the importance of each place relates to the importance and characteristics of other places, relates to how the characteristics of places change in predictable ways, and relates to how all this knowledge affects his well-being. To understand why, when, and how the animal lives its life within its home range, I need to know more than the "area traversed by the individual in its normal activities of food gathering, mating, and caring for young" (Burt 1943: 351). I need to learn what I can about what is in the animal's cognitive map, that is, what is his home range.

Related to the human-caused changes in our world, new research shows that air pollution reduces hippocampus volume in humans (Balboni et al. 2022), while hippocampus volume is related to memory loss, cognitive decline, and dementia (Drew 2025). Animals that live in cities and are, therefore, exposed to air pollution may suffer from reduced hippocampus function. Squirrels, rabbits, raccoons, coyotes, and, of course, commensal rats and mice have faster brain development and shorter lives than humans. Yet, the hypothesis that air pollution changes home range characteristics is worth testing.

New technologies are making recording brain activities of free-living animals actually possible. Rattenborg et al. (2008) monitored the sleep–wake cycles of free-living brown-throated three-toed sloths via electrodes on the sloths' skulls leading to recorders mounted on their backs that led to electroencephalograms and electromyograms. Different research groups have installed electrodes into hippocampi of Egyptian fruit bats in flight cages that broadcast to recorders mounted externally on the bats (Sarel et al. 2017, Liberti et al. 2022, Forli and Yartsev 2023). I anticipate that such systems will be mounted on free-living bats before long. MRIs have recorded brain function of unrestrained dogs and sheep, documenting emotional states (Berns 2017, Pluchot et al. 2024, Mugnai et al. 2024). Such technology opens a door for the future when field biologists might monitor the hippocampus complexes, and other brain functions, of free-living, wild animals. So, before too, too long, we shall be able to crawl, figuratively, inside our study animals' brains to understand them better.

13.3 Developing and Testing Home Range Hypotheses

How can a biologist make sense of all the information that a study animal knows but the biologist knows only in its rudiments? I start at the beginning with what I do know. Given my background knowledge, what new questions and hypotheses do new research results raise? Can I envision the new information and questions from the study animals' perspectives? How can I form that perspective into a hypothesis that I can test?

Evaluating field data inductively and looking for patterns can generate hypotheses. I can develop hypotheses by plotting animals' movements on every sort of map I can find: topography, human construction, habitat from the animal's perspective and habitat from my perspective, densities of potential predators and competitors, densities of foods, and so forth. Other maps are surely possible. I like to plot movements using a color gradient going from the start of a location data file to the end (arranged by time). These plots often suggest hypotheses related to, for example, foraging routes between resource patches and how an animal incorporates potentially limiting resources into her movements. A biologist can calculate optimal travel routes to balance diet with and without consideration of predators, leading to the hypothesis that predator avoidance affects travel significantly. Routes actually used will show the relative importance of food vs safety, or may show that the biologist has missed something important in her hypotheses.

Deriving hypotheses inductively has limits. Field data show what an animal has done but, if the field data were not collected within research designed to test hypotheses, then the data generally will, at best, provide only hints regarding why the animal did what it did. Consequently, results generated by such field data are usually limited by the conditions that existed while the data were collected. To generate hypotheses related to very different field conditions (or to thinking about field conditions from different perspectives), a biologist needs to derive logical hypotheses for why an animal behaved as it did under the original field conditions and, therefore, how the new conditions are expected to change behavior.

Hypotheses derived deductively based on results of tests of previous hypotheses, on theory, and on other a priori ideas of how the world works have more potential to produce reproducible results that can set a biologist on another round of the scientific method (Figure 1.3). Hypotheses built from results of models have the greatest potential to produce unanticipated results, which is, perhaps, the most consequential reward for modelling. Models should be as simple as needed to capture the basics of the biological processes of importance and as needed to collect field data that can be used to test the hypotheses. High levels of model complexity run the risk of exceeding the precision of the biological processes (May 1974). All biologists are capable of building theoretical models: verbal theoretical models, graphical theoretical models, and at least simple mathematical theoretical models. Biology is most productive when biologists with diverse perspectives contribute to areas of inquiry (Messeri and Crockett 2024). Accepting hypotheses (tentatively) is not necessary for learning new biology. Often, the most useful hypotheses are the ones rejected, potentially leading to completely new directions for research.

I want to keep track of everything that a study animal might be tracking with her cognitive map. I have realized while writing this book that I have (regretfully) missed innumerable possibilities to track aspects of the environment that my study animals were tracking. Being more mindful of data as it was produced and being more curious about "why" the animals behaved as they did would have led me to ask more creative questions in the field. Those questions would have stimulated me to collect new data, to do some new modelling, and to pursue potentially game-changing questions. I want to be curious and creative, which stimulates good hypotheses. Understanding cognitive maps begins with imagination, curiosity, and creativity.

This book provides introductions to many approaches to modelling that can generate hypotheses that would not have been conceived inductively. These approaches can, potentially, even be used to understand habitat choice, competition, and cooperation among plants (Wuest et al. 2022).

Nonetheless, good hypotheses can not be generated and tested without good field data. Likewise, good design for field research depends on good hypotheses. Field research and theoretical research do not form a dichotomy: they complement each other and lead to truly productive research when used together.

In many models, especially optimal foraging models, energy is used as a substitute for fitness. Nonetheless, even though one can replace each location in a data set with the rate of energy expenditure calculated for that location (Chapter 12.4), this approach will not model fitness. A mechanistic distribution built from energy kernels is strongly correlated with the standard time–location familiarity distribution (Powell and Mitchell 2012) and different activities having similar rates of energy expenditure have very different relationships to fitness. Consider energy expended to forage vs energy expended to elude a predator (the classic dinner–death trade-off). Calculations of fitness are not impossible but require serious thought. They have tremendous potential to provide new insights into why these animals develop the cognitive maps that they do. Sells et al. (2021, 2022a, b) compared the benefits of movements (food) to the costs of predation, competition, and travel to predict territories of wolf packs, making a first step toward testing the relative importance of benefits and costs on fitness (Mitchell and Powell 2004).

An alternative is to model energy gain. Energy gained at any particular site would have to share the gain proportionately with travel routes taken to obtain the energy gained. Energy gained could be subdivided by complementary food requirements (protein, carbohydrate, fat, water).

13.4 Machine Learning and Studying Home Ranges

Machine learning[1] has undoubted potential to generate hypotheses and, thereby, to provide novel insights into home ranges and research on home ranges. An animal's cognitive map is complex, anchored in multitudinous numbers of cells in her hippocampus and in the associated entorhinal cortex. Understanding will become more detailed in the future for how brains keep track of places, of the resources in patches, of behaviors of competitors and how they affect dependability of information, and of behaviors of predators and how predators affect foraging and other decisions. When biologists have gained such information, machine learning may be able to model an animal's cognitive map and, for example, to predict foraging plans in advance.

Machine learning and related fields are progressing with incredible speed and what they will be able to do even by the time this book is published can hardly be predicted. Parts of what I have written below will be out of date before any readers pick up this book.

On the down side, machine learning derives patterns from enormous data sets of observations and results of past research. The consequence is that machine learning can only generate hypotheses inductively (at least as I understand it). And, of course, the data sets need to be loaded into the cloud and made available. Both machine learning and the cloud have enormous energy costs, which have negative effects on biodiversity and climate.

To take advantage of the attributes of machine learning, however, biologists must learn to evaluate the accuracy of machine learning, to understand how pervasive the *illusions* of accuracy are, and to understand the ways that the illusions can mislead biologists and other scientists (Grossmann et al. 2023, Messeri and Crockett 2024, Skipper 2024a, b). Proposed solutions to problems obtained from machine learning have a history of leading researchers to believe that they (the researchers) have learned more than they really have. Machine learning consistently introduces unrealized biases, preventing researchers from knowing the shortcomings of proposed solutions. Large language models even introduce "hallucinations" (also called "confabulations") that can be missed. The problems of hallucinations are huge and difficult to solve (Jones 2025). And, when fed input that contains output from previous large learning model runs, the result can contain gibberish. During my decades as a biologist, I have experienced cycles of fashion in behavioral ecology, theoretical ecology, evolutionary ecology, and wildlife biology. Fashions that dominate alternative approaches limit innovation and productivity. Scientists have a long history of accepting foundational theories without question, and doing research within those foundational theories without even testing the foundational theories (Kuhn 1962). The prevailing approach can also limit what is considered to be science and what questions are worth asking (Baggott and Heilbron 2024, Baggott 2024, Shumailov et al. 2024).

If biases and hallucinations can be resolved, or at least identified, understood, and quantified (a big "if"), machine learning should be good for parameterizing a priori models, taking advantage of the ability of machine learning to digest and to organize past research results (in some unknown way). Machine learning may also be able to combine models in ways that biologists have not yet tried. And machine learning may be able to digest and to organize global data sets, making them available for testing models.

New approaches, such as machine learning now, are always productive in new ways but should not lead to ignoring or rejecting productive old approaches. New approaches are best incorporated into the existing suite of productive approaches on equal footing (Grossmann et al. 2023, Messeri and Crockett 2024, Skipper 2024b). Popularity of machine learning tools "risks introducing a phase of scientific enquiry in which we produce more but understand less" (Messeri and Crockett 2024, p. 49).

[1] I admit that "machine learning" and "artificial intelligence" apply to different but closely related entities. "Machine learning" is a more accurate term than "artificial intelligence", which is not intelligence but is artificial. "Artificial intelligence" is currently used to refer to a slough of technologies that work differently and have different applications (Narayanan and Kapoor 2024). Because "artificial intelligence" and machine learning can mean so many different things, my understanding is incomplete and may be in error on some points.

The best descriptive computer models based on machine learning failed to predict the record breaking weather and change of climate of 2023 (Schmidt 2024).

Modelling a cognitive map may, indeed, require black box computer methods available from machine learning but lack of transparency will always be a warning to be cautious. Scientists use the scientific method to minimize the effects of the biased black boxes between our ears. Biologists should also be able to use the scientific method to circumscribe the biases and limitations of any new technique. Machine learning is being used now in biomedicine, in industrial environmental programs, for identification of diverse species, and for species monitoring (Oldenburg et al. 2023, Bozeman et al. 2024, Hyde et al. 2024, Henrich et al. 2024, Rillig et al. 2024, Segura-Garcia et al. 2024, Villon et al. 2024), but its usefulness for ecological restoration has been seriously questioned (Woodrich and Pape 2024). I am cautiously eager to learn what machine learning will have to offer to field behavioral ecology.

13.5 So, Again, Why Home Ranges?

Animals' activities and movements vary across taxa, niches, landscapes, and scales but, nonetheless, animals do not wander aimlessly across those landscapes. Instead, home ranges emerge where animals concentrate their movements and activities and where animals learn the locations and characteristics of resources. Home ranges emerge because information about places and their relationships, stored as animals' cognitive maps, increases fitness. Thus, home ranges are central to animal behavior and ecology.

Once home ranges are established, foraging and other types of movements emerge because those movements depend on the animals knowing about their home ranges. An animal's cognitive map changes as resource distributions change, as the animal updates or loses information, and as the animal's resource requirements change.

Every animal uses his home range differently than other animals use their home ranges because the patches within each home range differ. Nonetheless, how each, individual animal uses his home range varies within the particular conformation of his home range, and can be learned by using the concepts and modelling approaches presented in Chapters 1 and 4 through 9. The variation in use of individual home ranges is presumed to be adaptive, which must be tested. At the ends of Chapters 5 and 6, I highlighted conditions caused by humans that can lead to maladaptive behavior. When an animal disperses or is introduced to an area where he has no prior experience, information about the landscape is critical for home range establishment. Following a period of apparent exploration, an animal settles into a home range: the animal finds a good place, gets to know it, and settles.

The consequence is that home ranges, animals' cognitive maps, contain an immense amount of information about the animals, their lives, and their livelihoods. Curiosity and creative, imaginative research will yield insight into this fascinating, exciting "mass" of information.

Acknowledgments

When live-trapping fishers in northern Minnesota one winter, I had trouble day after day with a weasel entering and springing one of my traps, hassling at the bait, then exiting the trap through the large mesh. One day I came to the trap and found it frozen open with its dead red squirrel bait gone. I followed weasel tracks only 10–15 m to where they entered a hollow log. I kneeled and looked into the log and there, close to the entrance, was my (or his) red squirrel carcass. I reached in and grabbed the squirrel. As I stood, the weasel climbed up my heavy wool pants looking from my face to the squirrel in my hand. A moment later, the weasel figured that he had bitten off more than he could chew and retreated. I admired the weasel's perseverance toward his goal. I hope, though, that *I* have not bitten off more than I can chew with writing this book. To help me avoid that fate, numerous people and animals have guided me.

Since 1970 I have studied diverse mammals and also some birds and herptiles. I have interrupted the lives of many and spied on others. They have provided the background and backbone of this book and I am indebted to them, each and every one.

Over the decades of field research, many graduate students and post docs helped to collect data and joined in discussions of diverse perspectives on home ranges, including Laura Brongo, Aaron Facka, Jorie Favreau, Mike Fritz, David Green, Peggy Horner, Michael Joyce, Adrienne Kovach, Tim Langer, Pete Laver, Sean Matthews, Brian Miller, John Noel, Jean Perry, Melissa Reynolds-Hogland, Erran Seaman, Jennifer Sevin, Julia Sizemore, Vanessa Sorenson, Wayne Spencer, Dennis Stewart, Rob Swiers, Linda Thomasma, Gordy Warburton, Bill Zielinski, and John Zimmerman. Most have continued on to become colleagues. Hundreds of technicians, undergraduate interns, devoted volunteers, and EarthWatch volunteers contributed to my field research projects. I thank them all.

Roger Peters introduced me to the concept of cognitive maps in the early 1970s, stimulating good discussion back then that planted the first seed for this book. Steve Buskirk gave me the idea of collecting data on my own home range. Analyzing those data made me realize how inadequate conventional approaches for studying home ranges are for understanding what animals' real home ranges are. That was another seed for this book. Scott Mills let me throw my sleeping bag on the floor of his hotel room during an annual meeting of the International Society of Behavioral Ecology. I discussed my idea of writing a review article about different modeling approaches to home ranges. He misunderstood my goal and, when the meeting was over, wished me good luck with my book. That was yet another seed.

Bill Sparklin, Laura Hanson, Mike Mitchell, and James Grand graciously provided and shared data on feral pigs on Fort Benning, Georgia, USA. Nick Gould and Chris DePerno shared data on the black bears living in Asheville, North Carolina. Lynn Rogers shared data on black bears in northern Minnesota. Tom Gable, Steve Winders, Dave Mech, and Shannon Barber-Meyer shared data on wolves in northern Minnesota. Hall Sawyer shared data on mule deer in Wyoming. And Cameron McQuivey shared data on pronghorns in New Mexico.

Roland Kays and Wayne Spencer reviewed manuscript drafts for several chapters of this book each and provided extremely helpful feedback. Each caught important problems. I was especially pleased that Lynn Nadel agreed to review some chapters, especially Chapter 3 on Cognitive Maps.

Aaron Facka and I have had innumerable discussions of home ranges as cognitive maps, as have Mike Mitchell and I, and Wayne Spencer and I.

Mike Mitchell reviewed every chapter, some multiple times, and provided some of the most irritating positive feedback I have ever received. Time and again he wrote "This is good. But, you know, you really need such-and-such. Now, I do not know how you might do that but I *know* you can do it. Go for it!" Thanks, Mike! The book would not be what it is without you.

Animals are the heroes of this book. Consie's (my wife) header drawings set the tone for the book and reinforce the point that the book is, first and foremost, about animals. In addition, Consie's excellent proof reading found typographical errors, of course, but more importantly found places where my writing or my logic was simply not clear. She provided continuous support from the beginning of this project and helped immeasurably to keep my spirits up and mood positive. The book would not exist without her.

Reading through the manuscript for this book for my one last time, I realize that this book has become quite personal. It has become a book not just about animals' home ranges but also a book about my philosophy of science (Figure 1.3). In some places, the book may be too technical for some readers, while in other places it may be too elementary. I hope my readers accept my unevenness. I have made critical comments on using statistics in some places without, perhaps, making clear that I find statistics extremely helpful and use them all the time to help me test hypotheses. Readers will find many mustelid and bear examples; my data on these critters are extensive and I am intimately familiar with them. I wish I had extensive experience with many other animals.

In the end, I hope that the book has imparted my love for being in the field and for studying the animals I love right where they live. The book is for them.

Scientific Names

Mammals

baboon, yellow – *Papio cynocephalus*
bat, Egyptian fruit – *Rousettus aegyptiacus*
bat, frog-eating (fringe-lipped) – *Trachops cirrhosus*
bear, American black – *Ursus americanus*
bear, grizzly and brown – *Ursus arctos*
beaver – *Castor canadensis*
beaver, Euroasian – *Castor fiber*
badger, Eurasian – *Meles meles*
bat, New Zealand lesser short-tailed – *Mystacina tuberculata*
bison – *Bison* (*Bos*) *bison*
capuchin – *Cebus capucinus*
chimpanzee – *Pan troglodytes*
chipmunk, eastern – *Tamias striatus*
chipmunk, yellow-pine – *Neotamias amoenus*
coyote – *Canis latrans*
dog, domestic and feral – *Canis familiaris*
dunker, common – *Sylvicapra grimmia*
elk – *Cervus canadensis*
fisher – *Pekania* (formerly *Martes*) *pennanti*
fox, arctic – *Vulpes lagopus*
fox, Eurasian red – *Vulpes vulpes*
fox, North American red – *Vulpes (vulpes) fulva*
gerbil, Anderson's – *Gerbillus andersoni*
gerbil, greater Egyptian – *Gerbillus pyramidum*
ground squirrel, golden mantled – *Callospermophilus lateralis*
hare, European – *Lepus europaeus*
hare, snowshoe – *Lepus americanus*
hyaena, spotted – *Crocuta crocuta*
kinkajou – *Potos flavus*
marten, American – *Martes americana*
marten, green-backed – *Martes hypotheticus*
marten, Pacific – *Martes caurina*
marten, pine – *Martes martes*
meerkat – *Suricata suricatta*
mink – *Neogale vison*
monkey, samango – *Cercopithecus (nictitans) mitis erythrarchus*
moose (elk) – *Alces alces*
muskrat – *Ondatra zibethica*
otter, Eurasian – *Lutra lutra*
otter, North American – *Lontra canadensis*
porcupine, North American – *Erethizon dorsatum*
possum (Virginia opossum) – *Didelphis virginiana*
puma or mountain lion and other names – *Puma concolor*
rabbit, cottontail – *Sylvilagus floridanus*
sheep, bighorn – *Ovis canadensis*
skunk, striped – *Mephitis mephitis*
sloth, brown-throated three-toed – *Bradypus variegatus*
squirrel, California ground – *Otospermophilus beecheyi*
squirrel, douglas – *Tamiasciurus douglasii*
squirrel, eastern grey – *Sciurus carolinensis*
squirrel, fox – *Sciurus niger*
squirrel, North American red – *Tamiasciurus hudsonicus*
squirrel, western grey – *Sciurus griseus*
stoat – *Mustela erminea*
vole, meadow – *Microtus pennsylvanicus*
vole, red-backed – *Myodes* (formerly *Clethrionmys*) *gapperi*
vole, woodland or pine – *Microtus* (or *Pitymys*) *pinetorum*
weasel, common – *Mustela nivalis*
weasel, least – *Mustela (nivalis) rixosa*
weasel, long-tailed – *Neogale* or *Mustela frenata*
weasel, short-tailed (North America) – *Mustela richardsonii*
wolf – *Canis lupus*

Birds

albatross, black-footed – *Phoebastria nigripes*
albatross, Laysan – *Phoebastria immutabilis*
albatross, shy – *Thalassarche cauta*
barbet, D'Arnaud's – *Trachyphonus darnaudi*
blackbird, red-winged – *Agelaius phoeniceus*
crane, Siberian – *Leucogeranus leucogeranus*
hawk, red-tailed – *Buteo jamaicensis*
honeycreeper, Hawaiian – *Drepanis coccinea*
Jay, grey or Canada – *Perisoreus canadensis*
nutcracker, Clark's – *Nucifraga columbiana*
nutcracker, Eurasian – *Nucifraga caryocatactes*

osprey – *Pandion haliaetus*
warbler, prairie – *Setophaga discolor*
woodpecker, red-cockaded – *Leuconotopicus borealis*
wren, banded – *Thryophilus pleurostictus*

Reptiles

anole, yellow-chinned – *Anolis gundlachi*
lizard, Be'er Sheva fringe-fingered – *Acanthodactylus beershebensis*
lizard, fence – *Sceloporus undulatus*
tortoise, desert – *Gopherus agassizii*

Fish

damselfish, threespot – *Eupomacentrus planifrons*

Invertebrates

ant, Neotropical arboreal – *Azteca trigona*
ant, red harvester – *Pogonomyrmex barbatus*
caddisfly, netspinning – *Arctopsyche ladogensis*
damselfly, ebony jewelwing – *Calopteryx maculata*

Herbaceous Plants

bear corn – *Conopholis americana*
blueberries – *Vaccinium* spp.Trees and Shrubs
azalea – *Rhododendron calendulaceum, R. nudiflorum*, etc.
blackberry – *Rubus* spp.
blueberry – *Vaccinium* spp.
blueberry, highbush – *Vaccinium corymbosum*
cedar, incense – *Calocedrus decurrens*
cherry – *Prunus* spp.
fir, Douglas – *Pseudotsuga menziesii*
fir, Fraser – *Abies fraseri*
fir, white – *Abies concolor*
hemlock, Carolina – *Tsuga caroliniana*
hemlock, eastern – *Tsuga canadensis*
huckleberry – *Gaylusacchia* spp.
laurel, mountain – *Kalmia latifolia*
magnolia – *Magnolia* spp.
oak, California black – *Quercus kellogii*
oak, cañon live – *Quercus chrysolepis*
oak, chestnut – *Quercus prinus*
oak, northern red – *Quercus rubra*
oak, scarlet – *Quercus coccinea*
oak, white – *Quercus alba*
pine – *Pinus* spp.
pine, Ponderosa – *Pinus ponderosa*
pine, sugar – *Pinus lambertiana*
poplar, balsam – *Populus balsamifera*
poplar, tulip – *Liriodendron tulipifera*
raspberry – *Rubus* spp.
rhododendron – *Rhododendron* spp.
serviceberry – *Amelanchier* spp.
spruce, red – *Picea rubens*
tanoak – *Notholithocarpus densiflorus*

Bibliography

Abelia, J. A., P. Erden and F. Canto. 2024. Using proximity loggers in studies of sheep behavior: An estimation of the daily frequency of water consumption. Animal Biotelemetry 12: 19. https://doi.org/10.1186/s40317-024-00376-z

Aben, J., J. Signer, J. Heiskanen, P. Pellikka and J. M. J. Travis. 2021. What you see is where you go: Visibility influences movement decisions of a forest bird navigating a three-dimensional-structured matrix. Biology Letters 17: 20200478. https://doi.org/10.1098/rsbl.2020.0478

Abu Baker, M. A. and J. S. Brown. 2014. Foraging and habitat use of common duikers, *Sylvicapra grimmia*, in a heterogeneous environment within the Soutpansberg, South Africa. African Journal of Ecology 52: 318–327.

Ackerman, B. B., F.A. Leban, E.O. Garton and M. D. Samuel. 1988. User's manual for program home range, 2nd edition. Forestry, Wildlife and Range Experiment Station Technical Report. University of Idaho, Moscow, Idaho, USA. 78 pp.

Ackerman, J. D., M. R. Mesler, K. L. Lu and A. M. Montalvo. 1982. Food-foraging behavior of male Euglossini (Hymenoptera, Apidae): Vagabonds or trapliners. Biotropica 14: 241–248.

Adams, E. S. 1990. Boundary disputes in the territorial ant Azteca trigona: Effects of asymmetries in colony size. Animal Behaviour 39: 321–328.

Adams, E. S. 2001. Approaches to the study of territory size and shape. Annual Review of Ecology and Systematics 32: 277–303.

Adler, F. R. and D. M. Gordon. 2003. Optimization, conflict, and nonoverlapping foraging ranges in ants. American Naturalist 162: 529–543.

Adolph, K. E. 2019. An ecological approach to learning in (not and) development. Human Development 63: 180–201.

Adolph, K. E. and J. W. Young. 2021. Learning to move in the real world. Science 373: 620–621.

Agarwal, A., A. Sarel, D. Derdikman, N. Ulanovsky and Y. Gutfreund. 2023. Spatial coding in the hippocampus and hyper pallium of flying owls. Proceedings of the National Academy of Sciences 120(5): e2212418120 https://doi.org/10.1073/pnas.2212418120

Aho, K. A. and R. T. Bowyer. 2015. Confidence intervals for a produce of proportions: Application to importance values. Ecosphere 6: 230. http://dx.doi.org/10.1890/ES15-00420.1

Akagi, M. 2018. Rethinking the problem of cognition. Synthese 195: 3547–3570.

Alexander, R. D. 1961. Aggressiveness, territoriality, and sexual behavior in field crickets (Orthoptera: Gryllidae). Behaviour 17: 130–223.

Alif, Z., J. Dunning, H. Y. J. Chik, T. Burke and J. Schroeder. 2022. What is the best fitness measure in wild populations? A case study on the power of short-term fitness proxies to predict reproductive value. Plos One 17. https://doi.org/10.1371/journal.pone.0260905

Allard, C. A. H., G. Kang, J. J. Kim, W. A. Valencia-Montoya, Ryan E. Hibbs and N. W. Bellono. 2023. Structural basis of sensory receptor evolution octopus. Nature 616: 373–377. https://doi.org/10.1038/s41586-023-05822-1

Allen, A. 1982. Habitat suitability index models: Marten. U.S. Fish and Wildlife Service, FWS/OBS-82/10.11.

Allen, A. W. 1983. Habitat suitability index models: Fisher. U.S. Fish and Wildlife Service, FWS/OBS-82/10.45.

Allen, D. L. 1938. Notes on the killing techniques of the New York weasel. Journal of Mammalogy 19: 225–229.

Almog, N., G. Tocker, T. Bonnevie, E. I. Moser, M. B. Moser and D. Derdikman. 2019. During hippocampal inactivation, grid cells maintain synchrony, even when the grid pattern is lost. eLife. doi: https://doi.org/10.7554/eLife.47147.001

Alt, G. L., G. J. Matula, F. W. Alt and J. S. Lindzey. 1977. Dynamics of home range and movements of adult black bears in northeastern Pennsylvania. International Conference on Bear Research and Management 4: 131–136.

Altmann, S. A. 1998. Foraging for survival. University of Chicago Press, Chicago. 609 pp.

Altmann, S. A. and J. Altmann. 1970. Baboon Ecology: African Field Research. University of Chicago Press, Chicago. 220 pp.

Amano, T., K. Ushiyama, G. Fujita and H. Higuchi. 2004. Alleviating grazing damage by white-fronted geese: An optimal foraging approach. Journal of Applied Ecology 41: 675–688.

Amstrup, S. C. and J. Beecham. 1976. Activity patterns of radio collared black bears in Idaho. Journal of Wildlife Management 40: 340–348.

Amsler, S. J. 2010. Energetic costs of territorial boundary patrols by wild chimpanzees. American Journal of Primatology 72: 93–103.

Amstrup, S. C., G. M. Durner, I. Stirling, N. J. Lunn, F. Messler and F. Messier. 2000. Movements and distribution of polar bears in the Beaufort Sea. Canadian Journal of Zoology 78: 948–966.

Andersen, D. E. and O. J. Rongstad. 1989. Home-range estimates of red-tailed hawks based on random and systematic relocations. Journal of Wildlife Management 53: 802–807.

Anderson, C. R. and F. G. Lindzey. 2003. Estimating cougar predation rates from GPS location clusters. Journal of Wildlife Management 67: 307–316.

Anderson, D. J. 1982. The home range: A new nonparametric estimation technique. Ecology 63: 103–112.

Anderson, D. J. 1983. Optimal foraging and the travelling salesman. Theoretical Population Biology 24: 145–159.

Anderson J. R. 1995. Learning and Memory: An Integrated Approach. Wiley and Sons, New York.

Andrews, E. F., R. Pascalau, A. Horowitz, G. M. Lawrence and P. J. Johnson. 2022. Extensive connections of the canine olfactory pathway revealed by tractography and dissection. Journal of Neuroscience 42: 6392–6407.

Andrews, K., J. Birch and J. Sebo. 2025. Evaluating animal consciousness. Science 387: 822–824.

Anthony, H. E. 1928. Field Book of North American Mammals. G. P. Putnam's Sons, New York. 627 pp.

Arimond, S. R. 1979. Fruit production in black bear *Ursus americanus* habitat of northeastern Minnesota. Plan B MS paper, University of Minnesota.

Aromaa, S., J. J. Ilvonen and J. Suhonen. 2019. Body mass and territorial defence strategy affect territgory size of odonate species. Proceedings of the Royal Society B: Biological Sciences 286: 20192398.

Aronov, D., R. Nevers and D. W. Tank. 2017. Mapping of a non-spatial dimension by the hippocampal-entorhinal circuit. Nature 543: 719–722.

Arraut, E. M., J. L. Arraut, M. Marmontel, J. E. Mantovani and E. Novo. 2017. Bottlenecks in the migration routes of Amazonian manatees and the threat of hydroelectric dams. Acta Amazonica 47: 7–17.

Asgharnia, A., H. Schwartz and M. Atia. 2022. Learning deception using fuzzy multi-level reinforcement learning in a multi-defender one-invader differential game. International Journal of Fuzzy Systems 24: 3015–3038. https://doi.org/10.1007/s40815-022-01352-6

Aspect, A. 2024. Why quantum theory is just like magic (and Einstein deserves extra credit). Nature 634: 1039.

Aubry, K. B. and D. B. Houston. 1992. Distribution and status of the fisher (*Martes pennanti*) in Washington. Northwestern Naturalist 73: 69–79.

Ausband, D. E., M. S. Mitchell, S. B. Bassing and C. White. 2013. No trespassing: Using a biofence to manipulate wolf movements. Wildlife Research 40: 207–216.

Ausband, D. E., M. S. Mitchell, C. R. Stansbury, J. L. Stenglein and L. P. Waits. 2017. Harvest and group effects on pup survival in a cooperative breeder. Proceedings of the Royal Society B: Biological Sciences 284: 20170580. https://doi.org/10.1098/rspb.2017.0580.

Baggott, J. 2024. How Einstein lost the battle to explain quantum theory. Nature 629L: 29–32.

Baggott J. and J. L. Heilbron. 2024. Quantum Drama: From the Einstein–Bohr Debate to the Riddle of Entanglement. Oxford University Press, Oxford. 352 pp.

Baker, R. 1946. A study of rodent populations on Guam, Mariana Islands. Ecological Monographs 16: 393–408.

Balboni, E., T. Filippini, M. Crous-Bou, M. Guxens, L. D. Erickson and M. Vinceti. 2022. The association between air pollutants and hippocampal volume from magnetic resonance imaging: A systematic review and meta-analysis. Environmental Research 204: 111976.

Balda, R. P. 1980. Recovery of cached seeds by a captive Nucifraga caryocatactes. Zeitschrift fűr Tierpsychologie - Journal of Comparative Ethology 52: 331–346.

Baluška, F., S. Mancuso, D. Volkmann and P. W. Barlow. 2009. The 'root-brain' hypothesis of Charles and Francis Darwin: Revival after more than 125 years. Plant Signaling & Behavior 4: 1121–1127.

Barcia, P., M. N. Bugalho, M. L. Campagnolo and J. O. Cerdeira. 2007. Using n-alkanes to estimate diet composition of herbivores: A novel mathematical approach. Animal 1: 141–149.

Bartlam-Brooks, H., S. Wilshin, T. Hubel, S. Hailes, E. Bennitt and A. M. Wilson. 2020. There and back again: A zebra's tale. Journal of Experimental Biology 223: jeb232140.

Bascompte, J. and C. Vilà. 1997. Fractals and search paths in mammals. Landscape Ecology 12: 213–221.

Bastos, A. P. M., Z. N. Houghton, L. Naranjo and F. Rossano. 2024. Soundboard-trained dogs produce non-accidental, non-random and non-imitative two-button combinations. Scientific Reports 14: 28771.

Bateman, A. W., M. A. Lewis, G. Gall, M. B. Manser and T. H. Clutton-Brock. 2015. Territoriality and home-range dynamics in meerkat, *Suricata suricatta*: A mechanistic modelling approach. Journal of Animal Ecology 84: 260–271. doi: 10.1111

Bhattacharyya, A. 1943. On a measure of divergence between two statistical populations defined by their probability distributions. Bulletin of the Calcutta Mathematical Society 35: 99–109.

Beardsell, A., D. Berteaux, F. Dulude-De Broin, G. Gauthier, J. Clermont, D. Gravel and J. Bety. 2023. Predator-mediated interactions through changes in predator home range size can lead to local prey exclusion. Proceedings of the Royal Society B: Biological Sciences 290: 20231154.

Beck, D. E. 1977. Twelve-year acorn yield in southern Appalachian oaks. USDA Forest Service Research Note. SE-244, 8 pp. Southeast Forest Experiment Station, Asheville, North Carolina.

Beck, D. E. and D. F. Olson, Jr. 1968. Seed production in southern Appalachian oak stands. USDA Forest Service Research Note. SE-91, 7 pp. Southeast Forest Experiment Station. Asheville, North Carolina.

Beesley, D. 2007. Crow's Range: An Environmental History of the Sierra Nevada. University of Nevada Press, Reno.

Bekoff, M. 2004. The great divide. American Scientist 92: https://www.americanscientist.org/article/the-great-divide.

Bekoff, M. and L. D. Mech. 1984. Computer simulation: Simulation analyses of space use: Home range estimates, variability, and sample size. Behavior Research Methods, Instruments, & Computers 16: 32–37.

Bekoff, M. and M. C. Wells. 1981. Behavioural budgeting by wild coyotes: The influence of food resources and social organization. Animal Behaviour 29: 794–801.

Belovsky, G. E. 1978. Diet optimization in a generalist herbivore: The moose. Theoretical Population Biology 14: 105–134.

Belovsky, G. E. 1984a. Diet optimization by beaver. American Midland Naturalist 111: 209–222.

Belovsky, G. E. 1984b. Herbivore optimal foraging: A comparative test of 3 models. American Naturalist 124: 97–115.

Belovsky, G. E. and P. A. Jordan. 1978. The time-energy budget of a moose. Theoretical Population Biology 14: 76–104.

Benhamou, S. 2004. How to reliably estimate the tortuosity of an animal's path: Straightness, sinuosity, or fractal dimension? Journal of Theoretical Biology 229: 209–220.

Benhamou, S. 2011. Dynamic approach to space and habitat use based on biased random bridges. Plos One 6(1): e14592. doi:10.1371/journal.pone.0014592

Benhamou, S. and P. Bovet. 1989. How animals use their environment: A new look at kinesis. Animal Behaviour 39: 375–383.

Berger, J., P. B. Stacey, L. Bellis and M. P. Johnson. 2001. A mammalian predator–prey imbalance: Grizzly bear and wolf extinction affect avian neotropical migrants. Ecological Applications 11: 947–960.

Berger, K. M. and E. M. Gese. 2007. Does interference competition with wolves limit the distribution and abundance of coyotes? Journal of Animal Ecology 76: 1075–1085.

Berger-Tal, O. and D. Saltz. 2019. Invisible barriers: Anthropogenic impacts on inter- and intra-specific interactions as drivers of landscape-independent fragmentation. Philosophical Transactions of the Royal Society B: Biological Sciences 374: 20180049. http://dx.doi.org/10.1098/rstb.2018.0049.

Berns, G. 2017. What it's Like to be a Dog: And Other Adventures in Animal Neuroscience. Basic Books, New York. 301 pp.

Berryman, A. A., M. Lima Arce and B. A. Hawkins. 2002. Population regulation, emergent properties, and a requiem for density dependence. Oikos 99: 600–606.

Beschta, R. L. and R. J. Ripple. 2009. Large predators and trophic cascades in terrestrial ecosystems of the western United States. Biological Conservation 142: 2401–2414. 10.1016/j.biocon.2009.06.015

Bingham, B. B. and B. R. Noon. 1997. Mitigation of habitat "take": Application to habitat conservation planning. Conservation Biology 11: 127–139.

Bjerknes, T. L., N. C. Dagslott, E. I. Moser and M. B. Moser. 2018. Path integration in place cells of developing rats. Proceedings of the National Academy of Sciences of the United States of America 115: E1637–E1646.

Blackwell, P. G. 1997. Random diffusion models for animal movement. Ecological Modelling 100: 87–102.

Blair, W. F. 1940. Notes on home ranges of populations of the short-tailed shrew. Ecology 21: 284–288.

Blanco, J. C. and Y. Cortés. 2007. Dispersal patterns, social structure and mortality of wolves living in agricultural habitats in Spain. Journal of Zoology 273: 114–124.

Blaser, N., G. Dell'Omo, G. Dell'Ariccia, D. P. Wolfer and H. P. Lipp. 2013. Testing cognitive navigation in unknown territories: Homing pigeons choose different targets. Journal of Experimental Biology 216: 3123–3131.

Boccara, C. N., M. Nardin, F. Stella, J. O'Neill and J. Csicsvari. 2019. The entorhinal cognitive map is attracted to goals. Science 363: 1443–1447.

Boitani, L., F. Francisci, P. Ciucci, and G. Andreoli 1995. Population biology and ecology of feral dogs in central Italy. Pp. 217–244 in Serpell, J. (editor), The Domestic Dog: Its Evolution, Behaviour and Interactions with People. Cambridge University Press, Cambridge.

Bolker, J. A. 2014. Models in context: biological and epistemological niches. Pp. 153–166 in Barker, G., E. Desjardins and T. Pearce (editors), Entangled Life: Organism and Environment in the Biological and Social Sciences. Springer, Dordrecht.

Bonanni, R., P. Valsecchi and E. Natoli. 2009. Pattern of individual participation and cheating in conflicts between groups of free-ranging dogs. Animal Behaviour 79: 957–968.

Botkin, D. B. 1995. Our Natural History: The Lessons of Lewis and Clark. Putnam, New York City.

Botton, M. L., R. E. Loveland and T. R. Jacobsen. 1994. Site selection by migratory shorebirds in Delaware Bay, and its relationship to beach characteristics and abundance of horseshoe crab (*Limulus polyphemus*) eggs. Auk 111: 605–616.

Boutin, S. and S. Schweiger. 1988. Manipulation of intruder pressure in red squirrels (*Tamiasciurus hudsonicus*): Effects on territory size and acquisition. Canadian Journal of Zoology 66: 2270–2274.

Bovet, P. and S. Benhamou. 1988. Spatial analysis of animals' movements using a correlated random-walk model. Journal of Theoretical Biology 131: 419–433.

Bovet, P. and S. Benhamou. 1991. Optimal sinuosity in central place for aging movements. Animal Behavior 42: 57–62.

Box, G. E. P. 1976. Science and statistics. Journal of the American Statistical Association 71: 791–799.

Bowen, W. D. and I. McTaggart-Cowan. 1980. Scent marking in coyotes. Canadian Journal of Zoology 58: 473–480.

Bozeman, J. F., III, C. Hollauer, A. T. Ramshankar, S. Nakkasunchi, J. Jambeck, A. Hicks, M. Bilec, D. McCauley and O. Heidrich. 2024. Embed systemic equity throughout industrial ecology applications: How to address machine learning unfairness and bias. Journal of Industrial Ecology 28(6): 1362–1376. doi: 10.1111/jiec.13509

Brandell, E. E., N. M. Fountain-Jones, M. L. J. Gilbertson, P. C. Cross, P. J. Hudson, D. W. Smith, D. R. Stahler, C. Packer and M. E. Craft. 2020. Group density, disease, and season shape territory size and overlap of social carnivores. Journal of Animal Ecology 90(1): 87–101. doi:10.1111/1365-2656.13294

Breed, G. A., E. A. Golson and M. T. Tinker. 2017. Predicting animal home-range structure and transitions using a multistate Ornstein–Uhlenbeck biased random walk. Ecology 98: 32–47.

Brent, L. J. N. 2015. Friends of friends: Are indirect connections in social networks important to animal behaviour? Animal Behaviour 103: 211–222.

Bridges, A. D., H. MaBouDi, O. Procenko, C. Lockwood, Y. Mohammed, A. Kowalewska, J. E. R. González, J. L. Woodgate and L. Chittka. 2023. Bumblebees acquire alternative puzzle-box solutions via social learning. Plos Biology 21. https://doi.org/10.1371/journal.pbio.3002019

Broadbooks, H. E. 1970. Home ranges and territorial behaviour of yellow-pine chipmunk, *Eutamias amoenus*. Journal of Mammalogy 51: 310–326.

Brongo, L. L., M. S. Mitchell and J. B. Grand. 2005. Long-term analysis of survival, fertility, and population growth rate of black bears in North Carolina. Journal of Mammalogy 86: 1029–1035.

Brosi, B. J. and E. G. Biber. 2009. Statistical inference, Type II error, and decision making under the US Endangered Species Act. Frontiers in Ecology and the Environment 7: 487–494.

Brown, J. L. 1964. The evolution of diversity in avian territorial systems. Wilson Bulletin 76: 160–169.

Brown, J. L. 1969. Territorial behavior and population regulation in birds: A review and re-evaluation. Wilson Bulletin 81: 293–329.

Brown, J. S. 1999. Vigilance, patch use and habitat selection: Foraging under predation risk. Evolutionary Ecology Research 1: 49–71.

Brown, J. S. and B. P. Kotler. 2004. Hazardous duty pay and the foraging cost of predation. Ecology Letters 7: 999–1014.

Brown, J. S., R. A. Morgan and B. D. Dow. 1992. Patch use under predation risk: II. A test with fox squirrels, *Sciurus niger*. Annales Zoologici Fennici 29: 311–318.

Brown, J. S., B. P. Kotler and A. Bouskila. 2001. Ecology of fear: Foraging games between predators and prey with pulsed resources. Annales Zoologici Fennici 38: 71–87.

Brown, J. S., J. W. Laundré and M. Gurung. 1999. The ecology of fear: Optimal foraging, game theory, and trophic interactions. Journal of Mammalogy 80: 385–399.

Bryant, J. P., F. D. Provenza, J. Pastor, P. B. Reichardt, T. P. Clausen and J. T. du Toit. 1991. Interactions between woody plants and browsing mammals mediated by secondary metabolites. Annual Review of Ecology and Systematics 22: 431–446.

Buehlmann, C. 2023. Animal homing: Ants build their own landmarks when they need them. Current Biology. https://doi.org/10.1016/j.cub.2023.05.048

Buderman, F. E., P. J. Helm, J. D. Clark, R. H. Williamson, J. Yarkovich and J. M. Mullinax. 2023. A multi-level modeling approach to guide management of female feral hogs in Great Smoky Mountains National Park. Biological Invasions. doi: 10.1007/s10530-023-03086-4

Bullard, F. 1999. Estimating the home range of an animal: the Brownian bridge approach. MS thesis, University of North Carolina, Chapel Hill, North Carolina.

Burnham, K. P. and D. R. Anderson. 1998. Model Selection and Multi Modal Inference: A Practical Information-Theoretic Approach. Springer, New York.

Burns, J. G., J. Foucaud and F. Mery. 2011. Costs of memory: Lessons from 'mini' brains. Proceedings of the Royal Society B: Biological Sciences 278: 923–929.

Burt, W. H. 1940. Territorial behavior and populations of some small mammals in southern Michigan. Miscellaneous Publications of the Museum of Zoology, University of Michigan 45: 1–58.

Burt, W. H. 1943. Territoriality and home range concepts as applied to mammals. Journal of Mammalogy 24: 346–352.

Buckner, C. 2015. A property cluster theory of cognition. Philosophical Psychology 28: 307–336.

Bush, J. M. and D. Simberloff. 2018. A case for anole territoriality. Behavioral Ecology and Sociobiology 72: 4.

Butler, W. N., K. Hardcastle, and L. M. Giocomo. 2019. Remembered reward locations restructure entorhinal spatial maps. Science 363: 1447–1452.

Byers, J., E. Hebets and J. Podos. 2010. Female mate choice based upon male motor performance. Animal Behaviour 79: 771–778.

Byers, J. A. 2003. Built for speed: A year in the life of pronghorn. Harvard University Press, Cambridge, Massachusetts. 230 pp.

Byers, J. A. and D. W. Kitchen. 1988. Mating system shift in a pronghorn population. Behavioral Ecology and Sociobiology 22: 355–360.

Byers, J. A. and L. Waits. 2006. Good genes sexual selection in nature. Proceedings of the National Academy of Sciences of the United States of America 103: 16343–16345.

Börger, L. and J. Fryxell. 2012. Quantifying individual differences in dispersal using net squared displacement. Pp. 222–247 in Clobert, J., M. Baguette, T. M. Benton and J. M. Bullock (editors), Dispersal Ecology and Evolution. Oxford University Press, Oxford.

Börger, L., B. D. Dalziel and J. M. Fryxell. 2008. Are there general mechanisms of animal home range behaviour? A review and prospects for future research Ecology Letters 11: 637–650. doi: 10.1111/j.1461

Börger, L., N. Franconi, G. De Michele, A. Gantz, F. Meschi, A. Manica, S. Lovari and T. Coulson. 2006. Effects of sampling regime on the mean and variance of home range size estimates. Journal of Animal Ecology 75: 1393–1405.

Cabrera, D., J. R. Nilsson and B. D. Griffen. 2021. The development of animal personality across ontogeny: A cross-species review. Animal Behaviour 173: 137–144.

Calabrese, J. M., C. H. Fleming and E. Gurarie. 2016. ctmm: an R package for analyzing animal relocation data as a continuous-time stochastic process. Methods in Ecology and Evolution 7: 1124–1132.

Calhoun, J. B. and J. U. Casby. 1958. Calculation of home range and density of small mammals. Public Health Monographs No. 55. USDA, Washington, DC. 24 pp.

California Department of Forestry and Fire Protection. 2013. California forest practice rules 2013. Resource Management, Forest Practice Program, Sacramento, California, USA. http://www.bof.fire.ca.gov/regulations/ca_forest_practice_rules_other_title_14_codes/california_forest_practice_rules/2013%20FPR%20and%20FPA.pdf.

Calvo, P. 2023. *Planta sapiens*: The New Science of Plant Intelligence. Norton, New York. 304 pp.

Cao, L., B. Wang, C. Yan, Z. Y. Wang, H. M. Zhang, Y. Z. Geng, J. Chen and Z. B. Zhang. 2018. Risk of cache pilferage determines hoarding behavior of rodents and seed fate. Behavioral Ecology 29: 984–991.

Čapek, F., J. Prucha, V. Socha, V. Hart and H. Burda. 2017. Directional orientation of pheasant chicks at the drinking dish and its potential for research on avian magnetoreception. Folia Zoologica 66: 175–182.

Caro, T. M. 1976. Observations on ranging behavior and daily activity of lone silverback mountain gorillas (Gorilla gorilla beringei). Animal Behaviour 24: 889–897.

Carpenter, F. L. and R. E. MacMillen. 1976. Threshold model of feeding territoriality and test with a Hawaiian honeycreeper. Science 194: 634–642.

Carpenter, F. L. and R. E. MacMillen. 1976. Energetic cost of feeding territories in a Hawaiian honeycreeper. Oecologia 26: 213–223.

Carscadden, K. A., N. C. Emery, C. A. Arnillas, M. W. Cadotte, M. E. Afkhami, D. Gravel, S. W. Livingstone and J. J. Wiens. 2020. Niche breadth: Causes and consequences for ecology, evolution, and conservation. Quarterly Review of Biology 95: 179–214.

Carter, J. H., J. R. Walters, S. H. Everhart and P. D. Doerr. 1989. Restrictors for red-cockaded woodpecker cavities. Wildlife Society Bulletin 17: 68–72.

Casey, C., I. Charrier, N. Mathevon, C. Nasr, P. Forman and C. Reichmuth. 2020. The genesis of giants: Behavioural ontogeny of male northern elephant seals. Animal Behaviour 166: 247–259.

Cassidy, K. A., D. R. MacNulty, D. R. Stahler, D. W. Smith and L. D. Mech. 2015. Group composition effects on aggressive interpack interactions of gray wolves in Yellowstone National Park. Behavioural Ecology 26: 1352–1360. doi:10.1093/beheco/arv081

Castles, M., R. Heinsohn, H. H. Marshall, A. E. G. Lee, G. Cowlishaw and A. J. Carter. 2014. Social networks created with different techniques are not comparable. Animal Behaviour 96: 59–67.

Catania, K. 2020. Great Adaptations: Star-nosed Moles, Electric Eels and Other Tales of Evolution's Mysteries Solved. Princeton University Press, Princeton, New Jersey.

Cattet, M. 2017. Capture, handling and release of bears standard operating procedure. Government of the Northwest Territories. https://www.enr.gov.nt.ca/sites/enr/files/resources/bears_care_sop.pdf

Cattet, M., J. Boulanger, G. Stenhouse, R. A. Powell and M. J. Reynolds-Hogland. 2008. An evaluation of long-term capture effects in ursids: Implications for wildlife welfare and research. Journal of Mammalogy 89: 973–990.

Chambert, T., L. Imberdis, C. Couloumy, R. Bonet and A. Besnard. 2020. Density dependence in Golden Eagle *Aquila chrysaetos* fecundity better explained by individual adjustment than territory heterogeneity. Ibis 162: 1312–1323.

Charnov, E. L. 1976a. Optimal foraging, attack strategy of a mantid. American Naturalist 110: 141–151.

Charnov, E., L. 1976b. Optimal foraging, the marginal value theorem. Theoretical Population Biology 9: 129–136.

Charnov, E. L., G. H. Orians and K. Hyatt. 1976. Ecological implications of resource depression. American Naturalist 110: 247–259.

Chase, J. M. and M. A. Liebold. 2003. Ecological Niches: Linking Classical and Contemporary Approaches. University of Chicago Press, Chicago.

Chatterjee, N., A. Kölzscg, S. C. Davidson, G. Borer and J. Fieberg. 2023. Tracking predators by algorithm. Wildlife Professional 17(4): 48–51.

Cheeseman, J. F., C. D. Millar, U. Greggers, K. Lehmann, M. D. M. Pawley, C. R. Gallistel, G. R. Warman and R. Menzel. 2014. Way-finding in displaced clock-shifted bees proves bees use a cognitive map. Proceedings of the National Academy of Sciences of the United States of America 111: 8949–8954.

Chettih, S. N., E. L. Mackevicius, S. Hale and D. Aronov. 2024. Bar coding of episodic memories in the hippocampus of a food-caching bird. Cell 187: 1–14.

Chirima, G. J. and N. Owen-Smith. 2017. Comparison of kernel density and local convex hull methods for assessing distribution ranges of large mammalian herbivores. Transactions in GIS 21: 359–375.

Chung, A., C. Jou, A. Grau-Perales, E. R. J. Levy, D. Dvorak, N. Hussain and A. A. Fenton. 2021. Cognitive control persistently enhances hippocampal information processing. Nature 600: 484–488.

Clayton, N. S., J. C. Reboreda and A. Kacelnik. 1997. Seasonal changes of hippocampus volume in parasitic cowbirds. Behavioural Processes 41: 237–243.

Clermont, J., C. Couchoux, D. Garant and D. Reale. 2017. Assessing anti-predator decisions of foraging eastern chipmunks under varying perceived risks: The effects of physical and social environments on vigilance. Behaviour 154: 131–148.

Cochran, W. W. 1972. Long-distance tracking of birds. Pp. 39–59 in Galler, S. R., K. Schmidt-Koenig, G. J. Jacobs and R. E. Belleville (editors), Animal Orientation and Navigation. National Aeronautics and Space Administration Special Publication 262, NASA, Washington, DC.

Cochran, W. W., G. G. Montgomery and R. R. Graber. 1967. Migratory flights of *Hylocichla* thrushes in spring: A radiotelemetry study. Living Bird 6: 213–225.

Cody, M. L. 1971. Finch flocks in the Mohave Desert. Theoretical Population Biology 2: 142–158.

Cody, M. L. 1974. Optimization in ecology. Science 183: 1156–1164.

Cohen, A. O., N. G. Matese, A. Filimontseva, X. Shen, T. C. Shi, E. Livne and C. A. Hartley. 2019. Aversive learning strengthens episodic memory in both adolescents and adults. Learning & Memory 26: 272–279. doi 10.1101/lm.048413.118

Cohen, N. J. 2015. Navigating life. Hippocampus 25: 704–708.

Collet, J., N. Pettorelli, A. Baniel, A. J. Carter, E. Huchard, A. J. King, A. E.G. Lee, H. H. Marshall and G. Cowlishaw. 2022. Immigrant males' knowledge influences baboon troop movements to reduce home range overlap and mating competition. Behavioral Ecology 33: 398–407.

Colombo, P. E., L. S. Elinder, E. P. A. Nykänen, E. Patterson, A. K. Lindroos and A. Parlesak. 2023. Developing a novel optimisation approach for keeping heterogeneous diets healthy and within planetary boundaries for climate change. European Journal of Clinical Nutrition 78(3): 193–201. doi: 10.1038/s41430-023-01368-7

Conrad, C. D., L. A. M. Galea, Y. Kuroda and B. S. McEwen. 1996. Chronic stress impairs rat spatial memory on the Y maze, and this effect is blocked by tianeptine pretreatment. Behavioral Neuroscience 110: 1321–1334.

Corsi, F., J. De Leeuw and A. Skidmore. 2000. Modeling species distributions with GIS. Pp. 389–434 in Boitani, L. and T. K. Fuller (editors), Research Techniques in Animal Ecology. Columbia University Press, New York.

Costanza, R., R. dArge, R. deGroot, S. Farber, M. Grasso, B. Hannon, K. Limburg, S. Naeem, R. V. Oneill, J. Paruelo, R. G. Raskin, P. Sutton and M. vandenBelt. 1997. The value of the world's ecosystem services and natural capital. Nature 387: 253–260.

Courellis, H. S., J. Minha, A. R. Cardenas, D. L. Kimmel, C. M. Reed, T. A. Valiante, C. D. Salzman, A. N. Mamelak, S. Fusi and U. Rutishauser. 2024. Abstract representations emerge in human hippocampal neurons during inference. Nature 632: 841–849.

Coutrot, A., E. Manley, S. Goodroe, C. Gahnstrom, G. Filomena, D. Yesiltepe, R. C. Dalton, J. M. Wiener, C. Holscher, M. Hornberger and H. J. Spiers. 2022. Entropy of city street networks linked to future spatial navigation ability. Nature 604: 104–110.

Coues, E. 1877. Furbearing Animals: A Monograph of North American Mustelidae. Department of the Interior Miscellaneous Publication 8. US Government Printing Office, Washington, DC.

Coulter, M. E. and C. Kemere. 2023. The neural basis of mental navigation in rats. Science 382: 517–518.

Craig, R. 1979. Parental manipulation, kin selection, and the evolution of altruism. Evolution 33: 319–334.

Craighead, J. J., J. S. Sumner and J. A. Mitchell. 1995. The Grizzly Bears of Yellowstone: Their Ecology in the Yellowstone Ecosystem, 1959–1992. Island Press. Washington, DC. 535 pp.

Crabtree, R. L. 1989. Sociodemography of an unexploited coyote population. PhD thesis, University of Idaho. 98 pp.

Crabtree, R. L., F. G. Burton, T. R. Garland, D. A. Cataldo and W. H. Rickard. 1989. Slow-release radioisotope implants as individual markers for carnivores. Journal of Wildlife Management 53: 949–954.

Creel, S., J. A. Winnie and D. Christianson. 2013. Underestimating the frequency, strength, and cost of antipredator responses with data from GPS collars: An example with wolves and elk. Ecology and Evolution 3: 5189–5200.

Crofoot, M. C., I. C. Gilby, M. C. Wikelski and R. W. Kays. 2008. Interaction location outweighs the competitive advantage of numerical superiority in *Cebus capucinus* intergroup contests. Proceedings of the National Academy of Sciences USA 105: 577–581.

Crofoot, M. C. and I. C. Gilby. 2012. Cheating monkeys undermine group strength in enenmy territory. Proceedings of the National Academy of Sciences USA 109: 501–505.

Cronk, N. E. and N. Pillay. 2021. Home range and use of residential gardens by yellow mongoose Cynictis penicillata in an urban environment. Urban Ecosystems 24: 127–139.

Cross, S. L., S. Tomlinson, M. D. Craig and P. W. Bateman. 2019. The time local convex hull method as a tool for assessing responses of fauna to habitat restoration: A case study using the perentie (Varanus giganteus: Reptilia: Varanidae). Australian Journal of Zoology 67: 27–37.

Croze, H. 1970. Searching image in carrion crows: Hunting strategy in a predator and some anti-predator devices in camouflaged prey. Zeitschrift fűr Tierpsychologie 5: 1–86.

Curio, E. 1976. The Ethology of Predation. Springer-Verlag, Berlin.

Currie, W. and E. D. Bellis. 1969. Home range and movements of bullfrog, Rana catesbeiana Shaw, in an Ontario pond. Copeia: 688–692.

Dacke, M., E. Baird, B. el Jundi, E. J. Warrant and M. Byrne. 2021. How dung beetles steer straight. Annual Review of Entomology 66: 243–256.

Dalke, P. D. and P. R. Sime. 1938. Home and seasonal ranges of the eastern cottontail in Connecticut. Transactions of the North American Wildlife Conference 3: 659–669.

Dalziel, B. D., J. M. Morales and J. M. Fryxell. 2008. Fitting probability distributions to animal movement trajectories: Using artificial networks to link distance, resources, and memory. American Naturalist 172: 248–258.

Danjo, T., T. Toyoizumi and S. Fujisawa. 2018. Spatial representations of self and other in the hippocampus. Science 359: 213–218.

Darwin, C. 1858. On the tendencies of species to form varieties and on the perpetuation of varieties and species by natural means of selection. Journal of the Proceedings of the Linnean Society, Zoology 3: 46–52.

Darwin, C. 1859. On the Origin of Species. Murray, London.

Darwin, C. 1871. Descent of Man and Selection in Relation to Sex. Murray, London.

Davies, J. R., E. Garcia-Pelegrin and N. S. Clayton. 2024. Eurasian jays (*Garrulus glandarius*) show episodic-like memory through the incidental encoding of information. Plos One 19(5): e0301298. https://doi.org/10.1371/journal.pone.0301298

Davies, N. B. and A. I. Houston. 1981. Owners and satellites: The economics of territory defence in the pied wagtail, *Motacilla alba*. Journal of Animal Ecology 50: 157–180.

Davis, A. 1970. Let's Eat Right to Keep Fit. New American Library, New York.

Davis S., B. Abbasi, S. Shah, S. Telfer and M. Begon. 2015. Spatial analyses of wildlife contact networks. Journal of the Royal Society Interface 12: 20141004. http://dx.doi.org/10.1098/rsif.2014.1004

DeBose, J. L. and G. A. Nevitt. 2008. The use of odors at different spatial scales: Comparing birds with fish. Journal of Chemical Ecology 34: 867–881.

De Montaigne, M. 1865. The language of animals: An apology of Taymond Sebond. In: The Works of Michel de Montaigne, 1865 (Hazlitt, W., translator). http://www.animal-rights-library.com/texts-c/montaigne01.htm

deVos, A. 1951. Recent findings in fisher and marten ecology and management. Transactions of the North American Wildlife Conference 16: 498–507.

deVos, A. 1952. Ecology and management of fisher and marten in Ontario. Technical Bulletin, Ontario Department of Lands and Forests.

Dice, L. R. and P. J. Clark. 1953. The statistical concept of home range and a consideration of the recapture radius of the prairie deermouse (*Peromyscus*). Anatomical Record 113: 545–545.

Diggle, P. J., J. A. Tawn and R. A. Moyeed. 1998. Model-based geostatistics. Journal of the Royal Statistical Society Series C: Applied Statistics 47: 299–326.

Dill, L. M. 1978. Energy-based model of optimal feeding-territory size. Theoretical Population Biology 14: 396–429.

Dixon, K. R. and J. A. Chapman. 1980. Harmonic mean measure of animal activity areas. Ecology 61: 1040–1044.

Dixon, M. M., P. L. Jones, M. J. Ryan, G. G. Carter and R. A. Page. 2022. Long-term memory in frog-eating bats. Current Biology 32: R555–R558.

Dimitri, L. A. and W. S. Longland. 2022. Pilfering of western juniper seed caches by scatter-hoarding rodents varies by microsite and canopy type. Integrative Zoology 17: 192–205.

Ditchkoff, S. S., D. B. Jolley, B. D. Sparklin, L. B. Hanson, M. S. Mitchell and J. B. Grand. 2012. Reproduction in a population of wild pigs (*Sus scrofa*) subjected to lethal control. Journal of Wildlife Management 76: 1235–1240.

Domenech, P., S. Rheims and E. Koechlin. 2020. Neural mechanisms resolving exploitation–exploration dilemmas in the medial prefrontal cortex. Science 369: 1076–1084.

Doncaster, C. P. 1990. Nonparametric estimates of interaction from radio-tracking data. Journal of Theoretical Biology 143: 431–443.

Doncaster, C. P. and D. W. Macdonald. 1991. Drifting territoriality in the red fox *Vulpes vulpes*. Journal of Animal Ecology 60: 423–439.

Doncaster, C. P. and D. W. Macdonald. 1992. Optimum group size for defending heterogeneous distributions of resources: A model applied to red foxes, *Vulpes vulpes*, in Oxford City. Journal of Theoretical Biology 159: 189–198.

Doncaster, C. P. and R. Woodroffe. 1993. Den site can determine shape and size of badger territories: Implications for group-living. Oikos 66: 88–93.

Donkor, L., E. Essien and N. S. Affrifah. 2023. A systematic review of linear programming techniques as applied to diet optimisation and opportunities for improvement. Journal of Optimization 2023(2): 1–17.

Dougherty, E. R., P. de Valpine, C. J. Carlson, J. K. Blackburn and W. M. Getz. 2018. Commentary to: A cross-validation-based approach for delimiting reliable home range estimates. Movement Ecology 6: 5.

Drew, L. 2025. Air pollution and brain damage: What the science says. Nature 637: 536–538.

Drewe, J. A. 2010. Who infects whom? Social networks and tuberculosis transmission in wild meerkats. Proceedings of the Royal Society B: Biological Sciences 277: 633–642.

Drewe, J. A., J. R. Madden and G. P. Pearce. 2009. The social network structure of a wild meerkat population: 1. Inter-group interactions. Behavioral Ecology and Sociobiology 63: 1295–1306.

Drupp, M. A., M. C. Hänsel, E. P. Fenichel, M. Freeman, C. Gollier, B. Groom, G. M. Heal, P. H. Howard, A. Millner, F. C. Moore, F. Nesje, M. F. Quaas, S. Smulders, T. Sterner, C. Traeger and F. Venmans. 2024. Accounting for the increasing benefits from scarce ecosystems. Science 383: 1062–1064.

Dubois, F. and L. A. Giraldeau. 2007. Food sharing among retaliators: Sequential arrivals and information asymmetries. Behavioral Ecology and Sociobiology 62: 263–271.

Dukas, R. and L. A. Real. 1993. Effects of nectar variance on learning by bumblebees. Animal Behaviour 45: 37–41.

Duncan, C., E. B. Nilsen, J. D. C. Linnell and N. Pettorelli. 2015. Life-history attributes and resource dynamics determine intraspecific home-range sizes in Carnivora. Remote Sensing in Ecology and Conservation 1: 39–50.

Dunkel, S. W. 2000. Dragonflies Through Binoculars. Oxford University Press, New York.

Dunn, J. E. and P. S. Gipson. 1977. Analysis of radio telemetry data in studies of home range. Biometrics 33: 85–101.

Dunning, J. B., B. J. Danielson and H. R. Pulliam. 1992. Ecological processes that affect populations in complex landscapes. Oikos 65: 169–175.

Eason, P. K., G. A. Cobbs and K. G. Trinca. 1999. The use of landmarks to define territorial boundaries. Animal Behaviour 58: 85–91.

Eaton, E. R. 2021. Wasps: The Astonishing Diversity of a Misunderstood Insect. Princeton University Press, Princeton, New Jersey, USA. 256 pp.

Ebbinghaus, H. 1885. Memory: A Contribution to Experimental Psychology. Republished 1964 by Dover, New York.

Ebersole, M. A. 1981. An experimental-analysis of territoriality in the California reef fish *Embiotoca jacksoni* (Embiotocidae). Copeia 3: 653–665.

Ebersole, J. P. 1980. Food density and territory size: An alternative model and a test on the reef fish *Eupomacentrus leucostictus*. American Naturalist 115: 492–509.

Edelhoff, H., J. Signer and N. Balkenhol. 2016. Path segmentation for beginners: An overview of current methods for detecting changes in animal movement patterns. Movement Ecology 4: 21.

Ehrlich, P. R. 1975. Population biology of coral-reef fishes. Annual Review of Ecology and Systematics 6: 211–246.

Eichenbaum, H. 2017. The role of the hippocampus in navigation is memory. Journal of Neurophysiology 117: 1785–1796.

Eichenbaum, H., D. G. Amaral, E. A. Buffalo, G. Buzsáki, N. Cohen, L. Davachi, L. Frank, S. Heckers, R. G. M. Morris, E. I. Moser, L. Nadel, J. O'Keefe, A. Preston, C. Ranganath, A. Silva and M. Witter. 2016. Hippocampus at 25. Hippocampus 26: 1238–1249.

Ekstrom, A. D., D. J. Huffman and M. Starrett. 2017. Interacting networks of brain regions underlie human spatial navigation: A review and novel synthesis of the literature. Journal of Neurophysiology 118: 3328–3344.

Ellison, N., B. J. Hatchwell, S. J. Biddiscombe and C. J. Napper. 2019. Mechanistic home range analysis reveals drivers of space use patterns for a non-territorial passerine. Journal of Animal Ecology 89: 2763–2776.

Ellison, N., B. J. Hatchwell, S. J. Biddiscombe, C. J. Napper and J. R. Potts. 2020. Mechanistic home range analysis reveals drivers of space use patterns for a non-territorial passerine. Journal of Animal Ecology 89: 2763–2776.

Ellner, S. and L. Real. 1989. Models for foraging in stochastic environments: Relationships and tests. Comments on Theoretical Biology 1: 129–158.

Emerson, S. E. and J. S. Brown. 2015. The influence of food chemistry on food-safety tradeoffs in samango monkeys. Journal of Mammalogy 96: 237–244.

Emlen, J. M. 1966. The role of time and energy in food preference. American Naturalist 100: 611–617.

Englund, G. and T. I. Olsson. 1990. Fighting and assessment in the net-spinning caddis larva Arctopsyche ladogensis: A test of the sequential assessment game. Animal Behaviour 39: 55–62.

Enquist, M. and O. Leimar. 1983. Evolution of fighting behavior: Decision rules and assessment of relative strength. Journal of Theoretical Biology 102: 387–410.

Enquist, M. and O. Leimar. 1987. Evolution of fighting behavior: The effect of variation in resource value. Journal of Theoretical Biology 127: 187–205.

Epling, B. P. and J. H. Powers. 2022. Cefiderocol and the need for higher-quality evidence: Methods matter for patients. Antimicrobial Agents and Chemotherapy 66(8): e0076622. doi: 10.1128/aac.00766-22

Epsztein, J. 2022. Mental replays enable flexible navigation. Nature 605: 35–36.

Erlinge, S. and M. Sandell. 1986. Seasonal changes in the social organization of male stoats, *Mustela erminea*. An effect of shifts between two decisive resources. Oikos 47: 57–62.

Evans, F. C. and R. Holdenried. 1943. A population study of the beechy ground squirrel in central California. Journal of Mammalogy 24: 231–260.

Evans, L., H. Cameron-Whytock and C. Ijichi. 2024. Whoa, no-go: Evidence consistent with model-based strategy use in horses during an inhibitory task. Applied Animal Behavior Science 277: 106339.

Ewer, R. F. 1968. Ethology of Mammals. Logos Press, London.

Facka, A. N, 2017. Conservation translocations as opportunities for scientific advancement: A case study with fishers (*Pekania pennanti*). PhD thesis, North Carolina State University, Raleigh, North Carolina. 299 pp.

Facka, A. N., J. C. Lewis, P. Happe, K. Jenkins, R. Callas and R. A. Powell. 2016. Timing of translocation influences birth rate and population dynamics in a forest carnivore. Ecosphere 7(1): e01223. doi:10.1002/ecs2.1223

Facka, A. N. and R. A. Powell. 2021. Intraspecific competition, habitat quality, niche partitioning, and causes of intrasexual territoriality for a reintroduced carnivoran. Frontiers in Ecology and Evolution 9: 734155. doi: 10.3389/fevo.2021.734155

Fagen, R. M. and D. Y. Young. 1978. Temporal patterns of behavior: durations, intervals, latencies, and sequences. Pp. 79–114 in Colgen, P. (editor), Quantitative Ethology. John Wiley and Sons, New York.

Fairbanks, B. and F. S. Dobson. 2007. Mechanisms of the group-size effect on vigilance in Columbian ground squirrels: Dilution versus detection. Animal Behaviour 73: 115–123.

Falcon-Cortes, A., D. Boyer, E. Merrill, J. L. Frair and J. M. Morales. 2021. Hierarchical, memory-based movement models for translocated elk (*Cervus canadensis*). Frontiers in Ecology and Evolution 9: 13.

Favreau, J. M. 2006. The effects of food abundance, foraging rules and cognitive abilities on local animal movements. PhD Dissertation, North Carolina State Univeristy, Raleigh. 219 pp.

Fawcett, T. W., S. Hamblin and L. A. Giraldeau. 2013. Exposing the behavioral gambit: The evolution of learning and decision rules. Behavioral Ecology 24: 2–11.

Fedorenko, E., S. T. Piantadosi and E. A. F. Gibson. 2024. Language is primarily a tool for communication rather than thought. Nature 630: 575–586.

Ferguson, E. S. 1977. Minds eye: Nonverbal thought in technology. Science 197: 827–836.

Fieberg, J. 2007a. Utilization distribution estimation with weighted kernel density estimators. Journal of Wildlife Management 71: 1669–1675.

Fieberg, J. 2007b. Kernel density estimators of home range, smoothing and the autocorrelation red herring. Ecology 88: 1059–1066.

Fieberg, J. and L. Börger. 2012. Could you please phrase "home range" as a question? Journal of Mammalogy 93: 890–902.

Fieberg, J. and C. Kochanny. 2005. Quantification of home range overlap: The importance of the utilization distribution. Journal of Wildlife Management 69: 1346–1359.

Finkel, E. 2023. Consciousness hunt yields results but not clarity. Science 380: 1309–1310.

Firth, J. A., B. C. Sheldon and L. J. N. Brent. 2017. Indirectly connected: Simple social differences can explain the causes and apparent consequences of complex social network positions. Proceedings of the Royal Society B: Biological Sciences 284. 20171939. http://dx.doi.org/10.1098/rspb.2017.1939

Fleming, C. H. and J. M. Calabrese. 2017. A new kernel density estimator for accurate home-range and species-range area estimation. Methods in Ecology and Evolution 8: 571–579.

Fleming, C. H., J. M. Calabrese, T. Mueller, K. A. Olson, P. Leimgruber and W. F. Fagan. 2014a. From fine-scale foraging to home ranges: A semivariance approach to identifying movement modes across spatiotemporal scales. American Naturalist 183: E154–E167.

Fleming, C. H., J. M. Calabrese, T. Mueller, K. A. Olson, P. Leimgruber and W. F. Fagan. 2014b. Non-Markovian maximum likelihood estimation of autocorrelated movement processes. Methods in Ecology and Evolution 5: 462–472.

Fleming, C. H., W. F. Fagan, T. Mueller, K. A. Olson, P. Leimgruber and J. M. Calabrese. 2015. Rigorous home range estimation with movement data: A new autocorrelated kernel density estimator. Ecology 96: 1182–1188.

Fleming, C. H., W. F. Fagan, T. Mueller, K. A. Olson, P. Leimgruber and J. M. Calabrese. 2016. Estimating where and how animals travel: An optimal framework for path reconstruction from autocorrelated tracking data. Ecology 97: 576–582.

Fleming, C. H., M. J. Noonan, E. P. Medici and J. M. Calabrese. 2019. Overcoming the challenge of small effective sample sizes in home-range estimation. Methods in Ecology and Evolution 10: 1679–1689.

Fleming, C. H., D. Sheldon, W. F. Fagan, P. Leimgruber, T. Mueller, D. Nandintsetseg, M. J. Noonan, K. A. Olson, E. Setyawan, A. Sianipar and J. M. Calabrese. 2018. Correcting for missing and irregular data in home-range estimation. Ecological Applications 28: 1003–1010.

Fletcher, R. J., C. W. Maxwell, J. E. Andrews and W. L. Helmey-Hartman. 2013. Signal detection theory clarifies the concept of perceptual range and its relevance to landscape connectivity. Landscape Ecology 28: 57–67.

Folse, L. J., J. M. Packard and W. E. Grant. 1989. AI modelling of animal movements in a heterogeneous habitat. Ecological Modelling 46: 57–72.

Ford, R. G. 1983. Home range in a patchy environment: Optimal foraging predictions. American Zoologist 23: 315–326.

Foresman, K. R. 2012. Carnivores in hand. Pp. 130–151 in Boitani, L. and R. A. Powell (editors), Carnivore Ecology and Conservation: A Handbook of Techniques. Oxford University Press, Oxford.

Forman, R. T. T. and M. Godron. 1986. Landscape Ecology. John Wiley and Sons, New York.

Forli, A. and M. M. Yartsev. 2023. Hippocampal representation during collective spatial behavior in bats. Nature. https://doi.org/10.1038/s41586-023-06478-7

Forrester, T. D., D. S. Casady and H. U. Wittmer. 2015. Home sweet home: Fitness consequences of site familiarity in female black-tailed deer. Behavioral Ecology and Sociobiology 69: 603–612.

Fortin, D., H. L. Beyer, M. S. Boyce, D. W. Smith, T. Duchesne and J. S. Mao. 2005. Wolves influence elk movements: Behavior shapes a trophic cascade in Yellowstone National Park. Ecology 86: 1320–1330.

Frafjord, K. and P. Prestrud. 1992. Home range and movements of arctic foxes *Alopex lagopus* in Svalbard. Polar Biology 12: 519–526.

Franckowiak, G. A., M. Perdicas and G. A. Smith. 2019. Spatial ecology of coyotes in the urbanizing landscape of the Cuyahoga Valley, Ohio. Plos One 14. https://doi.org/10.1371/journal.pone.0227028

Frank, P. W. 1964. On home range of limpets. American Naturalist 98: 99–104.

Freire, M., A. Bollix and M. Knaden. 2023. Absence of visual cues motivates desert ants to build their own landmarks. Current Biology 33: 2802–2805.

Fretwell, S. D. 1972. Populations in a Seasonal Environment. Princeton University Press, Princeton, NJ.

Fretwell, S. D. and H. L. Lucas., Jr. 1970. On territorial behavior and other factors influencing habitat distribution in birds. I. Theoretical development. Acta Biotheoretica 19: 16–36.

Fryxell, J. M., M. Hazell, L. Börger, B. D. Dalziel, D. T. Haydon, J. M. Morales, T. McIntosh and R. C. Rosatte. 2008. Multiple movement modes by large herbivores at multiple spatiotemporal scales. Proceedings of the National Academy of Sciences of the United States of America 105: 19114–19119.

Fuller, A. K., B. C. Augustine, D. J. Morin, K. Pigeon, J. Boulanger, D. C. Lee, F. Bisi and D. L. Garshelis. 2022. The occupancy-abundance relationship and sampling designs using occupancy to monitor populations of Asian bears. Global Ecology and Conservation 35: e02075.

Fuller, T.c K., L. D. Mech, and J. Fitts-Cochran. 2003. Wolf population dynamics. Pp. 161–191 in Mech, L. D. and L. Boitani (editors), Wolves: Behavior, Ecology, and Conservation. University of Chicago Press, Chicago, Illinois.

Fyhn, M., S. Molden, M. P. Witter, E. I. Moser and M. B. Moser. 2004. Spatial representation in the entorhinal cortex. Science 305: 1258–1264.

Gable, T. D., S. M. Johnson-Bice, A. T. Homkes, S. K. Windels and J. K. Bump. 2020. Outsized effect of predation: Wolves alter wetland creation and decolonization by killing ecosystem engineers. Science Advances 6: eabc5439.

Gable, T. D., S. K. Windels and A. T. Homkes. 2018a. Do wolves hunt freshwater fish in spring as a food source? Mammalian Biology 91: 30–33.

Gable, T. D., S. K. Windels, M. C. Romanski and F. Rosell. 2018b. The forgotten prey of an iconic predator: A review of interactions between grey wolves *Canis lupus* and beavers *Castor* spp. Mammal Review 48: 123–138. doi: 10.1111/mam.12118.

Gagliardo, A., P. Ioalè, M. Savini, G. Dell'Omo and V. P. Bingman. 2009. Hippocampal-dependent familiar area map supports corrective re-orientation following navigational error during pigeon homing: A GPS-tracking study. European Journal of Neuroscience 29: 2389–2400.

Gardner, R. J., L. Lu, T. Wernle, M. B. Moser and E. Moser. 2019. Correlation structure of grid cells is preserved during sleep. Nature Neuroscience 22: 598–608.

Garey, M.R. and D. S. Johnson. 1979. Computers and Intractability: A Guide to the Theory of NP-Completeness. Freeman, San Fransisco.

Garshelis, D. L. 2000. Delusions in habitat evaluation: Measuring use, selection, and importance. Pp. 111–164 in Boitani, L. and T. K. Fuller (editors), Research Techniques in Animal Ecology. Columbia University Press, New York.

Garshelis, D. L. and M. R. Pelton. 1980. Activity of black bears in the Great Smoky Mountains National Park. Journal of Mammalogy 61: 8–19.

Garshelis, D. L. and M. R. Pelton. 1981. Movements of black bears in the Great Smoky Mountains National Park. Journal of Wildlife Management 45: 912–925.

Garstang, M. 2004. Long-distance, low-frequency elephant communication. Journal of Comparative Physiology A: Neuroethology, Sensory Neural and Behavioral Physiology 190: 791–805.

Gautestad, A. O. 1998. Site fidelity and scaling vcomplexity in animal movement and habitat use: The multiscale random walk model. PhD dissertation, University of Oslo, Norway. 106 pp.

Gautestad, A. O. and I. Mysterud. 1993. Physical and biological mechanisms in animal movement processes. Journal of Applied Ecology 30: 523–535.

Gautestad, A. O. and I. Mysterud 1995. The home range ghost. Oikos 74: 195–204.

Gautestad, A. O. and I. Mysterud 2010. The home range fractal: From random walk to memory-dependent space use. Ecological Complexity 7: 458–470.

Gautestad, A. O., I. Mysterud and M. R. Pelton. 1995. Complex movement and scale-free habitat use: Testing the multiscaled home-range model on black bear telemetry data. Ursus 10: 219–234.

Gautestad, A. O., I. Mysterud and M. R. Pelton. 1998. Complex movement and scale-free habitat use: Testing the multiscaled home-range model on black bear telemetry data. Ursus 10: 219–234.

Gehr, B., N. C. Bonnot, M. Heurich, F. Cagnacci, S. Ciuti, A. J. M. Hewison, J. M. Gaillard, N. Ranc, J. Premier, K. Vogt, E. Hofer, A. Ryser, E. Vimercati and L. F. Keller. 2020. Stay home, stay safe: Site familiarity reduces predation risk in a large herbivore in two contrasting study sites. Journal of Animal Ecology 89: 1329–1339.

Gehrt, S. D. 2003. Raccoon *Procyon lotor* and allies. Pp. 611–634 in Feldhamer, G. A., B. C. Thompson and J. A. Chapman (editors), Wild Mammals of North America. Johns Hopkins University Press, Baltimore, Maryland.

Genzel, D. and M. M. Yartsev. 2021. The fully automated bat (FAB) flight room: A human-free environment for studying navigation in flying bats and its initial application to the retrosplenial cortex. Journal of Neuroscience Methods 348:108970.

Gerking, S. D. 1953. Evidence for the concepts of home range and territory in stream fishes. Ecology 34: 347–365.

Gerell, R. 1967. Food selection in relation to habitat in mink (*Mustela vison* Schreber) in Sweden. Oikos 18: 233–246.

Gerell, R. 1969. Activity patterns of mink *Mustela vison* Schreber in southern Sweden. Oikos 20: 451–460.

Gerell, R. 1970. Home ranges and movements of mink *Mustela vison* Schreber in southern Sweden. Oikos 21: 160–173.

Gese, E. M. 2001. Territorial defense by coyotes (*Canis latrans*) Yellowstone National Park, Wyoming: Who, how, where, when, why. Canadian Journal of Zoology 79: 980–987.

Gese, E. M., D. E. Andersen and O. J. Rongstad. 1990. Determining home-range size of resident coyotes from point and sequential locations. Journal of Wildlife Management 54: 501–506.

Getty, T. 1981. Territorial behavior of eastern chipmunks (*Tamias striatus*): Encounter avoidance and spatial time-sharing. Ecology 62: 915–921.

Getz, W. M. 2022. A hierarchical path-segmentation movement ecology framework. Ecological Processes 11: 56.

Getz, W. M., S. Fortmann-Roe, P. C. Cross, A. J. Lyons, S. J. Ryan and C. C. Wilmers. 2007. LoCoH: Nonparametric kernel methods for constructing home ranges and utilization distributions. Plos One 2: 11.

Getz, W. M. and C. C. Wilmers. 2004. A local nearest-neighbor convex-hull construction of home ranges and utilization distributions. Ecography 27: 489–505.

Gibson, J. J. 1979. The Perception of the Visual World. Houghton Mifflin, Boston, Massachusetts.

Gilbert, B. K. 2019. One of Us: A Biologist's Walk among Bears. Friesen Press, Victoria, British Columbia. 249 pp.

Gilbert, J. H., J. L. Wright, D. J. Lauten and J. R. Probst. 1997. Den and rest-site characteristics of American marten and fisher in northern Wisconsin. Pp. 135–145 in Proulx, G., H. N. Bryant and P. M. Woodard (editors), *Martes*: Taxonomy, Ecology, Techniques, and Management. The Provincial Museum of Alberta, Edmonton.

Gill, F. B. and L. L. Wolff. 1975. Economics of feeding territoriality in the golden-winged sunbird. Ecology 56: 333–345.

Gillingham, B. J. 1986. Sensory aspects of foraging behavior in the least weasel (*Mustela nivalis*). PhD thesis, University of Illinois: Urbana-Champaign.

Giraldeau, L. A. and R. Ydenberg. 1987. The center-edge effect: The result of a war of attrition between territorial contestants. Auk 104: 535–538.

Giuggioli L., J. R. Potts and S. Harris. 2011. Animal interactions and the emergence of territoriality. Plos Computational Biology 7: e1002008. doi:10.1371/journal.pcbi.1002008

Givon, S., M. Samina, O. Ben-Shahar and R. Segev. 2022. From fish out of water to new insights on navigation mechanisms in animals. Behavioural Brain Research 419: 8.

Glass, D. M., P. R. Prentice, A. D. Evans and O. J. Schlitz. 2022. Local differences in maximum temperature determine water use among desert bighorn sheep populations. Journal of Wildlife Management 86. doi: 10.1002/jwmg.22313

Gleason, H. A. 1926. The individualist concept of plant association. Torrey Botanical Club Bulletin 53: 7–26.

Gleitman, G. 1971. Forgetting of long-term memories in animals. Pp. 1–44 in Honig, W. K. and P. H. R. James (editors), Animal Memory. Academic Press, New York.

Godfrey, S. S., A. Sih and C. M. Bull. 2013. The response of a sleepy lizard social network to altered ecological conditions. Animal Behaviour 86: 763–772.

Goss-Custard, J. D. 1970. Feeding dispersion in some overwintering wading birds. Pp. 1–35 in Crook, J. H. (editor), Social Behavior in Birds and Mammals. Academic Press, London.

Goss-Custard, J. D. 1977. Optimal foraging and the size selection of worms by redshank Tringa totanus. Animal Behaviour 25: 10–29.

Gosselink, T. E., T. R. van Deelen, R. E. Warner and P. C. Mankin. 2007. Survival and cause-specific mortality of red foxes in agricultural and urban areas of Illinois. Journal of Wildlife Management 71: 1862–1873.

Gough, J. 2023. Cognitive science meets the mark of the cognitive: Putting the horse before the cart. Biology & Philosophy 38: 1. doi.org/10.1007/s10539-022-09889-4

Gould, J. L. 1986. The locale map of honey-bees: Do insects have cognitive maps? Science 232: 861–863.

Gould, N. P. 2020. Ecology of American black bears in an urban environment. PhD Dissertation. North Carolina State University. Raleigh. 126 pp.

Grandin. T. 2022. Visual Thinking: The Hidden Gifts of People Who Think in Pictures, Patterns, and Abstractions. Riverhead Books, New York. 340 pp.

Grant, J. W. A., C. A. Chapman and K. S. Richardson. 1992. Defended versus undefended home range size of carnivores, ungulates and primates. Behavioral Ecology and Sociobiology 31: 149–161.

Grant, J. W. A., L. K. Weir and S. O. Steingrímsson. 2017. Territory size decreases minimally with increasing food abundance in stream salmonids: Implications for population regulation. Journal of Animal Ecology 86: 1308–1316.

Grant, P. R. and B. R. Grant. 2011. Causes of lifetime fitness of Darwin's finches in a fluctuating environment. Proceedings of the National Academy of Sciences of the United States of America 108: 674–679.

Griffin, D. R. 1976. A possible window in the minds of animals: The ability of animals to communicate may permit experimental testing of the hypothesis that they have mental experiences and intentions. American Scientist 64: 530–536.

Griffin, D. R. 1977. Anthropomorphism. BioScience 27: 446–447.

Griffin, J. R. and W. B. Critchfield. 1972. The Distribution of Forest Trees in California. Pacific Southwest Forest and Range Experiment Station, Missoula, Montana.

Grimm, D. 2023. What are farm animals thinking? Science 382: 1103–1107.

Grossmann, I., M. Fein Berg, D. C. Parker, N. A. Christakis, P. E. Fetlock and W. A. Cunningham. 2023. AI and the transformation of social science research. Science 380: 1108–1109.

Gunner, R. M., R. P. Wilson, M. D. Holton, R. Scott, P Hoskins and C. M. Duarte. 2020. A new direction for differentiating animal activity based on measuring angular velocity about the yaw axis. Ecology and Evolution 10: 7872–7886. DOI: 10.1002/ece3.6515

Gunner, R. M., M. D. Holton, M. D. Scantlebury, O. L. Van Schalkwyk, H. M. English, H. J. Williams, P. Hopkins, F. Quintana, A. Gómez-Laich, L. Börger, J. Redcliffe, K. Yoda, T. Yamamoto, S. Ferreira, D. Govender, P. Viljoen, A. Bruna, S. H. Bell, M. J. Marks, N. C. Bennett, M. H. Tonini, C. M. Duarte, M. C. Van Rooyen, M. F. Bertelsen, C. J. Tambling and R. P. Wilson. 2021. Dead-reckoning animal movements in R: A reappraisal using *Gundog.Tracks*. Animal Biotelemetry 9: 23. https://doi.org/10.1186/s40317-021-00245-z

Gurarie, E., R. D. Andrews and K. L. Laidre. 2009. A novel method for identifying behavioural changes in animal movement data. Ecology Letters 12: 395–408.

Gurarie, E., F. Cagnacci, W. Peters, C. H. Fleming, J. M. Calabrese, T. Mueller and W. F. Fagan. 2017. A framework for modelling range shifts and migrations: Asking when, whither, whether and will it return. Journal of Animal Ecology 86: 943–959.

Gutowsky, S. E., M. L. Leonard, M. G. Conners, S. A. Shaffer and I. D. Jonsen. 2015. Individual-level variation and higher-level interpretations of space use in wide-ranging species: An albatross case study of sampling effects. Frontiers in Marine Science 2: 17.

Haenel, G. J., L. C. Smith and H. B. John-Alder. 2003. Home-range analysis in *Sceloporus undulatus*. II. A test of spatial relationships and reproductive success. Copeia 2003: 113–123.

Hagen, C. A., N. C. Kenkel, D. J. Walker, R. K. Baydack and C. E. Braun. 2001. Fractal-based spatial analysis of radiotelemetry. Pp. 167–187 in Millspaugh, J. and J. M. Marzluff (editors), Radio tracking and animal populations. Academic Press, New York.

Hagan, J. M. and J. R. Walters. 1990. Foraging behavior, reproductive success, and colonial nesting in ospreys. The Auk 107: 506–521.

Haig, D. 2020. From Darwin to Derrida: Selfish Genes, Social Selves, and the Meaning of Life. MIT Press, Cambridge, Massachusetts. 477 pp.

Hall, A. S. 2016. Acute artificial light diminishes Central Texas anuran calling behavior. American Midland Naturalist 175: 183–193.

Hamelin, F. M. and M. A. Lewis. 2010. A differential game theoretical analysis of mechanistic models for territoriality. Journal of Mathematical Biology 61: 665–694.

Hamilton, R. J. 1978. Ecology of the black bear in southeastern North Carolina. MS thesis, University of Georgia, Athens. 214 pp.

Hamilton, W. D. 1971. Geometry for the selfish herd. Journal of Theoretical Biology 31: 295–311.

Hamilton, W. J., Jr. 1937. Activity and home range of the field mouse, *Microtus pennsylvanicus pennsylvanicus*. Ecology 18: 255–263.

Harder, L. D. and L. A. Real. 1987. Why are bumblebees risk averse. Ecology 68: 1104–1108.

Harestad, A. S. and F. L. Bunnell. 1979. Home range and body weight: A reevaluation. Ecology 60: 389–402.

Harrington, F. H. and C. S. Asa. 2003. Wolf communication. Pp. 66–103 in Mech, L. D. and L. Boitani (editors), Wolves: Behavior, Ecology and Conservation. University of Chicago Press, Chicago, Illinois.

Harrington, L. A., A. L. Harrington and D. W. Macdonald. 2009. The smell of new competitors: The responses of American mink, Mustela vison, to the odours of otter, Lutra lutra and polecat, M. putorius. Ethology 115: 421–428.

Harris, R. B., S. G. Fancy, D. C. Douglas, et al. 1990. Tracking wildlife by satellite: Current systems and performance. Technical Report 30. U. S. Fish and Wildlife Service, Washington, DC.

Harrison, J. L. 1958. Range of movement of some Malayan rats. Journal of Mammalogy 39: 190–206.

Hatton, I. A., O. Azzarisi, A. Uteri and M. Smerlak. 2024. Diversity begets stability: Sublinear growth and competitive coexistence across ecosystems. Science 383: 1196. https://doi.org/10.1126/science.adg8488

Haugen, A. O. 1942. Home range of the cottontail rabbit. Ecology 23: 354–367.

Hawlena, D., D. Saltz, Z. Abramsky and A. Bouskila. 2010. Ecological trap for desert lizards caused by anthropogenic changes in habitat structure that favor predator activity. Conservation Biology 24: 803–809.

Hayne, D. W. 1949. Calculation of size of home range. Journal of Mammalogy 30: 1–18.

Hayne, D. W. 1950. Apparent home range of *Microtus* in relation to distance between traps. Journal of Mammalogy 31: 26–39.

Hayne, D. W. 1978. Experimental design and statistical analyses. Pymatuning Laboratory of Ecology Special Publication 5: 3–10.

Heft, H. 2013. Wayfinding, navigation, and environmental cognition from a naturalist's stance. Pp. 265–294 in Waller, D. and L. Nadel (editors), Handbook of Spatial Cognition. American Psychological Association, Washington, DC.

Hefty, K. L. and K. M. Stewart. 2018. Novel location data reveal spatiotemporal strategies used by a central-place forager. Journal of Mammalogy 99: 333–340.

Henrich, M., M. Burgueño, J. Hoyer, T. Haucke, V. Steinhage, H. S. Kühl and M. Heurich. 2024. A semi-automated camera trap distance sampling approach for population density estimation. Remote Sensing in Ecology and Conservation 10: 156–171.

Henrique, E. P., M. A. de Oliveira, D. C. Paulo, P. D. C. Pereira, C. Dias, L. S. de Siqueira, C. M. de Lima, D. D. Miranda, P. S. do Rego, J. Araripe, M. A. D. de Melo, D. G. Diniz, N. G. D. Magalhaes, D. F. Sherry, C. W. P. Diniz and C. G. Diniz. 2021. Contrasting migratory journeys and changes in hippocampal astrocyte morphology in shorebirds. European Journal of Neuroscience 54: 5687–5704.

Henschel, J. R. and J. D. Skinner. 1991. Territorial behavior by a clan of spotted hyenas *Crocuta crocuta*. Ethology 88: 223–235.

Herculano-Houzel, S. 2020. Birds do have a brain cortex – and think. Science 369: 1367–1368.

Hernández-Blanco, M., R. Costanza, H. J. Chen, D. deGroot, D. Jarvis, I. Kubiszewski, J. Montoya, K. Sangha, N. Stoeckl, K. Turner and V. Van't Hoff. 2022.

Ecosystem health, ecosystem services, and the well-being of humans and the rest of nature. Global Change Biology 28: 5027–5040.

Heuck, C., C. Herrmann, D. G. Schabo, R. Brandl and J. Albrecht. 2017. Density-dependent effects on reproductive performance in a recovering population of White-tailed Eagles *Haliaeetus albicilla*. Ibis 159: 297–310.

Hinsch, M. and J. Komdeur. 2017. What do territory owners defend against? Proceeding of the Royal Society B: Biological Sciences 284: 20162356. http://dx.doi.org/10.1098/rspb.2016.2356

Hinsch, M., I. Pen and J. Komdeur. 2013. Evolution of defense against depletion of local food resources in a mechanistic foraging model. Behavioral Ecology 24: 245–252.

Hirsch, B. T., M. P. Tujague, Y. E. Di Blanco, M. S. Di Bitetti and C. H. Janson. 2013. Comparing capuchins and coatis: Causes and consequences of differing movement ecology in two sympatric mammals. Animal Behaviour 86: 331–338.

Hixon, M. A. 1980. Food production and competitor density as the determinants of feeding territory size. American Naturalist 115: 510–530.

Holgate, P. 1971. Random walk models for animal behavior. Pp. 1–12 in Patil, G., E. Pielou and W. Walters (editors), Statistical Ecology: Sampling and Modeling Biological Populations and Population Dynamics. Volume 2 of Pennsylvania State Statistics. Pennsylvania State University Press, University Park, Pennsylvania.

Holland, J. S. 2024. Dog Smart: Life-changing Lessons in Dog Intelligence. National Geographic, Washington, DC. 299 pp.

Hollen, L. I., M. B. V. Bell, H. M. Wade, R. Rose, A. Russell, F. Niven, A. R. Ridley and A. N. Radford. 2011. Ecological conditions influence sentinel decisions. Animal Behaviour 82: 1435–1441.

Holling, C. S. 1959. Some characteristics of simple types of predation and parasitism. Canadian Entomologist 91: 385–398.

Holling, C. S. 1965. The functional response of predators to prey density and its role in mimicry and population regulation. Memoirs of the Entomological Society of Canada 97: 5–60.

Holzman, S., M. J. Conroy and J. Pickering. 1992. Home range, movements, and habitat use by coyotes in south-central Georgia. Journal of Wildlife Management 56: 139–146.

Hooten, M. B., D. S. Johnson, B. T. McClintock and J. M. Morales. 2017. Animal Movement: Statistical Models for Telemetry Data. CRC Press, Boca Raton.

Horne, J. S., J. Fieberg, L. Börger, J. L. Rachlow, J. M. Calabrese and C. H. Fleming. 2020. Animal home ranges. Pp. 315–332 in Murray, D. L. and B. K. Sanderock (editors), Population Ecology in Practice. John Wiley & Sons, London.

Horne, J. S. and E. O. Garton. 2006a. Selecting the best home range model: An information-theoretic approach. Ecology 87: 1146–1152.

Horne, J. S. and Garton, E. O. 2006b. Likelihood cross-validation versus least squares cross-validation for choosing the smoothing parameter in kernel home-range analysis. Journal of Wildlife Management 70: 641–648.

Horne, J. S., E. O. Garton, S. M. Krone and J. S. Lewis. 2007. Analyzing animal movements using Brownian bridges. Ecology 88: 2354–2363.

Horne, J. S., E. O. Garton and J. L. Rachlow. 2008. A synoptic model of animal space use: Simultaneous estimation of home range, habitat selection, and inter/intra-specific relationships. Ecological Modelling 214: 338–348.

Horne, J. S., A. M. Haines, M. E. Tewes and L. L. Laack. 2009. Habitat partitioning by sympatric ocelots and bobcats: Implications for recovery of ocelots in southern Texas. Southwestern Naturalist 54: 119–126.

Horner, M. A. 1986. Internal structure of home ranges of black bears and analyses of home range overlap. MS thesis, North Carolina State University, Raleigh, North Carolina. 54 pp.

Horner, M. A. and R. A. Powell. 1990. Internal structure of home ranges of black bears and analyses of home range overlap. Journal of Mammalogy 71: 402–410.

Hoskinson, R. L. and L. D. Mech. 1976. White-tailed deer migration and its role in wolf predation. Journal of Wildlife Management 40: 429–441.

Houle, D., C. Pélabon, G. P. Wagner and T. F. Hansen. 2011. Measurement and meaning in biology. Quarterly Review of Biology 86: 3–34.

Houston, A. I. and J. M. McNamara. 2005. John Maynard Smith and the importance of consistency in evolutionary game theory. Biology & Philosophy 20: 933–950.

Howard, H. E. 1907–1914. The British Warblers. R. H. Porter, London.

Howard, H. E. 1920. Territory in Bird Life. Oliver and Boyd, Edinburgh, Scotland. 484 pp.

Høydal, O. A., E. R. Skytoen, S. O. Andersson, M. B. Moser and E. I. Moser. 2019. Object–vector coding in the medial entorhinal cortex. Nature 568: 400–404.

Hubbell, S. 2001. The unified neutral theory of biodiversity and biogeography. Monographs in Population Biology 32. Princeton University Press, Princeton, New Jersey. 392 pp.

Huck, M., J. Davison and T. J. Roper. 2008. Comparison of two sampling protocols and four home-range estimators using radio-tracking from urban badgers *Meles meles*. Wildlife Biology 14: 467–477.

Hunger, C. 2021. How Stella learned to talk. William Morrow, New York. 260 pp.

Hunt, N. H., J. Jinn, L. F. Jacobs and R. J. Full. 2021. Acrobatic squirrels learn to leap and land on tree branches without falling. Science 373: 697–700.

Hurlbert, S. H. 1978. Measurement of niche overlap and some relatives. Ecology 59: 67–77.

Hurlbert, S. H. 1984. Pseudoreplication and the design of ecological field experiments. Ecological Monographs 54: 187–211.

Hutchinson, G. E. 1957. Concluding remarks, Cold Spring Harbor Symposium. Quantitative Biology 22: 415–427.

Huxley, J. 1934. A natural experiment on the territorial instinct. British Birds 27: 270–277.

Hüttner, T., L. von Fersen, L. Miersch and G. Dehnhardt. 2023. Passive electroreception in bottlenose dolphins (*Tursiops truncatus*): Implication for micro- and large-scale orientation. Journal of Experimental Biology 226: jeb245845. doi:10.1242/jeb.245845

Hyde, K. D., P. Baldrian, Y. P. Chen, K. W. T. Chethana, S. De Hoog, M. Doilom, A. R. G. de Farias, M. F. M. Gonçalves, D. Gonkhom, H. Gui, S. Hilario, Y. W. Hu, R. S. Jayawardena, S. Khyaju, P. M. Kirk, P. Kohout, T. Luangharn, S. S. N. Maharachchikumbura, I. S. Manawasinghe, P. E. Mortimer, A. G. T. Niego, M. Phonemany, B. Sandargo, I. C. Senanayake, M. Stadler, F. Surup, N. Thongklang, D. N. Wanasinghe, A. H. Bahkali and A. Walker. 2024. Current trends, limitations and future research in the fungi? Fungal Diversity 125: 1–71.

Ims, R. A. 1987a. Male spacing systems in microtine rodents. American Naturalist 130: 74–484.

Ims, R.A. 1987b. Responses in spatial organization and behaviour to manipulations of the food resource in the vole *Clethrionomys rufocanus*. Journal of Animal Ecology 56: 585–596.

Ims, R. A. 1988a. Spatial clumping of sexually receptive females induces space sharing among male voles. Nature 335: 541–543.

Ims, R. A. 1988b. The potential for sexual selection in males: Effect of sex ratio and spatiotemporal distribution of receptive females. Evolutionary Ecology 2: 338–352.

Ims, R. A. 1990. Mate detection success of male *Clethrionomys rufocanus* in relation to the spatial distribution of sexually receptive females. Evolutionary Ecology 4: 57–61.

Ingold, T. 2011. From science to art and back again: Essays in livelihood, dwelling and skill. Routledge, London. 602 pp.

Jacques, C. N., J. A. Jenks and R. W. Klaver. 2009. Seasonal movements and home-range use by female pronghorns in sagebrush-steppe communities of western South Dakota. Journal of Mammalogy 90: 433–441.

Jacobs, L. F. and W. D. Spencer. 1994. Natural space-use patterns and hippocampal site in kangaroo rats. Brain Behavior and Evolution 44: 125–132.

Jacobson, B., J. W. A. Grant and P. R. Peres-Neto. 2015. The interaction between the spatial distribution of resource patches and population density: Consequences for intraspecific growth and morphology. Journal of Animal Ecology 84: 934–942.

Jędrzejewska, B. and W. Jędrzejewski. 1990. Effect of a predator's visit on the spatial distribution of bank voles: Experiments with weasels. Canadian Journal of Zoology 68: 660–666.

Jędrzejewska, B. and W. Jędrzejewski. 1998. Predation in vertebrate communities: The Białowieża Primeval Forest as a case study. Ecological Studies series 135 (Series editors: J. G. Canadell, G. Heldmaier, D. F. Levia, E.-D. Schulze, U. Sommer, and D. A. Wardle). Springer-Verlag, Berlin.

Jędrzejewski, W., L. Rychlik and B. Jedrzejewska. 1993. Responses of bank voles to odors of 7 species of predators: Experimental data and their relevance to natural predator–vole relationships. Oikos 68: 251–257.

Jenkins, S. H. 1981. Common patterns in home-range–body-size relationships of birds and mammals. American Naturalist 118: 126–128.

Jenkins, S. H. and P. E. Busher. 1979. *Castor canadensis*. Mammalian Species 120: 1–8.

Jennrich, R. I. and F. B. Turner. 1969. Measurement of non-circular home range. Journal of Theoretical Biology 22: 227–237.

Jetz, W., C. Carbone, J. Fulford and J. H. Brown. 2004. The scaling of animal space use. Science 306: 266–268.

Johnson, D., N. Pelland and J. Sterling. 2021. A continuous-time semi-Markov model for animals in a dynamic environment. Annals of Applied Statistics 15: 797–812.

Johnson, D. D. P., D. W. Macdonald and A. J. Dickman. 2000. An analysis and review of models of the sociobiology of the Mustelidae. Mammal Review 30: 171–196.

Johnson, D. H. 1980. The comparison of usage and availability measurements for evaluating resource preference. Ecology 61: 65–71.

Johnson, D. S., D. L. Thomas, J. M. Ver Hoef and A. Christ. 2008. A general framework for the analysis of animal resource selection from telemetry data. Biometrics 64: 968–976. doi: 10.1111/j.1541-0420.2007.00943.x

Johnson, K. and M. R. Pelton. 1980. Prebaiting and snaring techniques for black bears. Wildlife Society Bulletin 8: 46–54.

Johnstone, R. A. 1996. Multiple displays in animal communication: 'Backup signals' and 'multiple messages.' Philosophical Transactions of the Royal Society B: Biological Sciences 351: 329–338.

Jones, E. N. 1990. Effects of forage availability on home range and population density of *Microtus pennsylvanicus*. Journal of Mammalogy 71: 382–389.

Jones, N. 2025. AI: Making it up. Nature 637: 778–780.

Jones, T. B., J. A. Green, S. C. Patrick, J. C. Evans, M. R. Wells, M. A. Rodríguez-Malagón and J. P. Y. Arnould. 2020. Consistent sociality but flexible social associations across temporal and spatial foraging contexts in a colonial breeder. Ecology Letters 23: 1085–1096.

Jones, W. T. 1989. Dispersal distance and the range of nightly movements in Merriams kangaroo rats. Journal of Mammalogy 70: 27–34.

Joshi, A. R., J. L. D. Smith and F. J. Cuthbert. 1995. Influence of food distribution and predation pressure on spacing behavior of palm civets. Journal of Mammalogy 76: 1205–1212.

Kadakia, N., M. Demir, B. T. Michaelis, B. D. DeAngelis, M. A. Reidenbach, D. A. Clark and T. Emonet. 2022. Odour motion sensing enhances navigation of complex plumes. Nature 611: 754–761.

Kamath, A. and J. Losos. 2018. Reconsidering territoriality is necessary for understanding Anolis mating systems. Behavioral Ecology and Sociobiology 72: 3.

Kamil, A. C. 1978. Systematic foraging for nectar by amakihi (*Loxops virens*). Journal of Comparative Physiological Psychology 92: 388–396.

Kang, G., C. A. H. Allard, W. A. Valencia-Montoya, L. van Giesen, J. J. Kim, P. B. Kilian, X. Bai, N. W. Bellono and R. E. Hibbs. 2023. Sensory specializations drive octopus and squid behavior. Nature 616: 378–383. https://doi.org/10.1038/s41586-023-05808-z

Kareiva, P. 1990. Population dynamics in spatially complex environments: Theory and data. Philosophical Transactions of the Royal Society of London 330: 175–190.

Karelus, D. L., B. W. Geary, L. A. Harveson and P. M. Harveson. 2021. Movement ecology and space-use by mountain lions in West Texas. Global Ecology and Conservation 31: e01859.

Karpanty, S. M., J. D. Fraser, J. Berkson, L. J. Niles, A. Dey and E. P. Smith. 2006. Horseshoe crab eggs determine red knot distribution in Delaware Bay. Journal of Wildlife Management 70: 1704–1710.

Kashetsky, T., T. Avgar and R. Dukas. 2021. The cognitive ecology of animal movement: Evidence from birds and mammals. Frontiers in Ecology and Evolution 9: 724887.

Kaspar, C., B. J. Ravoo, W. G. van der Wiel, S. V. Wegner and W. H. P. Pernice. 2021. The rise of intelligent matter. Nature 594: 345–355.

Katajisto, J. and A. Moilanen. 2006. Kernel-based home range method for data with irregular sampling intervals. Ecological Modelling 194: 405–413.

Kaufmann, J. H. 1962. Ecology and social behavior of the coati, *Nasua narica*, on Barro Colorado Island, Panama. University of California Publications in Zoology 60: 95–222.

Kay, C. 2007. Were native people keystone predators? A continuous-time analysis of wildlife observations made by Lewis and Clark in 1804–1806. Canadian Field Naturalist 121: 1–16.

Kays, R., M. C. Crofoot, W. Jetz and M. Wikelski. 2015. Terrestrial animal tracking as an eye on life and planet. Science 348: 1222. doi: 10.1126/science.aaa2478

Kays, R., A. Q. Vining, B. Hirsch, B. Alavi, R. W. Havmøller, L. W. Havmøller and M. Crofoot. In press. Remarkably strict territoriality in kinkajous. In Page, R. A., A. Aiello and W. T. Wcislo. The First 100 Years of Research on Barro Colorado: Animal Ecology, Evolution, and Behavior. Smithsonian Institution Scholarly Press. Washington, DC.

Kelt, D. A. and D. H. Van Vuren. 2001. The ecology and macroecology of mammalian home range area. American Naturalist 157: 637–645.

Kemp, G. A. and L. B. Keith. 1970. Dynamics and regulation of red squirrel (*Tamiasciurus hudsonicus*) populations. Ecology 51: 763–779.

Kenkre V. M. and L. Giuggioli. 2021. Extraction of parameters from observations, emergence of home ranges, and applicability of the fisher equation for bacteria. Pp. 98–125 in Kenkre, V. M. and L. Giugglioli, Theory of the Spread of Epidemics and Movement Ecology of Animals: An Interdisciplinary Approach Using Methodologies of Physics and Mathematics. Cambridge University Press, Cambridge.

Kie, J. G., R. T. Bowyer, B. B. Boroski, M. C. Nicholson and E. R. Loft. 2002. Landscape heterogeneity at differing scales: Effects on spatial distribution of mule deer. Ecology 83: 530–544.

Kie, J. G., J. Matthiopoulos, J. Fieberg, R. A. Powell, F. Cagnacci, M. S. Mitchell, J.-M. Gaillard and P. R. Moorcroft. 2010. The home-range concept: Are traditional estimators still relevant with modern telemetry technology? Philosophical Transactions of the Royal Society B: Biological Sciences 365: 2221–2231.

Killeen, J., H. Thurfjell, S. Ciuti, D. Paton, M. Musiani and M. S. Boyce. 2014. Habitat selection during ungulate dispersal and exploratory movement at broad and fine scale with implications for conservation management. Movement Ecology 2: 15.

Kimler, W. C. 2000. Reading Morgan's Canon: Reduction and unification in the foraging of a science of the mind. American Zoologist 40: 853–861.

Kimmerer, R. W. 2017. Speaking of nature: Finding language that affirms our kinship with the natural world. Orion Magazine 36 (2).

Kjelstrup, K. B., T. Solstad, V. H. Brun, T. Hafting, S. Leutgeb, M. P. Witter, E. I. Moser and M. B. Moser. 2008. Finite scale of spatial representation in the hippocampus. Science 321: 140–143.

Kleiber, M. 1975. The Fire of Life. 2nd edition. John Wiley & Sons, New York.

KleinHeerenbrink, M., L. A. France, C. H. Brighton and G. K. Taylor. 2022. Optimization of avian perching manoeuvres. Nature 607: 91–96.

Klenova, A. V., M. V. Goncharova and T. A. Kashentseva. 2020. Long-term stability in the vocal duets of the endangered Siberian Crane *Leucogeranus leucogeranus*. Polar Biology 43: 813–823.

Klopfer, P. H. 1969. Habitats and Territories: A Study of the Use of Space by Animals. Basic Books, New York.

Klopfer, P. H. and Rubenstein, D. I. 1977. The concept of privacy and its biological basis. Journal of Social Issues 33: 52–65.

Kochanny, C. O., DelGiudice, G. D. and J. Fieberg. 2009. Comparing global positioning system and very high frequency telemetry home ranges of white-tailed deer. Journal of Wildlife Management 73: 779–787.

Kodric-Brown, A. and J. H. Brown. 1978. Influence of economics, interspecific competition, and sexual dimorphism on territoriality of migrant rufous hummingbirds. Ecology 59: 285–296.

Koeppl, J. W., N. A. Slade, K. S. Harris and R. S. Hoffmann. 1977a. 3-dimensional home range model. Journal of Mammalogy 58: 213–220.

Koeppl, J. W., N. A. Slade and R. S. Hoffmann. 1975. Bivariate home range model with possible application to ethological data-analysis. Journal of Mammalogy 56: 81–90.

Koeppl, J. W., N. A. Slade and R. S. Hoffmann. 1977b. Distance between observations as an index of average home range size. American Midland Naturalist 98: 476–482.

Kotler, B. P., J. S. Brown and O. Hasson. 1991. Factors affecting gerbil foraging behavior and rates of owl predation. Ecology 72: 2249–2260.

Kozhevnikov, M., O. Blazhenkova and M. Becker. 2010. Trade-off in object versus spatial visualization abilities: Restriction in the development of visual-processing resources. Psychonomic Bulletin & Review 17: 29–35. doi:10.3758/PBR.17.1.29

Kozhevnikov, M., M. Hegarty and R. E. Mayer. 2002. Revising the visualizer–verbalizer dimension: Evidence for two types of visualizers. Cognition and Instruction 20: 47–77.

Kozhevnikov, M., S. Kosslyn and J. Shephard. 2005. Spatial versus object visualizers: A new characterization of visual cognitive style. Memory & Cognition 33: 710–726.

Kranstauber, B. 2019. Modelling animal movement as Brownian bridges with covariates. Movement Ecology 7: 22.

Kranstauber, B., R. Kays, S. D. LaPoint, M. Wikelski and K. Safi. 2012. A dynamic Brownian bridge movement model to estimate utilization distributions for heterogeneous animal movement. Journal of Animal Ecology 81: 738–746.

Kranstauber, B., K. Safi and F. Bartumeus. 2014. Bivariate Gaussian bridges: Directional factorization of diffusion in Brownian bridge models. Movement Ecology 2: 5.

Krebs, C. J. 1966. Demographic changes in fluctuating populations of Microtus californicus. Ecological Monographs 36; 239–273.

Krebs, J. R. and A. Kacelnik. 1991. Decision-making. Pp. 105–136 in Krebs, J. R. and N B. Davies (editors), Behavioural Ecology: An Evolutionary Approach. Blackwell Scientific Publications, London.

Krebs, J. R., D. F. Sherry, S. D. Healy, V. H. Perry and A. L. Vaccarino. 1989. Hippocampal specialization of food-storing birds. Proceedings of the National Academy of Sciences of the United States of America 86: 1388–1392.

Krige, D. G. 1951. Statistical approach to some mine valuations and allied problems of Wiswatersrand. Unpublished master's degree thesis. University of Witwatersrand, South Africa.

Krohn, W. 2012. Distribution changes of American martens and fishers in eastern North America, 1699–2001. Pp. 58–77 in Aubry, K. B., W. Zielinski, M. G. Raphael, G. Proulx and S. W. Buskirk (editors), Biology and Conservation of Martens, Sables, and Fishers: A New Synthesis. Cornell University Press, Ithaca, New York.

Krohn, W. B. and C. L. Hoving. 2010. Early Maine Wildlife. University of Maine Press, Orono. 523 pp.

Kruuk, H. 1972a. Surplus killing by carnivores. Journal of Zoology 166: 233–244.

Kruuk, H. 1972b. The Spotted Hyaena: A Study of Predation and Social Behavior. University of Chicago Press, Chicago, Illinois.

Kruuk, H. 1989. The Social Badger. Oxford University Press, Oxford.

Kruuk, H. and T. Parish. 1982. Factors affecting population density, group size and territory size of the European badger, *Meles meles*. Journal of Zoology 196: 31–39.

Kubiszewski, I., C. Ward, K. E. Pickett and R. Costanza. 2024. The complex relationships between economic inequality and biodiversity: A scoping review. Anthropocene Review 11: 49–66.

Kuhn, T. S. 1962. The Structure of Scientific Revolutions. University of Chicago Press, Chicago, Illinois.

Lai, C. X., S. Tanaka, T. D. Harris and A. K. Lee. 2023. Volitional activation of remote place representations with a hippocampal brain-machine interface. Science 382: 566–573.

Laidre, M. E. and S. L. Vehrencamp. 2008. Is bird song a reliable signal of aggressive intent? Behavioral Ecology and Sociobiology 62: 1207–1211.

Lair, H. 1987. Estimating the location of the focal center in red squirrel home ranges. Ecology 68: 1092–1101.

Landers, J. L., R. J. Hamilton, A. S. Johnson and R. L. Marchinton. 1979. Foods and habitat of black bears in southeastern North Carolina. Journal of Wildlife Management 43: 143–153.

Langer, T. J. 2006. Population estimates with age and genetic structure of a harvested bear population in eastern North Carolina. PhD thesis, North Carolina State University, Raleigh, North Carolina. 208 pp.

Laundré, J. W., L. Hernández and K. B. Altendorf. 2001. Wolves, elk, and bison: Reestablishing the "landscape of fear" in Yellowstone National Park, USA. Canadian Journal of Zoology-Revue Canadienne De Zoologie 79: 1401–1409.

Laver, P. N. and M. Kelly. 2005. A critical review of home range studies. Journal of Wildlife Management 72: 290–298.

Laver, P. N. and M. J. Kelly. 2008. A critical review of home range studies. Journal of Wildlife Management 72: 290–298. doi: 10.2193/2005-589

Laverty, T. M. and J. Berger. 2020. Do bats seek clean water? A perspective on biodiversity from the Namib Desert. Biological Conservation 248. doi:10.1016/j.biocon.2020.108686

Lawler, E., J. Lenstra, A. Rinnooy Kan and D. Shmoys. (editors). 1985. The Traveling Salesman Problem: A Guided Tour of Combinatorial Optimization. Wiley, New York.

Lay, D. W. 1942. Ecology of the opossum in eastern Texas. Journal of Mammalogy 23: 147–159.

Lee, J. 2023. What is cognitive about 'plant cognition'? Biology & Philosophy 38: 18. doi.org/10.1007/s10539-023-09907-z

Lee, J. Y., H. Jun, S. Soma, T. Nakazono, K. Shiraiwa, A. Dasgupta, T. Nakagawa, J. L. Xie, J. Chavez, R. Romo, S. Yungblut, M. Hagihara, K. Murata and K. M. Igarashi. 2021. Dopamine facilitates associative memory encoding in the entorhinal cortex. Nature 598: 321–326.

Lemanski, N., M. Silk, N. Fefferman and O. Udiani. 2021. How territoriality reduces disease transmission among social insect colonies. Behavioral Ecology and Sociobiology 75: 164.

Lenharo, M. 2023. Philosopher wins consciousness bet with neuroscientist. Nature 619: 14–15.

Lenharo, M. 2024a. AI consciousness: Scientists say we need answers. Nature 625: 226–226.

Lenharo, M. 2024b. Consciousness: The future of an embattled field. Nature 625: 438–440.

Leopold, A. 1933. Game Management. Charles Scribner and Sons, New York.

Leopold, A. 1949. A Sand County Almanac and Sketches Here and There. Oxford University Press, New York.

Lesmerises, F., C. Dassault and M.-H. St-Laurent. 2012. Wolf habitat selection is shaped by human activities in a highly managed boreal forest. Forest Ecology and Management 276: 125–131. https://doi.org/10.1016/j.foreco.2012.03.025.

Leutgebz, S., J. K. Leutgeb, E. I. Moser and M. B. Moser. 2006. Fast rate coding in hippocampal CA3 cell ensembles. Hippocampus 16: 765–774.

Levins, R. 1966. Strategy of model building in population biology. American Scientist 54: 421–431.

Leviton, M. 2022. Gray matter: Daniel J. Levitin on why memory isn't so black and white. The Sun 2022(544): 4–13.

Lewis, J. C., R. A. Powell and W. J. Zielinski. 2012. Carnivore translocations and conservation: insights from population models and field data for fishers (*Martes pennanti*). Plos One. http://dx.plos.org/10.1371/journal.pone.0032726

Lewis, J. C., K. J. Jenkins, P. J. Happe, D. J. Manson and M. McCalmon. 2016. Landscape-scale habitat selection by fishers translocated to the Olympic Peninsula of Washington. Forest Ecology and Management 369: 170–183.

Lewis, M. A., W. F. Fagan, M. Auger-Methe, J. Frair, J. M. Fryxell, C. Gros, E. Gurarie, S. D. Healy and J. A. Merkle. 2021. Learning and animal movement. Frontiers in Ecology and Evolution 9: 20.

Lewis, M. A. and J. D. Murray. 1993. Modelling territoriality and wolf–deer interactions. Nature 366: 738–740.

Lewis, L. S., E. G. Wessling, F. Kano, J. M. G. Stevens, J. Call and C. Krupenye. 2023. Bonobos and chimpanzees remember familiar conspecifics for decades. Proceedings of the National Academy of Sciences 120: e2304903120. https://doi.org/10.1073/pnas.2304903120

Lewontin, R. C. 1961. Evolution and the theory of games. Journal of Theoretical Biology 1: 382–403.

Li, Z. H., B. L. Ding, F. Wu, T. K. H. Lei, R. Kays and M. C. Crofoot. 2013. Attraction and avoidance detection from movements. Proceedings of the Vldb Endowment 7: 157–168.

Liberti, W. A. III, T. A. Schmid, A. Forli, M. Snyder and M. M. Yartsev. 2022. A stable hippocampal code in freely flying bats. Nature 604: 98–103.

Lichti, N. I. and R. K. Swihart. 2011. Estimating utilization distributions with kernel versus local convex hull methods. Journal of Wildlife Management 75: 413–422.

Lilkendey, J., C. Barrelet, J. Zhang, M. Meares, H. Larbi, G. Subsol, M. Chaumont and A. Sabetian. 2024. Herbivorous fish feeding dynamics and energy expenditure on a coral reef: Insights from stereo-video and AI-driven 3D tracking. Ecology and Evolution 14(3): e11070. doi: 10.1002/ece3.11070

Lima, S. L. and L. M. Dill. 1990. Behavioral decisions made under the risk of predation: A review and prospectus. Canadian Journal of Zoology 68: 619–640.

Lin, S. 1965. Computer solutions of traveling salesman problem. Bell Systems Technical Journal 44: 2245–2269.

Lincoln, R. J., G. A. Boxshall and P. F. Clark (Compilers). 1982. A Dictionary of Ecology, Evolution and Systematics. Cambridge University Press, Cambridge. 298 pp.

Lindstedt, S. L., B. J. Miller and S. W. Buskirk. 1986. Home range, time, and body size in mammals. Ecology 67: 413–418.

Lindström, E. 1989. Food limitation and social regulation in a red fox population. Holarctic Ecology 12: 70–79.

Lindzey, F. G. and E. C. Meslow. 1977. Home range and habitat use by black bears in southwestern Washington. Journbal of Wildlife Management 41: 413–425.

Linsdale, J. M. 1946. The California Ground Squirrel. University of California Press, Berkeley and Los Angeles.

Litvaitis, J. A., J. A. Sherburne and J. A. Bissonette. 1986. Bobcat habitat use and home range size in relation to prey density. Journal of Wildlife Management 50: 110–117.

Liu, C., R. Todorova, W. Tang, A. Oliva and A. Fernandez-Ruiz. 2023. Associative and predictive hippocampal codes support memory-guided behaviors. Science 382: 283. doi.org/10.1126/science.adi8237

Liu, C. Q., Z. F. Ye, C. S. Chen, N. Axmacher and G. Xue. 2022. Hippocampal representations of event structure and temporal context during episodic temporal order memory. Cerebral Cortex 32: 1520–1534.

Lloyd, M. 1967. Mean crowding. Journal of Animal Ecology 36: 1–30.

Lloyd-Smith, J. O., P. C. Cross, C. J. Briggs, M. Daugherty, W. M. Getz, J. Lotto, M S. Sanchez, A. B. Smith and A. Swei. 2005. Should we expect population thresholds for wildlife disease? Trends in Ecology and Evolution. 20 (9): 511–519. doi:https://doi.org/10.1016/j.tree.2005.07.004

Lockie, J. D. 1966. Territory in small carnivores. Symposium of the Zoological Society of London 18: 143–165.

Loehle, C. 1990. Home range: a fractal approach. Landscape Ecology 5: 39–52.

Łomnicki, A. 1978. Individual differences between animals and the natural regulation of their numbers. Journal of Animal Ecology 47: 461–475.

Łomnicki, A. 1988. Population Ecology of Individuals. Monographs in Population Biology 25. Princeton University Press, Princeton, New Jersey. 240 pp.

Łomnicki, A. 1999. Individual-based models and the individual-based approach to population ecology. Ecological Modelling 115: 191–198.

López-Sepulcre, A. and H. Kokko. 2005. Territorial defense, territory size, and population regulation. American Naturalist 166: 317–329.

López-Bao, J. V., A. Rodríguez, M. Delibes, J. M. Fedriani, J. Calzada, P. Ferreras and F. Palomares. 2014. Revisiting food-based models of territoriality in solitary predators. Journal of Animal Ecology 83: 934–942.

Lourie, E., I. Schiffner, S. Toledo and R. Nathan. 2021. Memory and conformity, but not competition, explain partial partitioning between two neighboring fruit bat colonies. Frontiers in Ecology and Evolution. doi: 10.3389/fevo.2021.732514

Lyons, A. J., W. C. Turner and W. M. Getz. 2013. Home range plus: A space-time characterization of movement over real landscapes. Movement Ecology 1: 14.

MacArthur, R. H. and E. R. Pianka. 1966. On optimal use of a patchy environment. American Naturalist 100: 603–609.

Macdonald, D. W. 1979. 'Helpers' in fox society. Nature 282: 69–71.

Macdonald, D. W. 1981. Resource dispersion and the social organization of the red fox, *Vulpes vulpes*. Pp. 918–949 in Chapman, J. A. and D. Pursley (editors), Proceedings of the Worldwide Furbearer Conference. Worldwide Furbearer Conference, Baltimore, Maryland.

Macdonald, D. W. 1983. The ecology of carnivore social behavior. Nature 301: 379–384.

Macdonald, D. W. 2021. Foreword. pp. ix–xiii in Vucetich 2021.

Macdonald, D. W. and G. M. Carr. 1989. Food security and the rewards of tolerance. Pp. 75–99 in Standen, V. and R. A. Foley (editors), Comparative Socioecology: The Behavioural Ecology of Humans and Other Mammals. Blackwell, Oxford.

Macdonald, K. E., J. W. Bartlett, K. K. Leung, S. Ourselin and J. Barnes. 2013. The value of hippocampal and temporal horn volumes and rates of change in predicting future conversion to AD. Alzheimer Disease & Associated Disorders 27: 168–173.

MacNulty, D. R., D. W. Smith, L. D. Mech, J. A. Vucetich and C. Packer. 2012. Nonlinear effects of group size on the success of wolves hunting elk. Behavioral Ecology 23: 75–82.

Madden, J. R., J. A. Drewe, G. P. Pearce and T. H. Clutton-Brock. 2009. The social network structure of a wild meerkat population: 2. Intragroup interactions. Behavioral Ecology and Sociobiology 64: 81–95.

Magowan, E. A., I. E. Maquire, S. Smith, S. Redpath, N. J. Marks, R. P. Wilson, F. Menzies, M. O'Hagan and D. M. Scantlebury. 2022. Dead-reckoning elucidates fine-scale habitat use by European badgets *Meles meles*. Animal Biotelemetry 10: 10. https://doi.org/10.1186/s40317-022-00282-2

Mahoney, P. J. and J. K. Young. 2017. Uncovering behavioural states from animal activity and site fidelity patterns. Methods in Ecology and Evolution 8: 174–183.

Mansfield S.A. 2007. Effects of supplemental food on weights and reproductive success of black bears in northeastern Minnesota. Master's thesis, Antioch University, New England.

Mansfield, S. A., L. L. Rogers, S. Robison and R. A. Powell 2022. Bed site selection by female North American black bears (*Ursus americanus*). Journal of Mammalogy 103: 361–372.

Mao, Y. X., H. J. Liu, Y. Wang and E. D. Brenner. 2023. A deep learning approach to track Arabidopsis seedlings' circumnutation from time-lapse videos. Plant Methods 19: 18. doi: https://doi.org/10.1186/s13007-023-00984-5

Mares, R., A. J. Young and T. H. Clutton-Brock. 2012. Individual contributions to territory defence in a cooperative breeder: Weighing up the benefits and costs. Proceedings of the Royal Society B: Biological Sciences 279: 3989–3995.

Marlow, R. W. and K. Tollestrup. 1982. Mining and exploitation of natural mineral deposits by the desert tortoise, *Gopherus agassizii*. Animal Behaviour 30: 475–478.

Martin, J. K. and A. A. Martin. (2007). Resource distribution influences mating system in the bobuck (*Trichosurus cunninghami*: Marsupialia). Oecologia 154(1): 227–236. doi: 10.1007/s00442-007-0823-y

Martinez-Garcia, R., C. H. Fleming, R. Seppelt, W. F. Fagan and J. M. Calabrese. 2020. How range residency and long-range perception change encounter rates. Journal of Theoretical Biology 498:110267. doi: 10.1016/j.jtbi.2020.110267

Martinig, A. R., H. J. Karst, E. R. Siracusa, E. K. Studd, A. G. McAdam, B. Dantzer, D. M. Delaney, J. E. Lane, P. Pokharel and S. Boutin. 2022. Animal personality: A comparison of standardized assays and focal observations in North American red squirrels. Animal Behaviour 190: 221–232.

Mason, C., R. Alderman, J. McGowan, H. P. Possingham, A. J. Hobday, M. Sumner and J. Shaw. 2018. Telemetry reveals existing marine protected areas are worse than random for protecting the foraging habitat of threatened shy albatross (*Thalassarche cauta*). Diversity and Distributions 24: 1744–1755.

Mathis, A., P. Mamidanna, K. M. Cury, T. Abe, V. N. Murthy, M. W. Mathis and M. Bethge. 2018. DeepLabCut: Markerless pose estimation of user-defined body parts with deep learning. Nature Neuroscience 21: 1281–1289.

Matthiopoulos, J. 2003a. The use of space by animals as a function of accessibility and preference. Ecological Modelling 159: 239–268.

Matthiopoulos, J. 2003b. Model-supervised kernel smoothing for the estimation of spatial usage. Oikos 102: 367–377.

Mauritzen, M., A. E. Derocher and O. Wiig. 2001. Space-use strategies of female polar bears in a dynamic sea ice habitat. Canadian Journal of Zoology 79: 1704–1713.

May, R. 1973. Stability and Complexity in Model Ecosystems. Monographs in Population Biology 6. Princeton University Press, Princeton, New Jersey.

May, R. M. 1974. Biological populations with nonoverlapping generations: Stable points, stable cycles, and chaos. Science 186: 645–647.

May, R. M. 1989. Citation classic: Stability and complexity in model ecosystems. Current Contents/Agriculture, Biology and Environmental Sciences 50: 22.

Mayes, A. 1983. Memory in Animals and Humans. Cambridge University Press, Cambridge, United Kingdom. 338 pp.

Maynard Smith, J. 1974. The theory of games and the evolution of animal conflicts. Journal of Theoretical Biology 47: 209–222.

Maynard Smith, J. 1976. Evolution and the theory of games. American Scientist 64: 41–45.

Maynard Smith, J. 1982. Evolution and the Theory of Games. Cambridge University Press, Cambridge. 224 pp.

Mayor, S. J., D. C. Schneider, J. A. Schaefer and S. P. Mahoney. 2009. Habitat selection at multiple scales. Ecoscience 16: 238–247.

Mazurkiewicz, M. 1969. Elliptical modification of home range pattern. Bulletin De L Academie Polonaise Des Sciences-Serie Des Sciences Biologiques 17: 427–431.

Mazurkiewicz, M. 1970. Analysis of home range directions based on catch-mark-release method. Bulletin De L Academie Polonaise Des Sciences-Serie Des Sciences Biologiques 18: 465–468.

McAdam, A. G., Q. M. R. Webber, B. Dantzer, J. E. Lane and S. Boutin 2022. Social effects on annual fitness in red squirrels. Journal of Heredity 113: 69–78.

McKeown, B., Z. Walton and T. Willebrand. 2020. Does recursive use of resource locations shape a home range? Exploring the red fox's cognitive map. Wildlife Biology. doi: 10.2981/wlb.00602

McNab, B. K. 1963. Bioenergetics and the determination of home range size. American Naturalist 97: 133–140.

McNab, B. K. 1999. On the comparative ecological and evolutionary significance of total and mass-specific rates of metabolism. Physiological and Biochemical Zoology 72: 642–644.

McNamara, J. M. and O. Leimar. 2020. Game Theory in Biology. Oxford University Press, Oxford. 328 pp.

McNamara, J. M., E. M. K. Wilson and A. I. Houston. 2006. Is it better to give information, receive it, or be ignorant in a two-player game? Behavioral Ecology 17: 441–451.

McQuivey, C. 2015. Home range, habitat use, survival, and limiting factors of pronghorn ten years post-reintroduction. P. 74 in O'Shaughnessy, R. and S. S. Gray (editors), Proceedings of the 26th Biennial Pronghorn Workshop: Managing Pronghorn on Private Land. Texas Parks and Wildlife Department and Borderlands Research Institute, Alpine, Texas.

McRoberts, R. E. and L. D. Mech. 2014. Wolf population regulation revisited, again. Journal of Wildlife Management 78: 963–967.

Mech, L. D. 1970. The Wolf: Ecology and Behavior of an Endangered Species. Natural History Press, Garden City, New Jersey.

Mech, L. D. 1977a. Wolf pack buffer zones as prey reservoirs. Science 198: 320–321

Mech, L. D. 1977b. Population trend and winter deer consumption in a Minnesota wolf pack. Pp. 55–83 in Phillips, R. I. and C. Jonkel (editors), Proceedings of the 1975 Predator Symposium. Montana Forest and Conservation Experiment Station, University of Montana, Missoula.

Mech, L. D. 1994. Buffer zones of territories of gray wolves as regions of intraspecific strife. Journal of Mammalogy 75: 199–202.

Mech, L. D. and L. Boitani. 2003. Wolf social ecology. Pp. 1–34 in Mech, L. D. and L. Boitani (editors), Wolves: Behavior, Ecology and Conservation. University of Chicago Press, Chicago, Illinois.

Mech, S. G. and P. A. Zollner. 2002. Using body size to predict perceptual range. Oikos 98: 47–52.

Meier, T. J., J. W. Burch, L. D. Mech and L. G. Adams. 1995. Pack structure dynamics and genetics relatedness among wolf packs in a naturally regulated population. Pp. 293–302 in Carbyn, L. N., S. H. Fritts and D.R. Seip (editors), Ecology and Conservation of Wolves in a Changing World. Canadian Circumpolar Institute, Edmonton, Alberta.

Menzel, R., U. Greggers, A. Smith, S. Berger, R. Brandt, S. Brunke, G. Bundrock, S. Hulse, T. Plumpe, F. Schaupp, E. Schuttler, S. Stach, J. Stindt, N. Stollhoff and S. Watzl. 2005. Honey bees navigate according to a map-like spatial memory. Proceedings of the National Academy of Sciences of the United States of America 102(8): 3040–3045.

Menezes, J. F. S., B. P. Kotler and A. K. Dixon. 2019. Defense exploitation in Negev gerbils. Behavioural Processes 162: 97–103.

Merkle, J. A., B. Abrahms, J. B. Armstrong, H. Sawyer, D. P. Costa and A. D. Calhoun. 2022. Site fidelity as a maladaptive behavior in the Anthropocene. Frontiers in Ecology and the Environment 20: 187–194.

Merkle, J. A., S. G. Cherry and D. Fortin. 2015. Bison distribution under conflicting foraging strategies: site fidelity vs energy maximization. Ecology 96: 1793–1801.

Merkle, J. A., D. Fortin and J. M. Morales. 2014. A memory-based foraging tactic reveals an adaptive mechanism for restricted space use. Ecology Letters 17: 924–931.

Merkle, J. A., J. R. Potts and D. Fortin. 2017. Energy benefits and emergent space use patterns of an empirically parameterized model of memory-based patch selection. Oikos 126: 185–195.

Merkle, J. A., H. Sawyer, K. L. Monteith, S. P. H. Dwinnell, G. L. Fralick and M. J. Kauffman. 2019. Spatial memory shapes migration and its benefits: Evidence from a large herbivore. Ecology Letters 22: 1797–1805.

Messeri, L. and M. J. Crockett. 2024. Artificial intelligence and illusions of understanding in scientific research. Nature 627: 49–59.

Mesterton-Gibbons, M. and E. S. Adams. 2003. Landmarks in territory partitioning: A strategically stable convention? American Naturalist 161: 685–697.

Mesterton-Gibbons, M. and T. N. Sherratt. 2016. How residency duration affects the outcome of a territorial contest: Complementary game-theoretic models. Journal of Theoretical Biology 394: 137–148.

Metzgar, L. H. 1972. Measurement of home range shape. Journal of Wildlife Management 36: 643–645.

Metzgar, L. H. and A. L. Sheldon. 1974. Index of home range size. Journal of Wildlife Management 38: 546–551.

Michelot, T. and P. G. Blackwell. 2019. State-switching continuous-time correlated random walks. Methods in Ecology and Evolution 10: 637–649.

Michelot, T., P. G. Blackwell and J. Matthiopoulos. 2019. Linking resource selection and step selection models for habitat preferences in animals. Ecology 100: e02452.

Milinski, M. and G. A. Parker. 1991. Competition for resources. Pp. 137–168 in Krebs, J. R. and N. B. Davies (editors), Behavioural Ecology: An Evolutionary Approach, 3rd edition. Blackwell Science, Oxford.

Miller, C., M. C. Christman and I. Estevez. 2011. Movement in a confined space: Estimating path tortuosity. Applied Animal Behaviour Science 135: 13–23.

Milton, K. 1980. The Foraging Strategy of Howler Monkeys. Columbia University Press, New York.

Minta, S. C. 1992. Tests of spatial and temporal interaction among animals. Ecological Applications 2: 178–188.

Mishchanchuk, K., G. Gregorio, A. Qü, A. Kastler, Q. J. M. Huys, L. Wilbrecht and A. F. MacAskill. 2024. Hidden state inference requires abstract contextual representations in the ventral hippocampus. Science 386: 926–932.

Mitchell, M. S. 1997. Optimal home ranges: Models and application to black bears. PhD thesis, North Carolina State University, Raleigh, North Carolina.

Mitchell, M. S. and M. Hebblewhite. 2012. Carnivore habitat ecology: Integrating theory and application. Pp. 218–255 in Boitani, L. and R. A. Powell (editors), Carnivore Ecology and Conservation: A Handbook of Techniques. Oxford University Press. Oxford.

Mitchell, M. S. and R. A. Powell. 2002. Linking fitness landscapes with the behavior and distribution of animals. Pp. 93–123 in Bissonette, J. A. and I. Storch (editors), Landscape ecology and resource management: Linking theory with practice. Island Press, Washington, DC.

Mitchell, M. S. and R. A. Powell. 2003. Linking fitness landscapes with the behavior and distribution of animals. Pp. 93–124 in Bissonette, J. A. and I. Storch (editors), Landscape Ecology and Resource Management: Linking Theory with Practice. Island Press, Washington, DC.

Mitchell, M. S. and R. A. Powell. 2004. A mechanistic home range model for optimal use of spatially distributed resources. Ecological Modelling 177: 209–230.

Mitchell, M. S. and R. A. Powell. 2007. Optimal use of resources structures home ranges and spatial distribution of black bears. Animal Behaviour 74: 219–230.

Mitchell, M. S. and R. A. Powell. 2008. Estimated home ranges can misrepresent habitat relationships on patchy landscapes. Ecological Modelling 216: 409–414.

Mitchell, M. S. and R. A. Powell. 2012. Foraging optimally for home ranges. Journal of Mammalogy 93: 917–928.

Mitchell, M. S., J. W. Zimmerman and R. A. Powell. 2002. Test of a habitat suitability index for black bears. Wildlife Society Bulletin 30: 794–808.

Moen, A. N. and G. S. Boomer. 2005. Modeling annual energy metabolism rhythms in mammals. Ecological Modelling 184: 193–202.

Mohr, C. O. 1947. Table of equivalent populations of North American small mammals. American Midland Naturalist 37: 223–249.

Mohr, C. O. and W. A. Stumpf. 1964. Relation of tick and chigger infestations to home areas of California meadow mice. Journal of Medical Entomology 1: 73–77.

Mohr, C. O. and W. A. Stumpf. 1966. Comparison of methods for calculating areas of animal activity. Journal of Wildlife Management 30: 293–304.

Moiron, M., A. Charmantier and S. Bouwhuis. 2022. The quantitative genetics of fitness in a wild seabird. Evolution 76: 1443–1452.

Montagnese, C. M., J. R. Krebs and G. Meyer. 1996. The dorsomedial and dorsolateral forebrain of the zebra finch, *Taeniopygia guttata*: A Golgi study. Cell and Tissue Research 283: 263–282.

Moorcroft, P. R. and M. A. Lewis. 2006. Mechanistic Home Range Analysis. Princeton University Press, Princeton, New Jersey.

Moorcroft, P. R., M. A. Lewis and R. L. Crabtree. 1999. Home range analysis using a mechanistic home range model. Ecology 80: 1656–1665.

Moorcroft, P. R., M. A. Lewis, and R. L. Crabtree. 2006. Mechanistic home range models capture spatial patterns and dynamics of coyote territories in Yellowstone. Proceedings of the Royal Society B: Biological Sciences 273: 1651–1659.

Moorhouse, T. P. and D. W. Macdonald. 2005. Temporal patterns of range use in water voles: Do females' territories drift? Journal of Mammalogy 86: 655–661.

Mora, C. V., J. D. Ross, P. V. Gorsevski, B. Chowdhury and V. P. Bingman. 2012. Evidence for discrete landmark use by pigeons during homing. Journal of Experimental Biology 215: 3379–3387.

Morales, J. M., D. T. Haydon, J. Frair, K. E. Holsiner and J. M. Fryxell. 2004. Extracting more out of relocation data: Building movement models as mixtures of random walks. Ecology 85: 2436–2445.

Moran, P. A. P. 1950. Notes on continuous stochastic phenomena. Biometrika. 37: 17–23.

Morgan, C. Lloyd. 1894. An Introduction to Comparative Psychology. Walter Scott, Ltd., London.

Morrell, L. J. and H. Kokko. 2003. Adaptive strategies of territory formation. Behavioral Ecology and Sociobiology 54: 385–395.

Morrell, L. J. and H. Kokko. 2005. Bridging the gap between mechanistic and adaptive explanations of territory formation. Behavioral Ecology and Sociobiology 57: 381–390.

Morris, D. W. 2003. Toward an ecological synthesis: A case for habitat selection. Oecologia 136: 1–13.

Morris, D. W. and J. T. MacEachern. 2013. Respect for property rights: When does it pay to defend territory? Evolutionary Ecology Research 15: 757–768.

Moser, E. I., M. B. Moser and B. L. McNaughton. 2017. Spatial representation in the hippocampal formation: A history. Nature Neuroscience 20: 1448–1464.

Mudrik, L. 2023. The grand challenge of consciousness. Nature 623: 25–26.

Mugnai, M., E. Auriemma, B. Contiero, D. Franchini, E. Zini and F. Tirrito. 2024. Effect of gadolinium contrast medium administration on susceptibility-weighted imaging of the canine brain. Veterinary Radiology & Ultrasound 65(5): 539–546. doi: 10.1111/vru.13395

Murie, A. 1936. Following fox trails. Miscellaneous Publications of the Museum of Zoology, University of Michigan 32: 1–45.

Muth, F. 2023. Bumblebees show capacity for behavioral traditions. Learning & Behavior 52(2): 137–138. doi: 10.3758/s13420-023-00594-0

Nabe-Nielsen, J., J. Tougaard, J. Teilmann, K. Lucke and M. C. Forchhammer. 2013. How a simple adaptive foraging strategy can lead to emergent home ranges and increased food intake. Oikos 122: 1307–1316.

Nagel, T. 1974. What is it like to be a bat? Philosophical Review 83: 435–450.

Nagy, K. A. 1980. CO_2 production in animals: Analysis of potential errors in the doubly labeled water method. American Journal of Physiology 238: R466–R473.

Nagy, K. A. 1987. Field metabolic rate and food requirement scaling in mammals and birds. Ecology Monographs 57: 111–128.

Nams, V. O., D. M. Parker, F. J. Weise, B. D. Patterson, R. Buij, F. G. T. Radloff, A. T. Vanak, P. Tumenta, M. Hayward, L. H. Swanepoel, P. J. Funston, H. Bauer, R. J. Power, J. O'Brien, T. G. O'Brien, C. J. Tambling, H. H. de Iongh, S. M. Ferreira, N. Owen-Smith, J. W. Cain, J. Fattebert, B. M. Croes, G. Spong, A. J. Loveridge, A. M. Houser, K. A. Golabek, C. M. Begg, T. Grant, P. Trethowan, C. Musyoki, V. Menges, S. Creel, G. A. Balme, R. T. Pitman, C. Bissett, D. Jenny, P. Schuette, C. C. Wilmers, L. T. B. Hunter, M. F. Kinnaird, K. S. Begg, C. R. Owen, V. Steyn, D. Bockmuehl, S. J. Munro, G. K. H. Mann, B. D. du Preez, L. L. Marker, T. J. Huqa, G. Cozzi, L. G. Frank, P. Nyoni, A. B. Stein, S. M. Kasiki, D. W. MacDonald, Q. E. Martins, R. J. van Vuuren, K. J. Stratford, L. R. Bidner, A. Oriol-Cotteril, N. W. Maputla, N. Maruping-Mzileni, T. Parker, M. Van't Zelfde, L. A. Isbell, O. B. Beukes and M. Beukes. 2023. Spatial patterns of large African cats: A large-scale study on density, home range size, and home range overlap of lions *Panthera leo* and leopards *Panthera pardus*. Mammal Review 53: 49–64.

Narayanan A. and S. Kapoor. 2024. AI Snake Oil. Princeton University Press, Princeton, New Jersey. 360 pp.

Nash, J. F. 1950. Equilibrium points in n-person games. Proceedings of the National Academy of Sciences of the United States of America 36: 48–49.

Nathan, R. 2008. An emerging movement ecology paradigm. Proceedings of the National Academy of Sciences of the United States of America 105: 19050–19051.

Nathan, R., W. M. Getz, E. Revilla, M. Holyoak, R. Kadmon, D. Saltz and P. E. Smouse. 2008. A movement ecology paradigm for unifying organismal movement research. Proceedings of the National Academy of Sciences of the United States of America 105: 19052–19059.

Naumann, R. K., J. M. Ondracek, S. Reiter, M. Shein-Ldelson, M. A. Tosches, T. M. Yamawaki and G. Laurent. 2015. Primer: The reptilian brain. Current Biology 25: R317–R321.

Neft, D. S. Statistical analysis for areal distributions. Monograph Series Number 2. Regional Science Research Institute, Philadelphia, Pennsylvania, USA.

Nelson, R. K. 1973. Hunters of the Northern Forest: Designs for Survival among Alaskan Kutchin. University of Chicago Press, Chicago.

Nevitt, G. A. 2008. Sensory ecology on the high seas: The odor world of the procellariiform seabirds. Journal of Experimental Biology 211: 1706–1713.

Nevitt, G. A., M. Losekoot and H. Weimerskirch. 2008. Evidence for olfactory search in wandering albatross, *Diomedea exulans*. Proceedings of the National Academy of Sciences of the United States of America 105: 4576–4581.

Nevoux, M., O. Gimenez, D. Arlt, M. Nicoll, C. Jones and K. Norris. 2011. Population regulation of territorial species: Both site dependence and interference mechanisms matter. Proceedings of the Royal Society B: Biological Sciences 278: 2173–2181.

Nicholson, A. J. 1957. The Self-adjustment of Populations to Change. Cold Spring Harbor Symposia on Quantitative Biology 22: 153–173.

Nickel, B. A., J. P. Suraci, A. C. Nisi and C. C. Wilmers. 2021. Energetics and fear of humans constrain the spatial ecology of pumas. Proceedings of the National Academy of Sciences of the United States of America 118. https://doi.org/10.1073/pnas.2004592118

Nieh, E. H., M. Schottdorf, N. W. Freeman, R. J. Low, S. Lewallen, S. A. Koay, L. Pinto, J. L. Gauthier, C. D. Brody and D. W. Tank. 2021. Geometry of abstract learned knowledge in the hippocampus. Nature 595: 80–84.

Noble, G. K. 1939. The role of dominance in the life of birds. Auk 56: 263–273.

Noel, J. 1993. Food productivity and home range size in black bears. MS thesis, North Carolina State University, Raleigh, North Carolina.

Noonan, M. J., C. H. Fleming, T. S. Akre, J. Drescher-Lehman, E. Gurarie, A. L. Harrison, R. Kays and J. M. Calabrese. 2019a. Scale-insensitive estimation of speed and distance traveled from animal tracking data. Movement Ecology 7: 35. doi: 10.1186/s40462-019-0177-1

Noonan, M. J., C. H. Fleming, M. A. Tucker, R. Kays, A. L. Harrison, M. C. Crofoot, B. Abrahms, S. C. Alberts, A. H. Ali, J. Altmann, P. C. Antunes, N. Attias, J. L. Belant, D. E. Beyer, L. R. Bidner, N. Blaum, R. B. Boone, D. Caillaud, R. C. de Paula, J. A. de la Torre, J. Dekker, C. S. DePerno, M. Farhadinia, J. Fennessy, C. Fichtel, C. Fischer, A. Ford, J. R. Goheen, R. W. Havmoller, B. Hirsch, C. Hurtado, L. A. Isbell, R. Janssen,

F. Jeltsch, P. Kaczensky, Y. Kaneko, P. Kappeler, A. Katna, M. Kauffman, F. Koch, A. Kulkarni, S. LaPoint, P. Leimgruber, D. W. Macdonald, A. C. Markham, L. McMahon, K. Mertes, C. E. Moorman, R. G. Morato, A. M. Mossbrucker, G. Mourao, D. O'Connor, L. G. R. Oliveira-Santos, J. Pastorini, B. D. Patterson, J. Rachlow, D. H. Ranglack, N. Reid, D. M. Scantlebury, D. M. Scott, N. Selva, A. Sergiel, M. Songer, N. Songsasen, J. A. Stabach, J. Stacy-Dawes, M. B. Swingen, J. J. Thompson, W. Ullmann, A. T. Vanak, M. Thaker, J. W. Wilson, K. Yamazaki, R. W. Yarnell, F. Zieba, T. Zwijacz-Kozica, W. F. Fagan, T. Mueller and J. M. Calabrese. 2020. Effects of body size on estimation of mammalian area requirements. Conservation Biology 34: 1017–1028.

Noonan, M. J., R. Martinez-Garcia, G. H. Davis, M. C. Crofoot, R. Kays, B. Hirsch, D. Caillaud, E. Payne, A. Sih, D. L. Sinn, O. Spiegel, W. F. Fagan, C. H. Fleming and J. M. Calabrese. 2021. Estimating encounter location distributions from animal tracking data. Methods in Ecology and Evolution 12: 1158–1173.

Noonan, M. J., M. A. Tucker, C. H. Fleming, T. S. Akre, S. C. Alberts, A. H. Ali, J. Altmann, P. C. Antunes, J. L. Belant, D. Beyer, N. Blaum, K. Bohning-Gaese, L. Cullen, R. C. de Paula, J. Dekker, J. Drescher-Lehman, N. Farwig, C. Fichtel, C. Fischer, A. T. Ford, J. R. Goheen, R. Janssen, F. Jeltsch, M. Kauffman, P. M. Kappeler, F. Koch, S. LaPoint, A. C. Markham, E. P. Medici, R. G. Morato, R. Nathan, L. G. R. Oliveira-Santos, K. A. Olson, B. D. Patterson, A. Paviolo, E. E. Ramalho, S. Rosner, D. G. Schabo, N. Selva, A. Sergiel, M. X. da Silva, O. Spiegel, P. Thompson, W. Ullmann, F. Zieba, T. Zwijacz-Kozica, W. F. Fagan, T. Mueller and J. M. Calabrese. 2019b. A comprehensive analysis of autocorrelation and bias in home range estimation. Ecological Monographs 89(2): e01344.

Noonburg, E. G. and R. C. Anderson. 2021. Asymmetric competition and floater dynamics. Ecology 102: e03238.

Nouvellet, P., J. P. Bacon and D. Waxman. 2009. Fundamental insights into the random movement of animals from a single distance-related statistic. American Naturalist 174: 506–514.

Noyce, K. V. and D. L. Garshelis. 1997. Influence of natural food abundance on black bear harvests in Minnesota. Journal of Wildlife Management 61: 1067–1074.

Noy-Meir, I. 1973. Desert ecosystems: Environment and producers. Annual Review of Ecology and Systematics. 4: 25–51.

Nunn, C. L, P. H. Thrall, F. H. Leendertz and C. Boesch. 2011. The spread of fecally transmitted parasites in socially structured populations. Plos One 6(6): e21677. doi:10.1371/journal.pone.0021677

Nussbaum, M. C. 2023. Justice for Animals: Our Collective Responsibility. Simon & Schuster, New York.

Nyholm. E. 1959. Kärpästä ja lumikosta ja niiden talvisista elinpiireistä. Suomen Riista, 13, 106–116.

O'Connor, M-F. 2022. The Grieving Brain. Harper Collins, New York. 181 pp.

O'Connor, M. R. 2019. Wayfinding: The science and mystery of how humans navigate the world. St. Martin's Press, New York. 354 pp.

Ofiteru, I. D. and C. Picioreanu. 2022. No model is perfect, but some are useful. Science 376: 914–916.

Ofstad, E. G., I. Herfindal, E. J. Solberg and B-E. Sæther. 2016. Home ranges, habitat and body mass: Simple correlates of home range size in ungulates. Proceedings of the Royal Society B: Biological Sciences 283: 20161234. http://dx.doi.org/10.1098/rspb.2016.1234

Ogasawara, T., F. Sogukpinar, K. Zhang, Y-Y Feng, J. Pai, A. Jezzini and I. E. Monosov. 2022. A primate temporal cortex–zone incerta pathway for novelty seeking. Nature Neuroscience 25: 52–60.

O'Keefe, J. and J. Dostrovsky. 1971. The hippocampus as a spatial map. Brain Research 34: 171–175.

O'Keefe, J. and L. Nadel. 1978. The Hippocampus as a Cognitive Map. Oxford University Press, New York.

Oldenburg, E., R. M. Kronberg, B. Niehoff, O. Ebenhöh and O. Popa. 2023. DeepLOKI: a deep learning based approach to identify zooplankton taxa on high-resolution images from the optical plankton recorder LOKI. Frontiers in Marine Science 10. https://doi.org/10.3389/fmars.2023.1280510

Olsson, O., J. S. Brown and H. G. Smith. 2001. Gain curves in depletable food patches: A test of five models with European starlings. Evolutionary Ecology Research 3: 285–310.

Omer, D. B., S. R. Maimon, L. Las and N. Ulanovsky. 2018. Social place-cells in the bat hippocampus. Science 359: 218–224.

Ord, T. J. 2021. Costs of territoriality: A review of hypotheses, meta-analysis, and field study. Oecologia 197: 615–631.

Ormond, J. and J. O'Keefe. 2022. Hippocampal place cells have goal-oriented vector fields during navigation. Nature 607: 741–746.

Ostfeld, R. S. 1985. Limiting resources and territoriality in microtine rodents. American Naturalist 126: 1–15.

Ostfeld, R. S. 1986. Territoriality and mating system of California voles. Journal of Animal Ecology 55: 691–706.

Ostfeld, R. A. 1990. The ecology of territoriality in small mammals. Trends in Ecology & Evolution. 5: 411–415.

Ostro, L. E. T., S. C. Silver, F. W. Koontz, T. P. Young and R. H. Horwich. 1999a. Ranging behavior of translocated and established groups of black howler monkeys *Alouatta pigra* in Belize, Central America. Biological Conservation 87: 181–190.

Ostro, L. E. T., T. P. Young, S. C. Silver and F. W. Koontz. 1999b. A Geographic Information System method for estimating home range size. Journal of Wildlife Management 63: 748–755.

Oudman, T., V. Hin, A. Dekinga and J. A. van Gils. 2015. The effect of digestive capacity on the intake rate of toxic and non-toxic prey in an ecological context. Plos One 10. https://doi.org/10.1371/journal.pone.0136144

Owen-Smith, N. 1977. On territoriality in ungulates and an evolutionary model. The Quarterly Journal of Biology 52: 1–38.

Pacheco, X. P. 2020. How consistently do personality attributes relate to an individual's position within a social network: A comparison across groups of captive meerkats. Behavioral Ecology and Sociobiology 74. https://doi.org/10.1007/s00265-020-02880-7

Padget, O., G. Stanley, J. K. Willis, A. L. Fayet, S. Bond, L. Maurice, A. Shoji, B. Dean, H. Kirk, I. Juarez-Martinez, R. Freeman, M. Bolton and T. Guilford. 2019. Shearwaters know the direction and distance home but fail to encode intervening obstacles after free-ranging foraging trips. Proceedings of the National Academy of Sciences of the United States of America 116: 21629–21633.

Palomares, F. 1994. Site fidelity and effects of body mass on home-range size of Egyptian mongooses. Canadian Journal of Zoology 72: 465–469.

Pastalkova,E., V. Itskov, A. Amarasingham and G. Buzsáki. 2008. Internally generated cell assembly sequences in the rat hippocampus. Science 321: 1322–1327.

Parker, G. A. and D. I. Rubenstein. 1981. Role of assessment, reserve strategy, and acquisition of information in asymmetric animal conflicts. Animal Behaviour 29: 221–240.

Paton, D. C. and F. L. Carpenter. 1984. Peripheral foraging by territorial rufous hummingbirds: Defense by exploitation. Ecology 65: 1808–1819.

Payne, H. L., G. F. Lynch and D. Aronov. 2021. Neural representations of space in the hippocampus of a food-caching bird. Science 373: 348.

Peper, A. 2020. A general theory of consciousness I: Consciousness and adaptation. Communicative and Integrative Biology 13: 6–21. https://doi.org/10.1080/19420889.2020.1713967

Pereira, H. M., A. Bergman and J. Roughgarden. 2003. Socially stable territories: The negotiation of space by interacting foragers. American Naturalist 161: 143–152.

Peris, A., F. Closa, I. Marco, P. Acevedo, J. A. Barasona and E. Casas-Diaz. 2020. Towards the comparison of home range estimators obtained from contrasting tracking regimes: The wild boar as a case study. European Journal of Wildlife Research 66: 10.

Person, D. K. and D. H. Hirth. 1991. Home range and habitat use of coyotes in a farm region of Vermont. Journal of Wildlife Management 55: 433–441.

Peters, R. 1978. Communication, cognitive mapping, and strategy in wolves and hominids. Pp. 95–108 in Hall, R.L. and H. S. Sharp (editors), Wolf and Man: Evolution in Parallel. Academic Press, New York.

Peters, R. and L. D. Mech. 1975. Scent-marking in wolves. American Scientist 63: 628–637.

Peterson, R. O. 1977. Wold ecology and prey relationships on Isle Royale. U. S. National Park Service Monograph Series No. 11, Washington, DC. 210 pp.

Petroski, H. 2023. Building knowledge. American Scientist 111: 232–234.

Phillips, M. K. and D. W. Smith. 1996. The Wolves of Yellowstone. Voyager Press, Stillwater, Minnesota. 128 pp.

Picasso, P. 1923. Picasso speaks. The Arts 3: 315–326

Pinter-Wollman, N., E. A. Hobson, J. E. Smith, A. J. Edelman, D. Shizuka, S. de Silva, J. S. Waters, S. D. Prager, T. Sasaki, G. Wittemyer, J. Fewell and D. B. McDonald. 2014. The dynamics of animal social networks: Analytical, conceptual, and theoretical advances. Behavioral Ecology 25: 242–255.

Pluchot, C., H. Adriaensen, C. Parias, D. Dubreuil, C. Arnould, E. Chaillou and S. A. Love. 2024. Sheep (*Ovis aries*) training protocol for voluntary awake and unrestrained structural brain MRI acquisitions. Behavior Research Methods 56(7): 7761–7773. doi: 10.3758/s13428-024-02449-6

Poo, C., G. Agarwal, N. Bonacchi and Z. F. Mainen. 2022. Spatial maps in piriform cortex during olfactory navigation. Nature 601: 595–599.

Poole, K., G. M. Matson, M. A. Strickland, A. J. Magoun, R. P. Graf and L. M. Dix. 1994. Age and sex determination for American martens and fishers. Pp. 204–223 in Buskirk, S. W., A. Harestad, M. Raphael and R. A. Powell (editors), Martens, Fishers, and Sables: Biology and Conservation. Cornell University Press, Ithaca, New York.

Popper, K. 1934/1959. Logik der Forschung/The logic of scientific discovery. Julian Springer, Vienna/ Hutchinson & Company, London. 513 pp.

Possingham, H. P. and A. I. Houston. 1990. Optimal patch use by a territorial forager. Journal of Theoretical Biology 145: 343–353.

Potts, J. R. and M. A. Lewis. 2014. How do animal territories form and change? Lessons from 20 years of mechanistic modelling. Proceedings of the Royal Society B: Biological Sciences 281: 1–9.

Potts, J. R. and M. A. Lewis. 2016. How memory of direct animal interactions can lead to territorial pattern formation. Journal of the Royal Society Interface 13: 20160059. http://dx.doi.org/10.1098/rsif.2016.0059

Potts, J. R., K. Mokross, P. C. Stouffer and M. A. Lewis. 2014. Step selection techniques uncover the environmental predictors of space use patterns in flocks of Amazonian birds. Ecology and Evolution 4: 4578–4588.

Powell, R. A. 1977. Hunting behavior, biological energetics and predator–prey community stability of the fishers (*Martes pennanti*). PhD thesis, University of Chicago. 132 pp.

Powell, R. A. 1978. A comparison of fisher and weasel hunting behavior. Carnivore 1(1): 28–34.

Powell, R. A. 1979a. Mustelid spacing: Variations on a theme by *Mustela*. Zeitschrift für Tierpsychologie 50: 153–165.

Powell, R. A. 1979b. Ecological energetics and foraging strategies of the fisher (*Martes pennanti*). Journal of Animal Ecology 48: 195–212.

Powell, R. A. 1981. Fisher food requirements and hunting behavior. Pp. 883–917 in Chapman, J. A. and D. Pursley (editors), Proceedings of the First Worldwide Furbearer Conference. Worldwide Furbearer Conference, Inc., Baltimore, Maryland.

Powell, R. A. 1987. Black bear home range overlap in North Carolina and the concept of home range applied to black bears. International Conference on Bear Research & Management 7: 235–242.

Powell, R. A. 1989. Effects of resource productivity, patchiness and predictability on mating and dispersal strategies. Pp. 101–123 in Standen, V. and R. A. Foley (editors), Comparative Socioecology of Mammals and Humans. British Ecological Society Symposium, Blackwell, Oxford.

Powell, R. A. 1993. The Fisher: Life History, Ecology, and Behavior, 2nd edition. University of Minnesota Press, Minneapolis, Minnesota.

Powell, R. A. 1994. Structure and spacing of *Martes* populations. Pp. 101–121 in Buskirk, S. W., A. S. Harestad, M. G. Raphael and R. A. Powell (editors), Martens, Fishers, and Sables: Biology and Conservation. Cornell University Press, Ithaca, New York.

Powell, R. A. 2000. Animal home ranges and territories and home range estimators. Pp. 65–110 in Boitani, L. and T. K. Fuller (editors), Research Techniques in Animal Ecology: Controversies and Consequences. Columbia University Press, New York.

Powell, R. A. 2004. Home ranges, cognitive maps, habitat models and fitness landscapes for *Martes*. Pp. 135–146 in Harrison, D. J., A. K. Fuller and G. Proulx (editors), Martens and Fishers (*Martes*) in Human-altered Environments: An International Perspective. Springer Science+Business Media, New York.

Powell, R. A. 2005. Evaluating welfare of American black bears (*Ursus americanus*) captured in foot snares and in winter dens. Journal of Mammalogy 86: 1171–1177.

Powell, R. A. 2012a. Movements, home ranges, activity, and dispersal. Pp. 188–217 in Boitani, L. and R. A. Powell (editors), Ecology and Conservation of Carnivores: A Handbook of Techniques. Oxford University Press, London.

Powell, R. A. 2012b. Diverse perspectives on mammal home ranges or a home range is more than location densities. Journal of Mammalogy 93: 887–889.

Powell, R. A., D. Clifford, A. N. Facka, D. Green, S. Matthews and K. P. Smith. 2018. Understanding a fisher reintroduction in northern California from 2 perspectives. Final Report. Submitted to United States Fish and Wildlife Service, Yreka, California; California Department of Fish and Wildlife, Redding, California; and Sierra Pacific Industries, Anderson, California.

Powell, R. A., A. N. Facka, D. L. Clifford, K. P. Smith, S. M. Matthews, T. Engstrom, E. Murphy, L. Finley, J. S. Yaeger, P. Figura and R. Callas. Establishing a carnivoran of extensive forests on an intensively managed landscape: Habitat and population establishment. PLoS One 20(4): e0320594.

Powell, R. A., A. N. Facka, M. Gabriel, J. H. Gilbert, J. M. Higley, S. D. Lapoint, N. P. McCann, W. D. Spencer and C. Thompson. 2017. The fisher as a model organism. Pp. 299–313 in Macdonald, D., L. A. Harrington, and C. Newman (editors), Biology and Conservation of Wild Musteloids. Oxford University Press, Oxford.

Powell, R. A. and J. J. Fried 1992. Helping by juvenile pine voles (*Microtus pinetorum*), growth and survival of younger siblings, and the evolution of pine vole sociality. Behavioral Ecology 4: 325–333.

Powell, R.A., J. C. Lewis, B. G. Slough, S. M. Brainerd, N. R. Jordan, A. V. Abramov, V. Monakhov, P. A. Zollner and T. Murakami. 2012. Evaluating translocations of martens, sables, and fishers: Testing model predictions with field data. Pp. 93–137 in Aubry, K. B., W. J. Zielinski, M. G. Raphael, G. Proulx and S. W. Buskirk (editors), Biology and Conservation of Martens, Sables, and Fishers: A New Synthesis. Cornell University Press, Ithaca, New York.

Powell, R. A., S. A. Mansfield and L. L. Rogers. 2022. Comparison of behaviors of black bears with and without habituation to humans and supplemental research feeding. Journal of Mammalogy 104: 1350–1365.

Powell, R. A. and M. S. Mitchell. 1998. Topographic constraints and home range quality. Ecography 21: 337–341.

Powell, R. A. and M. S. Mitchell. 2012. What is a home range? Journal of Mammalogy 93: 948–958.

Powell, R. A. and D. E. Seaman. 1990. Production of important black bear foods in the Southern Appalachians. International Conference on Bear Research and Management 8: 183–187.

Powell, R. A. and W. J. Zielinski. 1994. Fisher. Pp. 38–73 in Ruggiero, L. F., K. B. Aubry, S. W. Buskirk and W.

J. Zielinski (editors), The Scientific Basis for Conserving Forest Carnivores: American Marten, Fisher, Lynx, and Wolverine. USDA Forest Service, Rocky Mountain Research Station, Missoula, Montana.

Powell, R. A., J. W. Zimmerman, D. E. Seaman and J. F. Gilliam. 1996. Demographic analyses of a hunted black bear population with a refuge. *Conservation Biology* 10: 224–234.

Powell, R. A., J. W. Zimmerman and D. E. Seaman. 1997. Ecology and Behaviour of North American Black Bears: Home Ranges, Habitat and Social Organization. Chapman & Hall, London.

Powers, D. R. and T. McKee. 1994. The effect of food availability on time and energy expenditure of territorial and non-territorial hummingbirds. Condor 96: 1064–1075.

Pretorius, Y., J. D. Stigter, W. F. de Boer, S. E. van Wieren, C. B. de Jong, H. J. de Knegt, C. C. Grant, I. Heitkönig, N. Knox, E. Kohi, E. Mwakiwa, M. J. S. Peel, A. K. Skidmore, R. Slotow, C. van der Waal, F. van Langevelde and H. H. T. Prins. 2012. Diet selection of African elephant over time shows changing optimization currency. Oikos 121: 2110–2120.

Prevedello, J. A., G. Forero-Medina and M. V. Vieira. 2010. Movement behaviour within and beyond perceptual ranges in three small mammals: Effects of matrix type and body mass. Journal of Animal Ecology 79: 1315–1323.

Price, K. and S. Boutin. 1993. Territorial bequeathal by red squirrel mothers. Behavioral Ecology 4: 144–150.

Price, K., S. Boutin and R. Ydenberg. 1990. Intensity of territorial defense in red squirrels: An experimental test of the asymmetric war of attrition. Behavioral Ecology and Sociobiology 27: 217–222.

Pulliainen, E. 1981. Winter habitat selection, home range, and movements of the pine marten (*Martes martes*) in a Finnish Lapland forest. Pp. 1068–1087 in Chapman, J. A. and D. Pursley (editors), Worldwide Furbearer Conference Proceedings. Worldwide Furbearer Conference, Inc., Baltimore, Maryland.

Pulliainen, E. 1982. Scent-marking in the pine marten (*Martes martes*) Finnish forest Lapland in winter. Zeitschrift für Saugetierkunde 47: 91–99.

Pulliainen, E. 1984. Use of the home range by pine martens (*Martes martes* L.). Acta Zoologica Fennica 171: 271–274.

Pulliam, H. R. 1974. Theory of optimal diets. American Naturalist 108: 59–74.

Purandare, C. S., S. Dhingra, R. Rios, C. Vuong, T. To, A. Hachisuka, K. Choudhary and M. R. Mehta. 2022. Moving bar of light evokes vectorial spatial selectivity in the immobile rat hippocampus. Nature 602: 461–467.

Pyke, G. H. 1978. Optimal foraging: Movement patterns of bumblebees between inflorescences. Theoretical Population Biology 13: 72–98.

Pyke, G. H. 1981. Optimal foraging in hummingbirds: Rule of movement between inflorescences. Animal Behaviour 29: 889–896.

Pyke, G. H. 1983. Animal movements: an optimal foraging approach. Pp. 7–31 in Swingland, I. R. and P. J. Greenland (editors), The Ecology of Animal Movement. Oxford University Press, Oxford.

Pyke, G. H. 1984. Optimal foraging theory: A critical review. Annual Review of Ecology and Systematics 15: 523–575.

Pyke, G. H., H. R. Pulliam and E. L. Charnov. 1977. Optimal foraging: Selective review of theory and tests. Quarterly Review of Biology 52: 137–154.

Rabinowitz, A. R. 1986. Jaguar predation on domestic livestock in Belize. Wildlife Society Bulletin 14: 170–174.

Ranc, N., F. Cagnacci and P. R. Moorcroft. 2020. Knowing your neighbours: How memory-mediated conspecific avoidance influences home ranges. Journal of Animal Ecology 89: 2746–2749.

Ranc, N., P. R. Moorcroft, F. Ossi and F. Cagnacci. 2021. Experimental evidence of memory-based foraging decisions in a large wild mammal. Proceedings of the National Academy of Sciences of the United States of America 118(15): e2014856118. doi: 10.1073/pnas.2014856118

Ranc, N., F. Cagnacci and P. R. Moorcroft. 2022. Memory drives the formation of animal home ranges: Evidence from a reintroduction. Ecology Letters 25: 716–728.

Rattenborg, N. C., B. Voirin, A. L. Vyssotski, R. W. Kays, K. Spoelstra, F. Kuemmeth, W. Heidrich and M. Wikelski. 2008. Sleeping outside the box: Electroencephalographic measures of sleep in sloths inhabiting a rainforest. Biology Letters 4: 402–405.

Real, L. A. 1981a. Nectar availability and bee-foraging on *Ipomoea* (Convolvulaceae). Biotropica 13: 64–69.

Real, L. A. 1981b. Uncertainty and pollinator–plant interactions: The foraging behavior of bees and wasps on artificial flowers. Ecology 62: 20–26.

Real, L. A. 1991. Animal choice behavior and the evolution of cognitive architecture. Science 253: 980–986.

Real, L. A. 1992. Information-processing and the evolutionary ecology of cognitive architecture. American Naturalist 140: S108–S145.

Real, L. A. 1996. Paradox, performance, and the architecture of decision-making in animals. American Zoologist 36: 518–529.

Reber, A. S., W. B. Miller and F. Baluska. 2023. Consciousness: Unicellular organisms know the secret. Nature 620: 37–37.

Reid, M. L. and P. J. Weatherhead. 1988. Topographic constraints on competition for territories. Oikos 51: 115–117.

Rennie, K. M. 2015. Home range overlap and spatial-temporal interactions in female fishers on the

Hoopa Indian Reservation in California. MS thesis, Humboldt State University, Humboldt, California. 67 pp.

Reser, J. E. 2014. Solitary mammals provide an animal model for autism spectrum disorders. Journal of Comparative Psychology 128: 99–113.

Rettler, S. J., A. N. Tri, V. St-Louis, J. D. Forester and D. L. Garshelis. 2021. Three decades of declining natural foods alters bottom-up pressures on American black bears. Forest Ecology and Management 493: 119276. https://doi.org/10.1016/j.foreco.2021.119267

Reynolds, T. D. and J. W. Laundré. 1990. Time intervals for estimating pronghorn and coyote home ranges and daily movements. Journal of Wildlife Management 54: 316–322.

Reynolds-Hogland, M. J. and M. S. Mitchell. 2007a. Three axes of ecological studies: Matching process and time in landscape ecology. Pp. 174–194 in Bissonette, J. A. and I. Storch (editors), Temporal Dimensions of Landscape Ecology: Wildlife Responses to Variable Resources. Springer, New York.

Reynolds-Hogland, M. J. and M. S. Mitchell. 2007b. Effects of roads on habitat quality for bears in the Southern Appalachians: A long-term study. Journal of Mammalogy 88: 1050–1061.

Reynolds-Hogland, M. J., L. B. Pacifici and M. S. Mitchell. 2007. Linking resources with demography to understand resource limitation for bears. Journal of Applied Ecology 44: 1166–1175.

Rheault, H., C. R. Anderson, M. Bonar, R. R. Marrotte, T. R. Ross, G. Wittemyer and J. M. Northrup. 2021. Some memories never fade: Inferring multi-scale memory effects on habitat selection of a migratory ungulate using step-selection functions. Frontiers in Ecology and Evolution 9: 15.

Rich, J. G. 1888. The panther's scream. Forest and Stream 31(2): 25.

Rich, L. N., M. S. Mitchell, J. A. Gude and C. A. Sime. 2012. Anthropogenic mortality, intraspecific competition, and prey availability influence territory sizes of wolves in Montana. Journal of Mammalogy 93, 722–731. https://doi.org/10.1644/11-MAMM-A-079.2

Richardson, K., W. Steffen, W. Lucht, J. Bendtsen, S. E. Cornell, J. F. Donges, M. Drüke, I. Fetzer, G. Bala, W. von Bloh, G. Feulner, S. Fiedler, D. Gerten, T. Gleeson, M. Hofmann, W. Huiskamp, M. Kummu, C. Mohan, D. Nogués-Bravo, S. Petri, M. Porkka, S. Rahmstorf, S. Schaphoff, K. Thonicke, A. Tobian, V. Virkki, L. Wang-Erlandsson, L. Weber and J. Rockström. 2023. Earth beyond six of nine planetary boundaries. Science Advances 9: eadh2458.

Richardson, T. O., L. Giuggioli, N. R. Franks and A. B. Sendova-Franks. 2017. Measuring site fidelity and spatial segregation within animal societies. Methods in Ecology and Evolution 8: 965–975.

Ries, L., R. J. Fletcher, Jr., J. Battin and T. D. Sisk. 2004. Ecological responses to habitat edges: Mechanisms, models, and variability explained. Annual Review of Ecology, Evolution and Systematics 35: 491–522.

Rillig, M. C., I. Mansour, S. Hempel, M. Bi, B. König-Ries and A. Kasirzadeh. 2024. How widespread use of generative AI for images and video can affect the environment and the science of ecology. Ecology Letters 27(3). doi:10.1111/ele.14397

Riotte-Lambert, L., S. Benhamou and S. Chamaillé-Jammes. 2015. How memory-based movement leads to nonterritorial spatial segregation. American Naturalist 185: E103–E116.

Riotte-Lambert, L., S. Benhamou, C. Bonenfant and S. Chamaille-Jammes. 2017. Spatial memory shapes density dependence in population dynamics. Proceedings of the Royal Society B: Biological Sciences 284: 8.

Ripley, B. D. 1976. The second-order analysis of stationary point processes. Journal of Applied Probability 13: 255–266.

Riris, P., J. R. Oliver and N. L. Mendieta. 2024. Monumental snake engravings of the Orinoco Rover. Antiquity 98: 724–742.

Robbins, C. T. 1983. Wildlife Feeding and Nutrition. Academic Press, New York. 343 pp.

Robertson, B. A., J. S. Rehage and A. Sih. 2013. Ecological novelty and the emergence of evolutionary traps. Trends in Ecology & Evolution 28: 552–560.

Robertson, S. M. 2016. Mammalian plasticity in response to resource availability. PhD thesis, North Carolina State University, Raleigh, North Carolina. 154 pp.

Rockhill, A. P., C. S. DePerno and R. A. Powell. 2013. The effect of illumination and time of day on movements of bobcats (*Lynx rufus*). Plos One 8. https://doi.org/10.1371/journal.pone.0069213

Rodriguez, F., J. C. Lopez, J. P. Vargas, C. Broglio, Y. Gomez and C. Salas. 2002. Spatial memory and hippocampal pallium through vertebrate evolution: Insights from reptiles and teleost fish. Brain Research Bulletin 57: 499–503.

Rogers, L. L. 1977. Social relationships, movements, and population dynamics of black bears in northeastern Minnesota. PhD thesis, University of Minnesota, Minneapolis, Minnesota.

Rogers, L. L. 1978. Interpretation of cementum annuli in first premolars of bears. Proc. Eastern Workshop on Black Bear Research and Management 4: 102–112.

Rogers, L. L. 1987. Effects of food supply and kinship on social behavior, movements, and population growth of black bears in northeastern Minnesota. Wildlife Monographs 97: 1–72.

Rohwäder, M. S. and F. Jeltsch. 2022. Foraging personalities modify effects of habitat fragmentation on biodiversity. Oikos 2022: e09056. doi: 10.1111/oik.09056

Rolland, E. and S. Trull. 2022. Spatial mapping memory: Methods used to determine the existence and type of cognitive maps in arboreal mammals. Mammal Review 52: 96–111.

Rongstad, O. J. and J. R. Tester. 1969. Movements and habitat use of white-tailed deer in Minnesota. Journal of Wildlife Management 33: 366–379.

Rood, J. P. (1986). Ecology and social evolution in the mongooses. Pp. 131–152 in Rubenstein, D. I. and R. W. Wrangham (editors), Ecological Aspects of Social Evolution. Princeton University Press, Princeton, New Jersey.

Rooney, S. M., A. Wolfe and T. J. Hayden. 1998. Autocorrelated data in telemetry studies: Time to independence and the problem of behavioural effects. Mammal Review 28: 89–98.

Rosenzweig, M. L. 1991. Habitat selection and population interactions: The search for mechanism. American Naturalist 137, Supplement: S5–S28.

Rossi, G. S. and P. A. Wright. 2021. Does leaving water make fish smarter? Terrestrial exposure and exercise improve spatial learning in an amphibious fish. Proceedings of the Royal Society B: Biological Sciences 288: 20210603.

Rossion, B. 2022. Twenty years of investigation with the case of prosopagnosia PS to understand human face identity recognition. Part I: Function. Neuropsychologia 173: 108278

Rozen-Rechels, D., P. Farigoule, S. Agostini, A. Badiane, S. Meylan and J. F. Le Galliard. 2020. Short-term change in water availability influences thermoregulation behaviours in a dry-skinned ectotherm. Journal of Animal Ecology 89: 2099–2110.

Rutz, C., M. Bronstein, A. Raskin, S. C. Vernes, K. Zacarian and D. E. Blasi. 2023. Using machine learning to decode animal communication. Science 381: 152–155.

Saarenmaa, H., N. D. Stone, L. J. Folse, J. M. Packard, W. E. Grant, M. E. Makela and R. N. Coulson. 1988. An artificial-intelligence modeling approach to simulating animal habitat interactions. Ecological Modelling 44: 125–141.

Saitoh, T. 1991. The effects and limits of territoriality on population regulation in grey red-backed voles, *Clethrionomys rufocanus bedfordiae*. Researches on Population Ecology 33: 367–386.

Saleem, A. B., E. M. Diamanti, J. Fournier, K. D. Harris and M. Carandini. 2018. Coherent encoding of subjective spatial position in visual cortex and hippocampus. Nature 562: 124–127.

Saltz, D. and G. C. White. 1990. Comparison of different measures of the error in simulated radiotelemetry locations. Journal of Wildlife Management 54: 169–174.

Samuel, M. D., D. J. Pierce and E. O. Garton. 1985. Identifying areas of concentrated use within the home range. Journal of Animal Ecology 54: 711–719.

Sandell, M. 1989. The mating tactics and spacing patterns of solitary carnivores. Pp. 164–182 in Gittleman, J. L. (editor), Carnivore Behavior, Ecology, and Evolution. Cornell University Press, Ithaca, New York.

Sanderson, G. C. 1966. Study of mammal movements: A review. Journal of Wildlife Management 30: 215–235.

Sanderson, G. C. and B.C. Sanderson. 1964. Radio-tracking rats in Malaysa: A preliminary study. Journal of Wildlife Management 28: 752–768.

Sangha, K. K., I. J. Gordon and R. Costanza. 2023. Editorial: Ecosystem services, policy, and human wellbeing. Frontiers in Ecology and Evolution 11. doi: 10.3389/fevo.2023.1174160

Saraiva Menezes, J. F. and L. G. R. Oliveira-Santos. 2021. Cautious individuals have non-invadable territories, according to an evolutionary mechanistic model. Ecological Modelling 449: 109551.

Sarel, A., A. Finkelstein, L. Las and N. Ulanovsky. 2017. Vectorial representation of spatial goals in the hippocampus of bats. Science 355: 176–180.

Sargeant, A. B., G. A. Swanson and H. A. Doty. 1973. Selective predation by mink, *Mustela vison*, on waterfowl. American Midland Naturalist 89: 208–214.

Sargolini, F., M. Fyhn, T. Hafting, B. McNaughton, M. P. Witter, M. B. Moser and E. I. Moser. 2006. Spatial navigation and entorhinal grid cells. Chemical Senses 31: E24.

Sauder, J. D. and J. L. Rachlow. 2014. Both forest composition and configuration influence landscape-scale habitat selection by fishers (*Pekania pennanti*) in mixed coniferous forests of the Northern Rocky Mountains. Forest Ecology and Management 314: 75–84.

Sawyer, H., C. W. LeBeau, T. L. McDonald, W. Xu and A. D. Middletown. 2019. All routes are not created equal: An ungulate's choice of migration route can influence its survival. Journal of Applied Ecology 56: 1860–1869. doi: 10.1111/1365-2664.13445

Sawyer, H., M. J. Kauffman, R. M. Nielson and J. S. Horne. 2009. Identifying and prioritizing ungulate migration routes for landscape-level conservation. Ecological Applications 19: 2016–2025.

Schaller, G. B. 1972. The Serengeti Lion: A Study of Predator–Prey Relations. University of Chicago Press, Chicago. 480 pp.

Schelling, T. C. 1960. The Strategy of Conflict. Harvard University Press, Cambridge, Massachusetts.

Schmidt, G. 2024. Why 2023's heat anomaly is worrying scientists. Nature 627: 467.

Schmidt-Nielsen, K. 1971. Locomotion: Energy costs of swimming, flying and running. Science 177: 222–228.

Schmuker, M., V. Bahr and R. Huerta. 2016. Exploiting plume structure to decode gas source distance using metal-oxide gas sensors. Sensors and Actuators B: Chemical 235: 636–646.

Schmutz, J. A. and G. C. White. 1990. Error in telemetry studies: Effects of animal movement on triangulation. Journal of Wildlife Management 54: 506–510.

Schmutz, J. A., K. G. Wright, C. R. DeSorbo, J. Fair, D. C. Evers, B. D. Uher-Koch and D. M. Mulcahy. 2014. Size and retention of breeding territories of yellow-billed loons (*Gavia adamsii*) in Alaska and Canada. Waterbirds 37: 53–63.

Schoener, T. W. 1981. An empirically based estimate of home range. Theoretical Population Biology 20: 281–325.

Schoener, T. W. 1983. Simple-models of optimal feeding-territory size: A reconciliation. American Naturalist 121: 608–629.

Schoener, T. W. and A. Schoener. 1982. Intraspecific variation in home-range size in some Anolis lizards. Ecology 63: 809–823.

Schumacher, S., G. Von der Emde and T. B. De Pereira. 2017. Sensory influence on navigation in weakly electric fish *Gnathonemus petersii*. Animal Behavior 132: 1–12.

Schwartz, C. W. 1941. Home range of the cottontail in central Missouri. Journal of Mammalogy 22: 386–392.

Seaman, D. E. 1993. Home range and male reproductive optimization in black bears. PhD thesis, North Carolina State University, Raleigh, North Carolina.

Seaman, D. E. and R. A. Powell. 1990. Identifying patterns and intensity of home range use. International Conference on Bear Research and Management 8: 243–249.

Seaman, D. E. and R. A. Powell. 1996. An evaluation of the accuracy of kernel density estimators for home range analysis. Ecology 77: 2075–2085.

Séchaud, R., K. Schalcher, B. Almasi, R. Buhler, K. Safi, A. Romano and A. Roulin. 2022. Home range size and habitat quality affect breeding success but not parental investment in barn owl males. Scientific Reports 12(1): 6516. doi: 10.1038/s41598-022-10324-7

Segura-Garcia, J., S. Sturley, M. Arevalillo-Herraez, J. M. Alcaraz-Calero, S. Felici-Castell and E. A. Navarro-Camba. 2024. 5G AI-IoT System for Bird Species Monitoring and Song Classification. Sensors 24(11): 3687. https://doi.org/10.3390/s24113687

Seibert, S. G and M. R. Pelton. 1994. Nutrient content of squaw root, *Conopholis americana*, and its importance to Southern Appalachian black bears, *Ursus americanus* (Carnivora: Ursidae). Brimleyana 21: 151–156.

Seidel, D. P. and M. S. Boyce. 2016. Varied tastes: Home range implications of foraging-patch selection. Oikos 125: 39–49.

Seigle-Ferrand, J., K. Atmeh, J. M. Gaillard, V. Ronget, N. Morellet, M. Garel, A. Loison and G. Yannic. 2021. A systematic review of within-population variation in the size of home range across ungulates: What do we know after 50 years of telemetry studies? Frontiers in Ecology and Evolution 8. https://doi.org/10.3389/fevo.2020.555429

Seigle-Ferrand, J., P. Marchand, N. Morellet, J. M. Gaillard, A. J. M. Hewison, S. Said, Y. Chaval, H. Santacreu, A. Loison, G. Yannic and M. Garel. 2022. On this side of the fence: Functional responses to linear landscape features shape the home range of large herbivores. Journal of Animal Ecology 91: 443–457.

Sells, S. N. 2019. Mechanisms driving territorial and social behavior in a cooperative carnivore. PhD thesis, The University of Montana, Missoula, Montana.

Sells, S. N. and M. S. Mitchell. 2020. The economics of territory selection. Ecological Modelling 438: 15. https://doi.org/10.1016/j.ecolmodel.2020.109329

Sells, S. N., M. S. Mitchell, D. E. Ausband, A. D. Luis, D. J. Emlen, K. M. Podruzny and J. A. Gude. 2022a. Economical defence of resources structures territorial space use in a cooperative carnivore. Proceedings of the Royal Society B: Biological Sciences 289. https://doi.org/10.1098/rspb.2021.2512

Sells, S. N., M. S. Mitchell, K. M. Podruzny, J. A. Gude, A. C. Keever, D. K. Boyd, T. D. Smucker, A. A. Nelson, T. W. Parks, N. J. Lance, M. S. Ross and R. M. Inman. 2021. Evidence of economical territory selection in a cooperative carnivore. Proceedings of the Royal Society B: Biological Sciences 288. https://doi.org/10.1098/rspb.2021.0108

Sells, S. N., K. M. Podruzny, J. J. Nowak, T. D. Smucker, T. W. Parks, D. K. Boyd, A. A. Nelson, N. J. Lance, R. M. Inman, J. A. Gude, S. B. Bassing, K. E. Loonam and M. S. Mitchell. 2022b. Integrating basic and applied research to estimate carnivore abundance. Ecological Applications 32(8): e2714. doi: 10.1002/eap.2714

Seton, E. T. 1909. Life-Histories of Northern Animals. Charles Scribner's Sons, New York.

Shen, Y., M. Zhou, D. Cai, D. Almeida Filho, G. Fernandes, Y. Cai, A. F. de Sousa, M. Tian, J. Lee, D. Necula, C. Zhou, S. Li, S. Salinas, A. Liu, X. Kang, M. Kamata, A. Lavi, S. Huang, T. Silva, W. Do Heo and A. J. Silva. 2022. CCR5 closes the temporal window for memory linking. Nature 606: 146–152.

Shepard, E. L. C., R. P. Wilson, W. G. Rees, E. Grundy, S. A. Lambertucci and S. B. Vosper. 2013. Energy landscapes shape animal movement ecology. American Naturalist 182: 298–312.

Sheppard, K., J. Gardin, G. S. Sabnis, A. Peer, M. Darrell, S. Deats, B. Geuther, C. M. Lutz and V. Kumar. 2022. Stride-level analysis of mouse open field behavior using deep-learning-based pose estimation. Cell Reports 38: 110231.

Sheridan, C. L., S. Lang, M. Knappenberger, C. Albers, R. Loper, B. Tillett, J. Sanchez, A. Wilcox, T. Harrison, D. Panoz-Brown and J. D. Crystal. 2024. Replay of incidentally encoded episodic memories in the rat. Current Biology 34: 1–7. https://doi.org/10.1016/j.cub.2023.12.043

Sherratt, T. N. and M. Mesterton-Gibbons. 2015. The evolution of respect for property. Journal of Evolutionary Biology 28: 1185–1202.

Shumailov, I., Z. Shumaylov, Y. Zhao, N. Papernot, R. Anderson and Y. Gal. 2024. AI models collapse when trained on recursively generated data. Nature 631: 755–759.

Signer, J. and J. R. Fieberg. 2021. A fresh look at an old concept: Home-range estimation in a tidy world. PeerJ 9: 11031.

Sih, A., S. F. Hanser and K. A. McHugh. 2009. Social network theory: New insights and issues for behavioral ecologists. Behavioral Ecology and Sociobiology 63: 975–988.

Sikes, R. S. and W. L. Gannon. 2011. Guidelines of the American Society of Mammalogists for the use of wild mammals in research. Journal of Mammalogy 92: 235–253.

Sikes, R. S. and the Animal Care and Use Committee of the American Society of Mammalogists. 2016. Guidelines of the American Society of Mammalogists for the use of wild mammals in research and education. Journal of Mammalogy 97: 663–668.

Silverman, B. W. 1990. Density Estimation for Statistics and Data Analysis. Chapman and Hall, London.

Siniff, D. B. and C. R. Jessen. 1969. A simulation model of animal movement patterns. Advances in Ecological Research 6: 186–219.

Siracusa, E. R., D. R. Wilson, E. K. Studd, S. Boutin, M. M. Humphries, B. Dantzer, J. E. Lane and A. G. McAdam. 2019. North American red squirrels mitigate costs of territory defence through social plasticity. Animal Behaviour 151: 29–42.

Skipper, M. 2024a. Prioritize study design in the era of big data. Nature 635: 786.

Skipper, M. 2024b. Why scientists trust AI too much – and what to do about it. Nature 627: 243.

Slater, P. J. B. and N. P. Lester. 1982. Minimising errors in splitting behaviour into bouts. Behaviour 79: 153–161.

Smereka, C. A., P. F. Frame, M. A. Edwards, O. M. Slater, D. D. Frame and A. E. Derocher. 2021. Space use of cougars at the northern edge of their range. Journal of Mammalogy 102: 1042–1053.

Smielak, M. K., G. Ballard, P. J. S. Fleming, G. Kortner, K. Vernes, and N. Reid. 2023. Brushtail possum terrestrial activity patterns are driven by climatic conditions, breeding and moonlight intensity. Mammal Research 68: 547–560.

Smith, C. C. 1968. Adaptive nature of social organization in genus of tree squirrels *Tamiasciurus*. Ecological Monographs 38: 31–63.

Smith, C. C. 1978. Structure and function of the vocalizations of tree squirrels (*Tamiasciurus*). Journal of Mammalogy 59: 793–808.

Smith, C. C. 1981. The indivisible niche of *Tamiasciurus*: An example of nonpartitioning of resources. Ecological Monographs 51: 343–363.

Smith, K. 2023. Neuroscience goes wild. Nature 618: 448–450.

Smith, K. P., A. N. Facka and R. A. Powell. 2020. Using reintroduction to elucidate breeding system, breeding phenology and infanticide for a solitary carnivore. Canadian Journal of Zoology 98: 433–447.

Smouse, P. E., S. Focardi, P. R. Moorcroft, J. G. Kie, J. D. Forester and J. M. Morales. 2010. Stochastic modelling of animal movement. Philosophical Transactions of the Royal Society B: Biological Sciences 365: 2201–2211.

Sober, E. 2008. Evidence and Evolution: The Logic Behind the Science. Cambridge University Press, Cambridge.

Socha, R. and R. Zemek. 2003. Wing morph-related differences in the walking pattern and dispersal in a flightless bug, Pyrrhocoris apterus (Heteroptera). Oikos 100: 35–42.

Solomon, N. G., B. Keane, L. R. Knoch, and P. J. Hogan. 2004. Multiple paternity in socially monogamous prairie voles (*Microtus ochrogaster*). Canadian Journal of Zoology-Revue Canadienne De Zoologie 82(10): 1667–1671. doi: 10.1139/z04-142

Solstad, T., C. N. Boccara, E. Kropff, M. B. Moser and E. I. Moser. 2008. Representation of geometric borders in the entorhinal cortex. Science 322: 1865–1868.

Solvi, C., S. G. Al-Khudhairy and L. Chittka. 2020. Bumblebees display cross-modal object recognition between visual and tactile senses. Science 367: 910–912.

Soma, M. and H. Brumm. 2020. Group living facilitates the evolution of duets in barbets. Biology Letters 16: 6.

Soulsbury, C. D., P. J. Baker, G. Iossa and S. Harris. 2008. Fitness costs of dispersal in red foxes. Behavioral Ecology and Sociobiology 62: 1289–1298.

South, A. 1999. Extrapolating from individual movement behaviour to population spacing patterns in a ranging mammal. Ecological Modelling 117: 343–360.

Sparklin, B. D., M. S. Mitchell, L. B. Hanson, D. B. Jolley and S. S. Ditchkoff. 2009. Territoriality of feral pigs in a highly persecuted population on Fort Benning, Georgia. Journal of Wildlife Management 73: 497–502.

Speakman, J. R. 1993. How should we calculate CO_2 production in doubly labelled water studies of animals? Functional Ecology 7: 746–750.

Speakman, J. R., K. S. Nair and M. I. Goran. 1993. Revised equations for calculating the CO_2 production from doubly labelled water in humans. American Journal of Physiology 264: E912–E917.

Speakman, J. R. and P. A. Racey. 1988. The doubly-labelled water technique for measurement of energy expenditure in free-living animals. Scientific Progress, Oxford 72: 227–237.

Spencer, S. R., G. N. Cameron and R. K. Swihart. 1990. Operationally defining home range: Temporal dependence exhibited by hispid cotton rats. Ecology 71: 1817–1822.

Spencer, W. D. 1992. Space in the lives of vertebrates: On the ecology and psychology of space use. PhD dissertation, University of Arizona, Tucson, Arizona.

Spencer, W. D. 2012. Home ranges and the value of spatial information. Journal of Mammalogy 93: 929–947.

Spencer W. D. and R. H. Barrett. 1984. An evaluation of the harmonic mean measure for defining carnivore activity areas. Acta Zoologica Fennica 171: 255–259.

Spitz, D. B., M. Hebblewhite and T. R. Stephenson. 2017. 'MigrateR': Extending model-driven methods for classifying and quantifying animal movement behavior. Ecography 40: 788–799.

Stacho. M., C. Herold, N. Rook, H. Wagner, M. Axer, K. Amunts and O. Güntürkün. 2020. A cortex-like canonical circuit in the avian forebrain. Science 369. doi:10.1126/science.abc5534

Stahlecker, D. W. and T. G. Smith. 1993. A comparison of home range estimates for a bald eagle wintering in New Mexico. Journal of Raptor Research 27: 42–45.

Stamps, J. 1995. Motor learning and the value of familiar space. American Naturalist 146: 41–58.

Stamps, J. A. and V. V. Krishnan. 1999. A learning-based model of territory establishment. Quarterly Review of Biology 74: 291–318.

Stangl, M., U. Topalovic, C. S. Inman, S. Hiller, D. Villaroman, Z. M. Aghajan, L. Christov-Moore, N. R. Hasulak, V. R. Rao, C. H. Halpern, D. Eliashiv, I. Fried and N. Suthana. 2021. Boundary-anchored neural mechanisms of location-encoding for self and others. Nature 589: 420–425.

Stearns, S. C. 1992. The Evolution of Life Histories. Oxford University Press, Oxford, England.

Stebbins, R. C. 1948. Additional observations on home ranges and longevity in the lizard *Sceloporus graciosus*. Copeia 1948: 20–22.

Steixner-Kumar, S. and J. Glascher. 2020. Strategies for navigating a dynamic world. Science 369: 1056–1057.

Stenger, J. 1958. Food habits and available food of ovenbirds in relation to territory size. The Auk 75: 335–346.

Stephan, P., J. P. M. Jäschke, E. Oberzaucher and K. Grammer. 2014. Sex differences and similarities in urban home ranges and in the accuracy of cognitive maps. Evolutionary Psychology 12: 814–826.

Stephens, D. W. and J. R. Krebs. 1986. Foraging Theory. Princeton University Press, Princeton, New Jersey.

Stephens, D. W., J. S. Brown and R. C. Ydenberg. 2007. Foraging Behavior and Ecology. University of Chicago Press, Chicago, Illinois. 576 pp.

Steudler, J. S. and H. F. Ólafsdóttir. 2023. Cracking the neuronal code of episodic memory. Science 382: 262–263.

Stevens, J. R. and D. W. Stephens. 2002. Food sharing: A model of manipulation by harassment. Behavioral Ecology 13: 393–400.

Stickel, L. F. 1946. The source of animals moving into a depopulated area. Journal of Mammalogy 27: 301–307.

Stickel, W. H. and J. B. Cope 1947. The home ranges and wanderings of snakes. Copeia 1947: 127–136.

Stiegler, J., A. Lins, M. Dammhahn, S. Kramer-Schadt, S. Ortmann and N. Blaum. 2022. Personality drives activity and space use in a mammalian herbivore. Movement Ecology 10: 33.

Stillman, R. A., R. W. G. Caldow, J. D. Goss-Custard and M. J. Alexander. 2000. Individual variation in intake rate: The relative importance of foraging efficiency and dominance. Journal of Animal Ecology 69: 484–493.

Storer, T. L., F. C. Evans and F. G. Palmer. 1944. Some rodent populations in the Sierra Nevada of California. Ecological Monographs 14: 165–192.

Strassmann, J. E. and D. C. Queller. 2014. Privatization and property in biology. Animal Behaviour 92: 305–311.

Striedter, G. F. 2016. Evolution of the hippocampus in reptiles and birds. Journal of Comparative Neurology 524: 496–517.

Ströckens, F., K. Neves, S. Kirchem, C. Schwab, S. Herculano-Houzel and O. Güntürkün. 2022. High associative neuron numbers could drive cognitive performance in corvid species. Journal of Comparative Neurology 530: 1588–1605.

Stumpf, W. A. and C. O. Mohr. 1962. Linearity of home ranges of California mice and other animals. Journal of Wildlife Management 26: 149–154.

Sugar, J. and M. B. Moser. 2019. Episodic memory: Neuronal codes for what, where, and when. Hippocampus 29: 1190–1205 (plus erratum).

Sugihara, G. and R. M. May. 1990. Applications of fractals in ecology. Trends in Ecology & Evolution 5: 79–86.

Suriyampola, P. S. and P. K. Eason. 2015. The effects of landmarks on territorial behavior in a convict cichlid, *Amatitlania siquia*. Ethology 121: 785–792.

Sutherland, W. J. 1996. Predicting the consequences of habitat loss for migratory populations. Proceedings of the Royal Society 263: 1325–1327.

Sutherland, W. J. 2006. Predicting the ecological consequences of environmental change: A review of the methods. Journal of Applied Ecology 43: 599–616.

Swihart, R. K. and J. P. Bryant. 2001. Importance of biogeography and ontogeny of woody plants in winter herbivory by mammals. Journal of Mammalogy 82: 1–21.

Swihart, R. K. and N. A. Slade. 1985. Testing for independence of observations in animal movements. Ecology 66: 1176–1184.

Swihart, R. K. and N. A. Slade. 1986. The importance of statistical power when testing for independence in animal movements. Ecology 67: 255–258.

Swihart, R. K., N. A. Slade and B. J. Bergstrom. 1988. Relating body size to the rate of home range use in mammals. Ecology 69: 393–399.

Tacikowski, P., G. Kalender, D. Ciliberti and I. Fried. 2024. Human hippocampal and entorhinal neurons encode the temporal structure of experience. Nature 635: 160–167.

Takagi, S., H. Chijiiwa, M. Arahori, A. Saito, K. Fujita and H. Kuroshima. 2021. Socio-spatial cognition in cats: Mentally mapping owner's location from voice. Plos One 16(11): e0257611. https://doi.org/10.1371/journal.pone.0257611

Tan, Z. J., X. W. Zou, S. Y. Huang, W. Zhang and Z. Z. Jin. 2002. Random walk with memory enhancement and decay. Physical Review E 65: 1–5.

Tao, Y., L. Börger and A. Hastings. 2016. Dynamic range size analysis of territorial animals: An optimality approach. American Naturalist 188: 460–474.

Tarter, M. E. and R. A. Kronmal. 1970. On multivariate density estimates based on orthogonal expansions. Annals of Methematical Statistics 41: 718–722.

Taylor, C. R., K. Schmidt-Nielsen and J. L. Rabb. 1970. Scaling the energetic cost of running to body size in mammals. American Journal of Physiology 219: 1104–1107.

Temeles, E. J. 2022. The effects of prey availability and capture success on the foraging and territory economics of a predatory bird, *Circus hudsonius*. Journal of Ornithology 163: 767–777.

Terada, S., T. Geiller, Z. R. Liao, J. O'Hare, B. Vancura and A. Losonczy. 2022. Adaptive stimulus selection for consolidation in the hippocampus. Nature 601: 240–244.

Terborgh, J. 1983. Five New World primates: A study in comparative ecology. Princeton University Press, Princeton.

Terceros, A. and P. Rajasethupathy. 2022. Closing the window on memory linking. Nature 606: 38–39.

Theng, M., C. Milleret, C. Bracis, P. Cassey and S. Delean. 2022. Confronting spatial capture–recapture models with realistic animal movement simulations. Ecology 103. https://doi.org/10.1002/ecy.3676

Thomson, J. D., M. Slatkin and B. A. Thomson. 1997. Trapline foraging by bumble bees .2. Definition and detection from sequence data. Behavioral Ecology 8: 199–210.

Thompson, E. S. 1898. Wild Animals I Have Known. Grosset & Dunlap, New York.

Tolman, E. C. 1948. Cognitive maps in rats and men. Psychological Review 55: 189–208.

Tórrez-Herrera, L. L., G. H. Davis and M. C. Crofoot. 2020. Do monkeys avoid areas of home range overlap because they are dangerous? A test of the risk hypothesis in white-faced capuchin monkeys (Cebus capucinus). International Journal of Primatology 41: 246–264.

Tran, V. T., A. C. Vitz and M. H. Bakermans. 2024. Evaluating habitat-specific interference in automated radio telemetry systems: Implications for animal movement studies. Animal Biotelemetry 12: 12. https://doi.org/10.1186/s40317-024-00369-y

Treinys, R., U. Bergmanis and Ü. Väli. 2017. Strong territoriality and weak density-dependent reproduction in Lesser Spotted Eagles *Clanga pomarina*. Ibis 159: 343–351.

Tsalyuk, M., W. Kilian, B. Reineking and W. M. Getz. 2019. Temporal variation in resource selection of African elephants follows long-term variability in resource availability. Ecological Monographs 89. https://doi.org/10.1002/ecm.1348

Tsao, A., J. Sugar, L. Lu, C. Wang, J. J. Knierim, M. B. Moser and E. I. Moser. 2018. Integrating time from experience in the lateral entorhinal cortex. Nature 561: 57–62. https://doi.org/10.1038/s41586-018-0459-6

Turchin, P. 1991. Translating foraging movements in heterogeneous environments into the spatial distribution of foragers. Ecology 72: 1253–1266.

Turchin, P. 1998. Quantitative analysis of movement: Measuring and modeling population redistribution in animals and plants. Sinauer Associates, Sunderland, Massachusetts. 396 pp.

Turchin, P. 2023. End Times: Elites, Counter-Elites and the Path of Political Disintegration. Penguin Press, New York. 352 pp.

Turner, C. H. 1907. The homing of ants: An experimental study of ant behavior. Journal of Comparative Neurology and Psychology 17: 367–434.

Turner, M. G., Y. A. Wu, L. L. Wallace, W. H. Romme and A. Brenkert. 1994. Simulating winter interactions among ungulates, vegetation, and fire in northern Yellowstone Park. Ecological Applications 4: 472–486.

Turner, W. C., S. Périquet, C. E. Goelst, K. B. Vera, E. Z. Cameron, K. A. Alexander, J. L. Belant, C. C. Cloete, P. du Preez, W. M. Getz, R. S. Hetem, P. L. Kamath, M. K. Kasaona, M. Mackenzie, J. Mendelsohn, J. K. E. Mfune, J. R. Muntifering, R. Portas, H. A. Scott, W. Maartin Strauss, W. Versfeld, B. Wachter, G. Wittemyer and J. W. Kilian. 2022. Africa's drylands in a changing world: Challenges for wildlife conservation under climate and land-use changes in the Greater Etosha Landscape. Global Ecology and Conservation 38: e02221. https://doi.org/10.1016/j.gecco.2022.e02221.

Twining, J. P., C. Sutherland, A. Zalerski, M. V. Cove, J. Birks, O. R. Wearn, J. Hansom, A. Wereszczuk, E. Manzo, P. Bartolommei, A. Mortelli, B. Evans, B. D. Gerber, T. J. McGreevy, Jr., L.S. Ganoe, J. Masseloux, A. Mayer, I. Wierzbowska, J. Loch, J. Akins, D. Drummey, W. McShea, R. Kays, S. Manke, L. Pardo, A. Boyce, S. Li, R. B. Ragai, Á. J. V. Trujillo, C. López-Gonzalez, N. E. Lara-Díaz, O. Cosby, C. W. Waggershauser, J. Bambeer, F. Stewart, J. Fisher, A. Fuller, K. Perkins and R. A. Powell. In press. Sociality in unexpected places: Using global remote camera data to evaluate the drivers of social complexity. Proceedings of the National Academy of Sciences of the United States of America.

Tyree, T. J., M. Metke and C. T. Miller. 2023. Cross-model representation of identity in the primate hippocampus. Science 382: 417–423.

Uhlenbeck, G. E. and L. S. Ornstein. 1930. On the theory of the Brownian motion. Physical Review 36: 823–841.

Valldeperes, M., R. Pascual-Rico, P. Fandos, R. C. Soriguer, J. M. Perez, F. J. Cano-Manuel León, P. P. Ferro, J. Ramon Lopez-Elvira and J. E. Granados. 2024. Home range in genus *Capra*: from polygons to Brownian bridges of scabietic and healthy Iberian ibexes (*Capra pyrenaica*). Journal of Mammalogy. https://doi.org/10.1093/jmammal/gyae013

van der Merwe, M. and J. S. Brown. 2008. Mapping the landscape of fear of the Cape ground squirrel (Xerus inauris). Journal of Mammalogy 89: 1162–1169.

van Moorter, B., C. M. Rolandsen, M. Basille and J. M. Gaillard. 2016. Movement is the glue connecting home ranges and habitat selection. Journal of Animal Ecology 85: 21–31.

van Moorter, B., D. Visscher, S. Benhamou, L. Borger. M. S. Boyce and J.-M. Gaillard. 2009. Memory keeps you at home: A mechanistic model for home range emergence. Oikos 118: 641–652.

Van Winkle, W., Jr., D. C. Martin and M. J. Sebetich. 1973. A home-range model for animals inhabiting and ecotone. Ecology 54: 205–209.

Van Winkle, W., Jr. 1975. Comparison of several probabilistic home-range models. Journal of Wildlife Management 39: 118–123.

Vander Wal, E. and A. R. Rodgers. 2012. An individual-based quantitative approach for delineating core areas of animal space use. Ecological Modelling 224: 48–53.

Vander Wall, S. B. 1982. An experimental analysis of cache recovery in Clark's nutcracker. Animal Behaviour 30: 84–94.

Vander Wall, S. B., J. S. Briggs, S. H. Jenkins, K. M. Kuhn, T. C. Thayer and M. J. Beck. 2006. Do food-hoarding animals have a cache recovery advantage? Determining recovery of stored food. Animal Behaviour 72: 189–197.

Vandermeer, J. 1981. A further note on community models. American Naturalist 117: 379–381.

Vangen, K. M., J. Persson, A. Landa, R. Andersen and P. Segerström. 2001. Characteristics of dispersal in wolverines. Canadian Journal of Zoology 79: 1641–1649.

Varga, T., J. Garay, J. Rychtář and M. Brown. 2020. A temporal model of territorial defence with antagonistic interactions. Theoretical Population Biology 134: 15–35.

Vehrencamp, S. L., J. M. Ellis, B. F. Cropp and J. M. Koltz. 2014. Negotiation of territorial boundaries in a songbird. Behavioral Ecology 25: 1436–1450.

Velazquez, J. L. P. 2024. Consciousness? It's all in the mind. Nature 627: 38.

Venditti, C., J. Baker and R. A. Barton. 2024. Co-evolutionary dynamics of mammalian brain and body size. Nature Ecology & Evolution. https://doi.org/10.1038/s41559-024-02451-3

Versace, E., L. Freeland and M. G. Emmerson. 2024. First-sight recognition of touched objects shows that chicks can solve Molyneux's problem. Biology Letters 20(4): 20240025. doi: 10.1098/rsbl.2024.0025

Villon, S., C. Iovan, M. Mangeas and L. Vigliola. 2024. Toward an artificial intelligence-assisted counting of sharks on baited video. Ecological Informatics 80(3): 102499. doi: 10.1016/j.ecoinf.2024.102499

Viskaitis, P., M. Arnold, C. Garau, L. T. Jensen, L. Fugger, D. Peleg-Raibstein and D. Burdakov. 2022. Ingested non-essential amino acids recruit brain orexin cells to suppress eating in mice. Current Biology 32: 1–10. https://doi.org/10.1016/j.cub.2022.02.067

von Bonin, G. 1937. Brain-weight and body-weight of mammals. Journal of General Psychology 16: 379–389.

von Neuman, J. and O. Morgenstern. 1953. Theory of Games and Economic Behaviour. Princeton University Press, Princeton, New Jersey.

Von Uexküll, J. 1934. Streifzüge durch die Umwelten con Tieren und Menschen. Verlag von Julius Springer, Vienna. O'Neill, J. D. (Translator). A foray into the Worlds of Animals and Humans with a Theory of Meaning. University of Minnesota Press, Minneapolis.

Vucetich, J. A. 2021. Restoring the Balance: What Wolves Tell Us About Our Relationship with Nature. Johns Hopkins University Press, Baltimore, Maryland. 382 pp.

Waage, J. K. 1988. Confusion over residency and the escalation of damselfly territorial disputes. Animal Behaviour 36: 586–595.

Wakefield, E. D., T. W. Bodey, S. Bearhop, J. Blackburn, K. Colhoun, R. Davies, R. G. Dwyer, J. A. Green, D. Grémillet, A. L. Jackson, M. J. Jessopp, A. Kane, R. H. W. Langston, A. Lescroël, S. Murray, M. Le Nuz, S. C. Patrick, C. Péron, L. M. Soanes, S. Wanless, S. C. Votier and K. C. Hamer. 2013. Space partitioning without territoriality in gannets. Science 341: 68–70.

Walters, J. R., C. K. Copeyon and J. H. Carter III. 1992. Test of the ecological basis of cooperative breeding in red-cockaded woodpecker. Auk 109: 90–97.

Walters, J. R., P. D. Doerr and J. H. Carter III. 1988. The cooperative breeding system of the red-cockaded woodpecker. Ethology 78: 275–305.

Walton, R. and V. Nolan, Jr. 1986. Imperfect information and the persistence of pretenders: Male prairie warblers contesting for territory. The American Naturalist 128: 427–432.

Wang, M. N., D. J. Foster and B. E. Pfeiffer. 2020. Alternating sequences of future and past behavior encoded within hippocampal theta oscillations. Science 370: 247–250.

Wang, M. and V. Grimm. 2007. Home range dynamics and population regulation: An individual-based model of the common shrew *Sorex araneus*. Ecological Modelling 205: 397–409.

Waser, P. M. 1976. *Cercocebus albigena*: Site attachment, avoidance, and intergroup spacing. American Naturalist 110: 911–935.

Wasserstein, R. L., A. L. Schirm and N. A. Lazar. 2019. Moving to a world beyond "$p < 0.05$." American Statistician 73: 1–19.

Watson, A. and R. Moss. 1970. Dominance, spacing behaviour and aggression in relation to population limitation in vertebrates. Symposium of the British Ecological Society 10: 167–218.

Watson, A., R. Moss, R. Parr, M. D. Mountford and P. Rothery. 1994. Kin landownership, differential aggression between kin and non-kin, and population fluctuations in red grouse. Journal of Animal Ecology 63: 39–50.

Weary, D. M., E. A. Pajor, B. K. Thompson and D. Fraser. 1996. Risky behaviour by piglets: A trade off between feeding and risk of mortality by maternal crushing? Animal Behaviour 51: 619–624.

Webber, Q. M. R., B. Dantzer, J. E. Lane, S. Boutin and A. G. McAdam. 2023. Density-dependent plasticity in territoriality revealed using social network analysis. Journal of Animal Ecology 92: 207–221.

Wehner, R. 2020. Desert Navigator: The Journey of an Ant. Harvard University Press, Cambridge, Massachusetts.

Weir, R. D. and F. B. Corbould 2010. Factors affecting landscape occupancy by fishers in north-central British Columbia. Journal of Wildlife Management 74: 405–410.

Weisstein, E. W. 2006. Traveling salesman problem. From MathWorld—A Wolfram Web Resource. https://mathworld.wolfram.com/TravelingSalesmanProblem.html – visited 27 November 2022.

Wells, M. C., and M. Bekoff. 1981. An observational study of scent-marking in coyotes, *Canis latrans*. Animal Behaviour 29: 332–350.

Westcott, D. A. and D. L. Graham. 2000. Patterns of movement and seed dispersal of a tropical frugivore. Oecologia 122: 249–257.

Westoby, M. 1974. Analysis of diet selection by large generalist herbivores. American Naturalist 108: 290–304.

White, C. R., L. A. Alton, C. L. Bywater, E. J. Lombardi and D. J. Marshall. 2022. Metabolic scaling is the product of life-history optimization. Science 377: 834–839.

White, G. C. and R. A. Garrott. 1990. Analysis of Wildlife Radio-Tracking Data. Academic Press, San Diego.

White, J. E. 1964. An index of the range of activity. American Midland Naturalist 71: 369–373.

White, L. A., S. VandeWoude, and M. E. Craft. 2020. A mechanistic, stigmergy model of territory formation in solitary animals: Territorial behavior can dampen disease prevalence but increase persistence. Plos Computational Biology 16. https://doi.org/10.1371/journal.pcbi.1007457

White, P. C. L. and S. Harris. 1994. Encounters between red foxes (*Vulpes vulpes*): Implications for territory maintenance, social cohesion and dispersal. Journal of Animal Ecology 63: 315–327.

Widloski, J. and D. J. Foster. 2022. Flexible rerouting of hippocampal replay sequences around changing barriers in the absence of global place field remapping. Neuron 110: 1547–1558. https://doi.org/10.1016/j.neuron.2022.02.002

Wiens, J. A. 1995. Landscape mosaics and ecological theory. Pp. 1–26 in Hansson, L., L. Fahrig and G. Merriam (editors), Mosaic Landscapes and Ecological Processes. Chapman and Hall, London, UK.

Wikelski, M., M. Quetting, Y. C. Cheng, W. Fiedler, A. Flack, A. Gagliardo, R. Salas, N. Zannoni and J. Williams. 2021. Smell of green leaf volatiles attracts white storks to freshly cut meadows. Scientific Reports 11: 12912. https://doi.org/10.1038/s41598-021-92073-7

Wilming, N., P. König, S. König and E. A. Buffalo. 2018. Entorhinal cortex receptive fields are modulated by spatial attention, even without movement. eLife. doi: https://doi.org/10.7554/eLife.31745.001

Wilson, E. O. 1975. Sociobiology. Harvard University Press, Cambridge, Massachusetts. 697 pp.

Wilson, R., M. Hooten, B. Strobel and J. Shivik. 2010. Accounting for individuals, uncertainty, and multiscale clustering in core area estimation. Journal of Wildlife Management 74: 1343–1352.

Winner, K., M. J. Noonan, C. H. Fleming, K. A. Olson, T. Mueller, D. Sheldon and J. M. Calabrese. 2018. Statistical inference for home range overlap. Methods in Ecology and Evolution 9: 1679–1691.

Wirth, S., P. Baraduc, A. Planté, S. Pinède and J. R. Duhamel. 2017. Gaze-informed, task-situated representation of space in primate hippocampus during virtual navigation. Plos Biology 15(2): e2001045. doi:10.1371/journal.pbio.2001045

With, K. A. and T. O. Crist. 1996. Translating across scales: Simulating species distributions as the aggregate response of individuals to heterogeneity. Ecological Modelling 93: 125–137.

Witter, M. P., H. J. Groenewegen, F. H. L. Dasilva and A. H. M. Lohman. 1989. Functional-organization of the extrinsic and intrinsic circuitry of the parahippocampal region. Progress in Neurobiology 33: 161–253.

Wolff, J. O. 1989. Social behavior. Pp. 271–291 in Kirkland, G. L. and J. N. Layne (editors), Advances in the Study of *Peromyscus* (Rodentia). Texas Tech University Press, Lubbock, Texas.

Wolff, J. O. 1993. Why are female small mammals territorial? Oikos 68: 364–369.

Wolff, J. O. and T. Van Horn. 2003. Vigilance and foraging patterns of American elk during the rut in habitats with and without predators. Canadian Journal of Zoology 81: 266–271.

Wood, E. R., P. A. Dudchenko and H. Eichenbaum. 1999. The global record of memory in hippocampal neuronal activity. Nature 397: 613–616.

Woodrich, S. T. and T. Pape. 2024. Ecological restoration and artificial intelligence: Whose values inform a project? Restoration Ecology 32. doi:10.1111/rec.14128

Woolfenden, G. E. and J. W. Fitzpatrick. 1984. The Florida Scrub Jay: Demography of a Cooperative-breeding Bird. Princeton University Press, Princeton, New Jersey.

Worton, B. 1987. A review of models of home range for animal movement. Ecology Modeling 38: 277–298.

Worton, B. J. 1989. Kernel methods for estimating the utilization distribution in home-range studies. Ecology 70: 164–168.

Wright, S. 1932. The roles of mutation, inbreeding, crossbreeding, and selection in evolution. Proceedings VI, International Congress of Genetics 1: 356–366.

Wright, S. 1978. Evolution and the Genetics of Populations. Volume 4: Variability Within and Among Natural Populations. University of Chicago Press, Chicago, Illinois.

Wuest, S. E., N. D. Pires, S. Luo, F. Vasseur, J. Messier, U. Grossniklaus and P. A. Niklaus. 2022. Increasing plant group productivity through latent genetic variation in cooperation. Plos Biology 20(11): e3001842. https://doi.org/10.1371/journal.pbio.3001842

Xu, J. J., L. E. Jarocha, T. Zollitsch, M. Konowalczyk, K. B. Henbest, S. Richert, M. J. Golesworthy, J. Schmidt, V. Déjean, D. J. C. Sowood, M. Bassetto, J. T. Luo, J. R. Walton, J. Fleming, Y. J. Wei, T. L. Pitcher, G. Moise, M. Herrmann, H. Yin, H. J. Wu, R. Bartölke, S. J. Käsehagen, S. Horst, G. Dautaj, P. D. F. Murton, A. S. Gehrckens, Y. Chelliah, J. S. Takahashi, K. W. Koch, S. Weber, I. A. Solov'yov, C. Xie, S. R. Mackenzie, C. R. Timmel, H. Mouritsen and P. J. Hore. 2021. Magnetic sensitivity of cryptochrome 4 from a migratory songbird. Nature 594: 535–540.

Xu, J. J., E. H. Merrill and M. A. Lewis. 2022. Spreading speed of chronic wasting disease across deer groups with overlapping home ranges. Journal of Theoretical Biology 547: 111135. doi: 10.1016/j.jtbi.2022.111135

Yamaguchi, N. and D. W. Macdonald. 2003. The burden of co-occupancy, intraspecific resource competition and spacing patterns in American mink, *Mustela vison*. Journal of Mammalogy 84: 1341–1355.

Yamaguchi, N., H. L. Dugdale and D. W. Macdonald. 2006. Female receptivity, embryonic diapause, and superfetation in the European badger (*Meles meles*): Implications for the reproductive tactics of males and females. Quarterly Review of Biology 82: 33–48.

Yamada, K. and M. Besom. 2013. What Do You Do with an Idea? Compendium, Inc., Seattle, Washington. 38 pp.

Yang, W., C. Sun, R. Huszár, T. Hainmueller, K. Kislev and G. Buzsáki. 2024. Selection of experience for memory by hippocampus sharp wave ripples. Science 383: 1478–1483.

Ydenberg, R. C., L. A. Giraldeau and J. B. Falls. 1988. Neighbors, strangers, and the asymmetric war of attrition. Animal Behaviour 36: 343–347.

Yerger, R. W. 1953. Home range, territoriality, and populations of the chipmunk in central New York. Journal of Mammalogy 34: 448–458.

Yi, S. J., M. H. Wang, M. Y. Ju and X. F. Yi. 2021. Olfaction alters spatial memory strategy of scatter-hoarding animals. Integrative Zoology 16: 128–135.

Ylönen, H. 1989. Weasels *Mustela nivalis* suppress reproduction in cyclic bank voles *Clethrionomys glareolus*. Oikos 55: 138–140.

Ylönen, H. 1994. Vole cycles and anti-predatory behaviour. Trends in Ecology & Evolution 9: 426–430.

Ylönen, H. and H. Ronkainen. 1994. Breeding suppression in the bank vole as antipredatory adaptation in a predictable environment. Evolutionary Ecology 8: 658–666.

York, J. E., K. J. Wells and A. J. Young. 2019. Dominance-related contributions to collective territory defence are adjusted according to the threat. Animal Behaviour 158: 25–34.

Young, B. F. and R. L. Ruff. 1982. Population dynamics and movements of black bears in east central Alberta. Journal of Wildlife Management 46: 845–860.

Young, P. J. 1999. Red squirrel *Tamiasciurus hudsonicus*. Pp. 460–461 in Wilson, D. E. and S. Ruff (editors), The Smithsonian Book of North American Mammals. The Smithsonian Institution, Washington, DC.

Yu, H., F. T. Muijres, J. S. Lindera, A. Hedenström and P. Henningsson. 2023. Accelerometer sampling requirements for animal behavior classification and estimation of energy expenditure. Animal Biotelemetry 11: 28. https://doi.org/10.1186/s40317-023-00339-w

Zehnder, A., S. Henley and R. Weibel. 2018. Home ranges of lions in the Kalahari, Botswana exhibit vast sizes and high temporal variability. Zoology 128: 46–54.

Zhou, G., J. K. Olofsson, M. Z. Koubeissi, G. Menelaou, J. Rosenow, S. U. Schuele, P. Xu, J. L Voss, G. Lane and C. Zelanoa. 2021. Human hippocampal connectivity is stronger in olfaction than other sensory systems. Progress in Neurobiology 201: 102027.

Zielinski, W. J., W. D. Spencer and R. D. Barrett. 1983. Relationship between food habits and activity patterns of pine martens. Journal of Mammalogy 64: 387–396.

Zimmerman, J. W. 1990. A critical review of the error polygon method. International Conference on Bear Research and Management 8: 251–256.

Zimmerman, J. W. 1992. A habitat suitability model for black bears in the southern Appalachian region, evaluated for location error. PhD thesis, North Carolina State University, Raleigh, North Carolina.

Zimmerman, J. W. and R. A. Powell. 1995. Radio telemetry error: Location error method compared to error polygons and error ellipses. Canadian Journal of Zoology 73: 1123–1133.

Ziv, Y., L. D. Burns, E. D. Cocker, E. O. Hamel, K. K. Ghosh, L. J. Kitch, A. El Gamal and M. J. Schnitzer. 2013. Long-term dynamics of CA1 hippocampal place codes. Nature Neuroscience 16: 264–266.

Zoellick, B. W. and N. S. Smith. 1992. Size and spatial organization of home ranges of kit foxes in Arizona. Journal of Mammalogy 73: 83–88.

Zollner, P. A. and S. L. Lima. 1997. Landscape-level perceptual abilities in white-footed mice: Perceptual range and the detection of forested habitat. Oikos 80: 51–60.

Zollner, P. A. and S. L. Lima. 1999. Illumination and the perception of remote habitat patches by white-footed mice. Animal Behaviour 58: 489–500.

Zollner, P. A. 2000. Comparing the landscape level perceptual abilities of forest sciurids in fragmented agricultural landscapes. Landscape Ecology 15: 523–533.

Index

Tables and figures are indicated by an italic *t* or *f*.